Geobotany Studies

Basics, Methods and Case Studies

Editor
Franco Pedrotti
University of Camerino
Via Pontoni 5
62032 Camerino
Italy

Editorial Board:
S. Bartha, Vácrátót, Hungary
F. Bioret, University of Brest, France
E. O. Box, University of Georgia, Athens, Georgia, USA
A. Čarni, Slovenian Academy of Sciences, Ljubljana, Slovenia
K. Fujiwara, Yokohama City University, Japan
D. Gafta, "Babes-Bolyai" University Cluj-Napoca, Romania
J. Loidi, University of Bilbao, Spain
L. Mucina, The University of Western Australia, Perth, Australia
S. Pignatti, Università degli Studi di Roma "La Sapienza", Italy
R. Pott, University of Hannover, Germany
A. Velázquez, Centro de Investigación en Sciéncias Ambientales, Morelia, Mexico
R. Venanzoni, University of Perugia, Italy

About the Series

The series includes outstanding monographs and collections of papers on a given topic in the following fields: Phytogeography, Phytosociology, Plant Community Ecology, Biocoenology, Vegetation Science, Eco-informatics, Landscape Ecology, Vegetation Mapping, Plant Conservation Biology and Plant Diversity. Contributions are expected to reflect the latest theoretical and methodological developments or to present new applications at broad spatial or temporal scales that could reinforce our understanding of ecological processes acting at the phytocoenosis and landscape level. Case studies based on large data sets are also considered, provided that they support refinement of habitat classification, conservation of plant diversity, or prediction of vegetation change. "Geobotany Studies: Basics, Methods and Case Studies" is the successor to the journal *Braun-Blanquetia*, which was published by the University of Camerino between 1984 and 2011 with the cooperation of the Station Internationale de Phytosociologie (Bailleul-France) and the Dipartimento di Botanica ed Ecologia (Università di Camerino, Italy) and under the aegis of the Société Amicale Francophone de Phytosociologie, the Société Française de Phytosociologie, the Rheinold Tüxen Gesellschaft, and the Eastern Alpine and Dinaric Society for Vegetation Ecology. This series aims to promote the expansion, evolution and application of the invaluable scientific legacy of the Braun-Blanquet school.

More information about this series at http://www.springer.com/series/10526

Andrew M. Greller • Kazue Fujiwara •
Franco Pedrotti
Editors

Geographical Changes in Vegetation and Plant Functional Types

Editors
Andrew M. Greller
Department of Biology
Queens College, CUNY
Flushing, New York
USA

Kazue Fujiwara
Graduate School in Nanobioscience
Yokohama City University
Yokohama, Kanagawa
Japan

Franco Pedrotti
Department of Botany and Ecology
University of Camerino
Camerino, Italy

ISSN 2198-2562 ISSN 2198-2570 (electronic)
Geobotany Studies
ISBN 978-3-319-68737-7 ISBN 978-3-319-68738-4 (eBook)
https://doi.org/10.1007/978-3-319-68738-4

Library of Congress Control Number: 2017964450

© Springer International Publishing AG, part of Springer Nature 2018

This work is subject to copyright. All rights are reserved by the Publisher, whether the whole or part of the material is concerned, specifically the rights of translation, reprinting, reuse of illustrations, recitation, broadcasting, reproduction on microfilms or in any other physical way, and transmission or information storage and retrieval, electronic adaptation, computer software, or by similar or dissimilar methodology now known or hereafter developed.

The use of general descriptive names, registered names, trademarks, service marks, etc. in this publication does not imply, even in the absence of a specific statement, that such names are exempt from the relevant protective laws and regulations and therefore free for general use.

The publisher, the authors and the editors are safe to assume that the advice and information in this book are believed to be true and accurate at the date of publication. Neither the publisher nor the authors or the editors give a warranty, express or implied, with respect to the material contained herein or for any errors or omissions that may have been made. The publisher remains neutral with regard to jurisdictional claims in published maps and institutional affiliations.

Printed on acid-free paper

This Springer imprint is published by the registered company Springer International Publishing AG part of Springer Nature.
The registered company address is: Gewerbestrasse 11, 6330 Cham, Switzerland

Dedication

Elgene Owen Box at least has a name, unlike his father and grandfather, who each had only the two letters E and O (with no periods, since they were not true initials). Although most of the family was originally from England, one great-grandmother was from Berlin; last year Elgene was able to see the dock in Bremerhaven from which she most probably embarked for the New World.

Elgene Owen Box at Muir Woods National Monument in California, in November, 2015

Elgene was born on 16 December 1945 to [later] Capt. and Mrs. E O Box jr, at the Naval Air Station in Corpus Christi, Texas (an old pirate hangout with a sanitized name). In spring 1946, they moved to Commerce, in the cotton country of northeast Texas, where his father had married the girl next door and where both sets of grandparents still lived. In August, they moved to Bartlesville (Oklahoma), which 40 years earlier was still an oil and cowboy town in the Indian Territory. After skipping the 4th grade, Elgene became the youngest boy in his class, graduating in 1963 from College High School (Bartlesville). Along the way, he learned to play baseball, to ride a horse, to make color fireworks (cf Photo 1), to play the piano and sing rock 'n' roll, and to love the woods. Since his mother was a Latin teacher, he also picked up four years of Latin and was ready in college to start learning modern languages.

Photo 1 Elgene liked fireworks and by middle-school age had learned to make them himself (shown here in his garage, with younger sister Diane)

At the University of Texas, Elgene studied mathematics and history, sang classical choral music, learned to program computers, and graduated in 1967. Then Duke University beckoned him to the many more opportunities available on the east coast, where he received his Master's degree (mathematics) in 1969 and discovered the holistic thinking of ecology and the environmental movement. While working in 1970 at the University of North Carolina Computer Center, he was thus in just the right place to be discovered by Helmut Lieth, who needed a programmer to help make the first model-driven world computer maps, in particular the "Miami Model" of net primary productivity (Photo 2, presented in 1971; see Lieth and Box 1972, Lieth 1975).

In those days, computing was done on mainframe computers, which cost money, even at universities. So, before his main doctoral time at UNC, Lieth sent Elgene to the Kernforschungsanlage in Jülich (Germany), where he had unlimited computer access. There he was able to compile a first detailed world climatic database, make a first computerized soil-water budgeting system (SOLWAT), and write a proto-GIS for world mapping and map quantification (MAPCOUNT, Box 1975, 1979a, b). Using data from the International Biological Program, Lieth and Box (1977) made a first global model of gross primary productivity (thus also respiration), which Elgene mapped and quantified, providing the first estimates of terrestrial gross production and the often stated 50% ratio of terrestrial net to gross primary production (Box 1978). In 1977, Elgene also lived for two months with the famous Prof. Heinrich Walter, where he translated the Eurasian portion of the "Ecosystems of the World" volume on Temperate Deserts and Semi-Deserts (Walter and Box 1983). Finally, also during the 1970s, he wrote the basic computer processors for climatic envelope modeling (ECOSIEVE) and mapping (FLEXVEG), which were used for his doctoral research (PhD in 1978) to demonstrate that the main natural vegetation types and regions of the world could be predicted and mapped reasonably from climatic data alone.

Beginning as an Assistant Professor at the University of Georgia in 1979, Elgene was confronted immediately with an unexpected teaching load of six courses per year, 50% more than at most US universities. Before his global modeling program was slowed, however, he was able to extend the envelope work to basic ecological plant types and to present the global-scale results, with model validation. This book, appropriately subtitled "an introduction to predictive modeling in phytogeography" (Box 1981), introduced the idea of combining a large global database (climate), envelope modeling, and basic ecological plant types to predict world vegetation composition and patterns in more detail. There were 90 plant types in that model, later expanded to about 125 (Box 1984, 1987, 1995a, 1996); these were the forerunners of what are today called plant functional types (cf Cramer and Leemans 1993, Cramer 1997, Foley 1995, Peng 2000, Wullschleger et al. 2014).

After predicting world vegetation patterns, it was time to see more of the world. In 1981–1982, Elgene managed to get a year's leave (Georgia has no sabbatical program) to study vegetation ecology in France and come home via tropical Asia, Australia, and Hawaii. This was followed in 1983 by his first IAVS [vegetation science] symposium and excursion, in Argentina, where he met Prof. Akira Miyawaki, host of the next year's IAVS meeting. Prof. Miyawaki said that Elgene

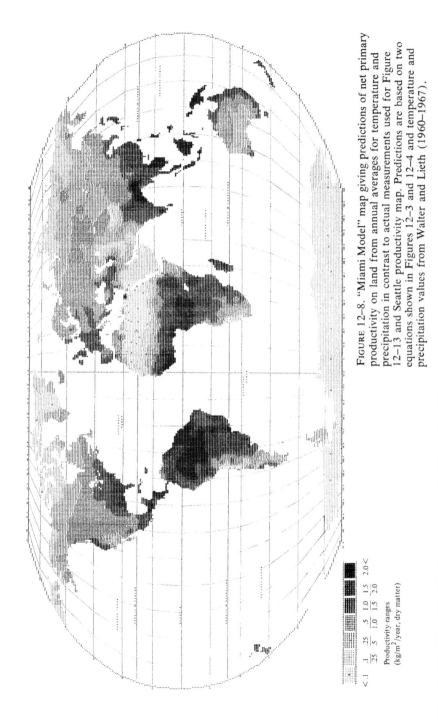

FIGURE 12–8. "Miami Model" map giving predictions of net primary productivity on land from annual averages for temperature and precipitation in contrast to actual measurements used for Figure 12–13 and Seattle productivity map. Predictions are based on two equations shown in Figures 12–3 and 12–4 and temperature and precipitation values from Walter and Lieth (1960–1967).

Photo 2 The "Miami Model" of net primary productivity predicted from temperature and precipitation, one of the first world computer maps (presented in November 1971; cf Lieth and Box 1972, Lieth 1973)

always sat in the back seat of the [German-speaking] bus and was very quiet. When people were in the field, though, he translated both ways between German (language of the field commentary) and English (language of the other busload of participants). The next year Elgene came to Japan for the IAVS symposium and excursions. He has worked with us ever since then and has visited Japan more than 30 times.

During the 1980s, Vernon Meentemeyer (also in the Georgia geography department) and Elgene gathered literature data, made geographically representative statistical models, projected the results globally, and quantified the totals, offering first quantitative estimates of additional components of the global carbon budget, including litterfall, decomposition, and potential accumulations (Meentemeyer et al. 1982); soil carbon (Meentemeyer et al. 1985); soil CO_2 (Brook et al. 1983); and soil CO_2 evolution (Box and Meentemeyer 1993). Completion of these models permitted the development of MONTHLYC, a first seasonal model of the carbon dynamics of a natural biosphere (Box 1988a, b). This provided values for inputs to global atmospheric models (e.g., Gillette and Box 1986)—but was soon superseded by better, more theoretical models of photosynthetic and respiratory fluxes, made by the new generation of well-funded modeling teams. During the 1980s, Elgene also worked with NASA to provide the first rigorous evaluation of the newly available NDVI "greenness index" as an estimator of primary productivity, leaf area, and standing biomass (Box et al. 1989). He always emphasized that all such large-area models are inherently geographic models and should be validated geographically, i.e., separately for the world's different biomes or other physiographic regions, not just by geographically blind statistics (Box and Meentemeyer 1991). Other validation methodologies were suggested in Box (1981) but also at regional scale (Box et al. 1993; cf Peters 1991).

For 1992–1994, Elgene was invited to spend three years as Guest Research Professor at Tokyo University, in a remote-sensing laboratory interested in global-scale work. There he had his first PC (and minicomputer access), learned new computer techniques such as color pixel mapping, and was able to produce a world map of actual net productivity (based on satellite data rather than climate) (Box and Bai 1993). He also produced a geographically complete world classification and "benchmark" map of 50 pheno-physiognomic "potential dominant vegetation" types (Box 1995b) and a first large-area attempt to estimate actual standing biomass amounts from satellite data (Hu et al. 1996).

During his first trips to Japan (1984) and China (1985), Elgene could see that East Asia was that part of the world most similar to eastern North America. But he also wondered why East Asia had evergreen broad-leaved forests at latitudes that are still deciduous in eastern North America. After some suggestions (Box 1988c), he demonstrated that lower absolute minimum temperatures are in fact the reason (Box 1995c) that these have a geography somewhat independent of mean minima and are important for warm-temperate to subtropical vegetation patterns in general. Work with us in East Asia also produced various field descriptions of East Asian vegetation (e.g., Box et al. 1989, 1991a, b, 1998, 2001; Box 2008), as well as comparisons (Box 1992, 2002; Box and Fujiwara 2012). Other comparative work involved more detailed treatments of Asian climates (e.g., Box 2000; Box and Choi 2003) and edited books (Box et al. 1995, 2002; Kolbek et al. 2003).

Comparative work with Asia also led to more fieldwork in eastern North America, including the Eastern North American Vegetation Survey (1988–1990, funded by Japan; see Miyawaki et al. 1994), which collected an extensive database from eastern Canada to south Florida and Louisiana (Box 1994, Box and Fujiwara 2010). A particular interest involved the position of evergreen broad-leaved forests, both in the southeastern USA (Box and Fujiwara 1988, Fujiwara and Box 1994) and worldwide (Box 1997, Fujiwara and Box 1999a). Work under an EPA grant, using climatic envelopes, looked at potential effects of global warming on the ranges of major woody plant species in Florida, suggesting unexpected sensitivity (Box et al. 1993, 1999; Crumpacker et al. 2001a, b, 2002). Work with a German postdoc produced comparisons between eastern North American and European deciduous forests, including cross-projections to study possible limiting factors (Box and Manthey 2005, 2006; Manthey and Box 2007; cf Box, in press). Finally, some results from this and newer work were brought together in an IAVS symposium on "warm-temperate deciduous forests" (Lyon and Lyon 2011) and subsequent book (Box and Fujiwara 2015). Other work on bioclimatic zonation has focused on the subtropical transition (Crumpacker et al. 2001c), northeast Asia (Box and Fujiwara 2012), uplands (Box 2014), summarization (Box and Fujiwara 2013, 2005), and a more detailed global system (Box, in press).

Elgene has been an IAVS member since 1983 (Photo 3), once attended 20 consecutive annual meetings (on all continents except Antarctica), and was President from 1994 to 2007 (three terms). In the 1980s, he was Secretary of the Plant Section of the International Society for Biometeorology. He has also organized or co-organized various special sessions for meetings of the IAVS, the International Botanical Congress, and the International Society for Ecology (INTECOL).

Photo 3 At the IAVS Symposium in Japan in 1984, which was the first trip to Japan for Elgene (right, in the front row), with Heinz Ellenberg (left), Akira Miyawaki (middle), and Hartmut Dierschke (behind Box)

Elgene has long had an interest in the effects of global warming on vegetation, first shown by the estimate that primary productivity might change by about 5% for each [Celsius] degree of temperature change (with moisture balance remaining constant; see Lieth 1976). The global envelope model (Box 1981) suggested which plant forms might decrease or increase with climate change. The Florida work continued this interest (e.g., Box et al. 1999; see also Fujiwara and Box 1999b, Lassiter et al. 2000). His pessimistic side came out in his portion of an otherwise upbeat book on *The Healing Power of Forests* (Miyawaki and Box 2006). Recently, he translated Franco Pedrotti's book *Cartografia Geobotanica* from Italian into "Plant and Vegetation Mapping." At the age of 70, he is still teaching, writing, and editing (e.g., Box ed., in press) and claims to have no intention to retire any time soon.

Elgene Owen Box is one of the well-known early biosphere modelers and still works on classification of plant life forms and modeling plant function, world vegetation distributions, and potential vegetation shift under climate change. But he is also a field scientist, with experience in about 75 countries. When traveling he usually carries printed climate data and modeling results with him, for use in the field. His memories are still sharp and he does not forget Latin plant names or important numbers, such as historical and conference dates. He loves history, not only of the United States but of all parts of the world. He can still calculate numbers and dates without a calculator. He also still likes to sing and sang until recently in a chorus at his University of Georgia. When people at meetings gather to sing, he will be in the group (Photo 4).

Photo 4 Quartet (from left to right: Elgene, Sandro Pignatti, Richard Pott and Franco Pedrotti) singing at the IAVS Symposium 2000 in Nagano

Throughout his career, Elgene's work has been global and often mathematical in nature, like his life history in general. He is very proud of the English language, still writes, and tries to speak it correctly, and his English is easy for foreign people to understand, perhaps because he has worked with people from so many countries. He has edited books, translated one and a half books (so far), reviewed books written in German and French, and reviewed and edited many papers.

He is very shy, but even so has worked well in the IAVS for more than 20 years. When Prof. Miyawaki first met him in Argentina (1983), it seemed that he hesitated to work except with friends.

From the mid-1960s, he has programmed mainframe computers and, later, minicomputers and personal computers. On the other hand, he hates the coercive nature of modern (i.e., commercial) computer software and is always fighting with new versions of Windows and web-ware. These contrasts make us think that Elgene is not only very sharp but also familiar.

We say thank you very much, sincerely, that he reviews and edits so much English for others. At this time, we would like to celebrate his 70th birthday. We hope that he will have the longevity that has blessed his father, who reached 100 in 2014, is still clear-headed, and was still driving a car in 2015. We like to work together to move toward clear theoretical vegetation systems and to help solve green-environment problems in the world.

Kazue Fujiwara

References

Box EO (1975) Quantitative evaluation of global productivity models generated by computers. In: Lieth H, Whittaker RH (eds) Primary productivity of the biosphere. Springer, New York, pp 265–283

Box EO (1978) Geographical dimensions of terrestrial net and gross primary productivity. Radiat Environ Biophys 15:305–322

Box EO (1979a) The MAPCOUNT series of programs for analysis and manipulation of SYMAP maps. In: Mapping software and cartographic data-bases. Harvard Library of Computer Mapping, vol 2. Harvard University, Cambridge, MA, pp 21–29

Box EO (1979b) Quantitative cartographic analysis: a summary (with geoenvironmental applications) of SYMAP auxiliary programs developed at the Jülich Nuclear Research Center. Reports of the Kernforschungsanlage Jülich GmbH, no. 1582, Jülich, Germany, 162 pp

Box EO (1981) Macroclimate and plant forms: an introduction to predictive modeling in phytogeography. Tasks for vegetation science, 1. Dr. W. Junk BV, Publisher, Den Haag, 258 pp, 25 world maps

Box EO (1984) Productivity and plant types—some thoughts on a synthesis. Port Acta Biol, Series A 17:129–148

Box EO (1987) Plant life forms in Mediterranean environments. Ann Bot 45(2):7–42

Box EO (1988a) Estimating the seasonal carbon source-sink geography of a natural steady-state terrestrial biosphere. J Appl Meteorol 27:1109–1124

Box EO (1988b) Effects of seasonal vegetation structure on the carbon dynamics of terrestrial biosphere models. In: Verhoeven J et al (eds) Effect of vegetation structure and composition on carbon and nutrient economy. SPB Academic Publishers, Den Haag, pp 19–27

Box EO (1988c) Some similarities in the climates and vegetations of central Honshu and central eastern North America. Veröff Geobot Inst ETH, Stiftung Rübel (Zürich) 98:141–168

Box EO, Fujiwara K (1988) Evergreen broad-leaved forests of the southeastern United States: preliminary description. Bull Inst Environ Sci Technol, Yokohama Natl Univ 15:71–93

Box EO, Fujiwara K et al (1989) A tropical montane evergreen forest and other vegetation on Hainan Island, southern China. Bull Inst Environ Sci Technol, Yokohama Natl Univ 16:75–94

Box EO, Holben BN, Kalb V (1989) Accuracy of the AVHRR Vegetation Index as a predictor of biomass, primary productivity, and net CO_2 flux. Vegetatio 80:71–89

Box EO, Song Y-C, Miyawaki A, Fujiwara K (1991a) An evergreen broad-leaved forest in transitional eastern China. Bull Inst Environ Sci Technol, Yokohama Natl Univ 17:63–84

Box EO, Fujiwara K, Qiu X-Z (1991b) Diversity and dissimilarity of three forest types in Xishuangbanna, tropical southern China. Bull Inst Environ Sci Technol, Yokohama Natl Univ 17:85–105

Box EO, Meentemeyer V (1991) Geographic modeling and modern ecology. In: Esser G, Overdieck D (eds) Modern ecology: basic and applied aspects. Elsevier, Amsterdam, pp 773–804

Box EO (1992) Comparing the natural montane vegetation types of East Asia and eastern North America. Braun-Blanquetia 5:22–29

Box EO, Bai X-M (1993) A satellite-based world map of current terrestrial net primary productivity. Seisan Kenkyuu (Tokyo) 45(9):666–672

Box EO, Crumpacker DW, Hardin ED (1993) A climatic model for plant species locations in Florida. J Biogeogr 20:629–644

Box EO, Meentemeyer V (1993) Climate and the world geography of soil carbon dioxide evolution. In: Breymeyer A (ed) Geography of organic matter production and decay. Conference papers, vol 18. SCOPE and Inst. of Geography, Polish Academy of Sciences, Warszawa, pp 21–50

Box EO (1994) Eastern North America: natural environment and sampling strategy. In: Miyawaki A et al (eds) Vegetation in Eastern North America. Tokyo University Press, Tokyo, pp 21–57

Box EO (1995a) Factors determining distributions of tree species and plant functional types. Vegetatio 121:101–116

Box EO (1995b) Global potential natural vegetation: dynamic benchmark in the era of disruption. In: Murai S (ed) Toward global planning of sustainable use of the earth – development of global eco-engineering. Elsevier, Amsterdam, pp 77–95

Box EO (1995c) Global and local climatic relations of the forests of East and Southeast Asia. In: Box EO et al (eds) Vegetation science in forestry, Handbook Vegetation Sci, vol 12/1. Kluwer, Dordrecht, pp 23–55

Box EO (ed), with Peet RK, Masuzawa T, Yamada I, Fujiwara K, Maycock PF (co-eds) (1995) Vegetation science in forestry: global perspective based on forest ecosystems of East and Southeast Asia. Kluwer, Dordrecht, 663 pp

Box EO (1996) Plant functional types and climate at the global scale. J Veg Sci 7:309–320

Box EO (1997) Bioclimatic position of evergreen broad-leaved forests. In: Island and high-mountain vegetation: biodiversity, bioclimate and conservation, pp 17–38. Proceedings annual meeting international association for vegetation science, Tenerife, April 1993. Universidad de La Laguna, Servicio de Publicaciones, Tenerife (Canary Islands, Spain)

Box EO, Chou C-H, Fujiwara K (1998) Richness, climatic position, and biogeographic potential of East Asian laurophyll forests, with particular reference to examples from Taiwan. Bull Inst Environ Sci Technol, Yokohama Natl Univ 24:61–95

Box EO, Crumpacker DW, Hardin ED (1999) Predicted effects of climatic change on distribution of ecologically important native tree and shrub species in Florida. Clim Change 41:213–248

Box EO (2000) Climate and climatology in Japan. In: 43rd IAVS Post-Symposium Excursion Committee (eds) A guide book of the post-symposium excursion in Japan, for the 43rd symposium of the international association for vegetation science, pp 144–169

Box EO, Hai-Mei Y, and Li D-L (2001) Climatic ultra-continentality and the abrupt boreal-nemoral forest boundary in northern Manchuria. In: Assn. Commemorate Retirement of Prof. Okuda (eds) Studies on the vegetation of alluvial plains, pp 183–200, Yokohama

Box EO (2002) Vegetation analogs and differences in the Northern and Southern Hemispheres: a global comparison. Plant Ecol 163:139–154 [appendix missing: ask author]

Box EO, Nakashizuka T, Fischer A (eds) (2002) Dynamics of temperate forests. Special features in vegetation science, vol 17. Opulus Press, Uppsala. 148 pp

Box EO, Choi JN (2003) Climate of Northeast Asia. In: Kolbek J et al (eds) Forest vegetation of Northeast Asia. Kluwer, Dordrecht, pp 5–31

Box EO, Manthey M (2005) Oak and other deciduous forest types of eastern North America and Europe. Botanika Chronika (Greece) 18(1):51–62

Box EO, Manthey M (2006) Conservation of deciduous tree species in Europe: projecting potential ranges and changes. In: Gafta D, Akeroyd J (eds) Nature conservation: concepts and practice. Springer, Berlin, pp 241–253

Box EO (2008) Sections in: "Integrated vegetation mapping in Asia" (K. Fujiwara, project leader). Final Report to Monbushô (Japanese Ministry of Science Education and Culture)

Box EO, Fujiwara K (2010) What else can one do with relevé data: Eastern North America? Braun-Blanquetia 46:139–142

Box EO, Fujiwara K (2012) A comparative look at bioclimatic zonation, vegetation types, tree taxa and species richness in Northeast Asia. Bot Pac 1:5–20

Box EO, Fujiwara K (2013, 2005) Vegetation types and their broad-scale distribution. In: van der Maarel E, Franklin J (eds) Vegetation ecology, 1st edn, pp 106–128. Blackwell Scientific, Oxford, pp 455–485

Box EO (2014) Uplands and global zonation. Contribṭtii Botanice, (Cluj-Napoca) 49:223–254

Box EO, Fujiwara K (eds) (2015) Warm-temperate deciduous forests around the Northern Hemisphere. Geobotany series. Springer, Heidelberg, 292 pp (including 4 chapters by Box)

Box EO (in press-a) Global bioclimatic zonation. In: Box EO (ed) Vegetation structure and function at multiple spatial, temporal and conceptual scales. Springer, Cham

Box EO (ed) (in press-b) Vegetation structure and function at multiple spatial, temporal and conceptual scales. Springer, Cham

Brook GA, Folkoff ME, Box EO (1983) A world model of soil carbon dioxide. Earth Surf Process 8:79–88

Cramer W, Leemans R (1993) Assessing impacts of climate change on vegetation using climate classification systems. In: Solomon AM, Shugart HH (eds) Vegetation dynamics and global change. Chapman & Hall, New York, pp 190–217

Cramer W (1997) Using plant functional types in a global vegetation model. In: Smith TM et al (eds) Plant functional types: their relevance to ecosystem properties and global change. Cambridge University Press, Cambridge, pp 271–288

Crumpacker DW, Box EO, Hardin ED (2001a) Implications of climatic warming for conservation of native trees and shrubs in Florida. Conserv Biol 15(4):1008–1020

Crumpacker DW, Box EO, Hardin ED (2001b) Potential breakup of Florida plant communities as a result of climatic warming. Fla Sci 64(1):29–43

Crumpacker DW, Box EO, Hardin ED (2001c) Temperate-subtropical transition areas for native trees and shrubs in Florida: present locations, predicted changes under climatic warming, and implications for conservation. Nat Areas J 21(2):136–148

Crumpacker DW, Box EO, Hardin ED (2002) Use of plant climatic envelopes to design a monitoring system for biotic effects of climatic warming. Fla Sci 65:159–184

Didukh Ya P (2011) The ecological scales for the species of Ukrainian flora and their use in synphytoindication [Vyd-vo Ukraïns'koho Fitosotsiolohichnoho Tsentru]. Phytosociocentre, Kyiv

Foley JA (1995) Numerical models of the terrestrial biosphere. J Biogeogr 22(4/5):837–842

Fujiwara K, Box EO (1994) Evergreen broad-leaved forest region of the southeastern United States. In: Miyawaki A et al (eds) Vegetation in Eastern North America. Tokyo University Press, Tokyo, pp 273–312

Fujiwara K, Box EO (1999a) Evergreen broad-leaved forests in Japan and eastern North America: vegetation shift under climatic warming. In: Klötzli F, Walther G-R (eds) conference on recent shifts in vegetation boundaries of deciduous forests, especially due to general global warming, Ascona. Birkhäuser, Basel, pp 273–300

Fujiwara K, Box EO (1999b) Climate change and vegetation shift. In: Farina A (ed) Perspectives in ecology, a glance from the VII international congress of ecology, Florence, 19–25 July 1998. Backhuys, Leiden, pp 121–126

Gillette DA, Box EO (1986) Modeling seasonal changes of atmospheric carbon dioxide and carbon-13. J Geophys Res 91(D4):5287–5304

Hu H, Shibasaki R, Box EO (1996) Generation of global terrestrial biomass map by integrating satellite data and carbon dynamics model. In: Proceedings, Asian conference on remote sensing, paper GLE 96-8. Asian Society for Photogrammetry and Remote Sensing

Kolbek J, Srůtek M, Box EO (eds) (2003) Forest vegetation of Northeast Asia. Geobotany, 28. Kluwer, Dordrecht, 472 pp

Lassiter RR, Box EO et al (2000) Vulnerability of ecosystems of the mid-Atlantic region, USA, to climate change. Environ Toxicol Chem 19(4):1153–1160

Lieth H, Box EO (1972) Evapotranspiration and primary productivity; C. W. Thornthwaite memorial model. Publ Climatol, Univ. of Delaware, Elmer, New Jersey 25(3):37–46

Lieth H (1975) Modeling the primary productivity of the world. In: Lieth H, Whittaker RH (eds) Primary productivity of the biosphere. Springer, New York, pp 237–263

Lieth H (1976) Possible effects of climate change on natural vegetation. In: Kopec RJ (ed) Atmospheric quality and climatic change. Univ. N Carolina, Chapel Hill, pp 150–159

Lieth H, Box EO (1977) The gross primary productivity pattern of the land vegetation: a first attempt. Trop Ecol 18:109–115

Lyon JG, Lyon LK (2011) Practical handbook for wetland identification and delineation. CRC Press, Boca Raton, FL

Manthey M, Box EO (2007) Realized climatic niches of deciduous trees: comparing western Eurasia and eastern North America. J Biogeogr 34:1028–1040

Meentemeyer V, Box EO, Thompson R (1982) World patterns and amounts of terrestrial plant litter production. BioScience 32:125–128

Meentemeyer V, Gardner J, Box EO (1985) World patterns and amounts of detrital soil carbon. Earth Surf Process 10:557–567

Miyawaki A, Iwatsuki K, Grandtner MM (eds) (1994) Vegetation in Eastern North America. University of Tokyo Press, Tokyo. 515 pp

Miyawaki A, Box EO (2006) The healing power of forests. Kôsei Publ, Tokyo. 286 pp

Peng C-H (2000) From static biogeographical model to dynamic global vegetation model: a global perspective on modeling vegetation dynamics. Ecol Model 135:33–54

Peters RH (1991) A critique for ecology. Cambridge University Press, Cambridge. 366 pp

Rivas Martinez S (1976) De plantis Hispaniae notulae systematicae, chorologicae et ecologicae: systematic, chorological and ecological notes on plants from Spain. Candollea 31(1):111–117

Walter H, Box EO (1983) Chapters 2–9 in. Temperate deserts and semi-deserts. In: West NE (ed) Ecosystems of the world, vol 5. Elsevier, Amsterdam

Wullschleger S, Epstein HE, Box EO et al (2014) Plant functional types in Earth system models: Past experiences and future directions for application of dynamic vegetation models in high-latitude ecosystems. Ann Bot 114:1–16

Preface/Overview

This book is a Festschrift honoring Prof. Dr. Elgene Owen Box on his 70th birthday. Elgene has many interests and has made many studies throughout the world, as outlined in the Dedication. Among these are long-standing interests in ecological plant forms (now commonly called "plant functional types"), temperate-zone forests around the world, and in particular evergreen broad-leaved forests. The chapters herein reflect some of these interests, including vegetation interpretation at two different spatial scales (global and regional).

Part I involves general problems of vegetation interpretation.

Jorge Capelo (Chap. 1) makes an important contribution to understanding seral relations of communities on disturbed land progressing toward the regional or local "climax" community. Capelo credits S. Rivas-Martinez (1976) with originating the concept of "vegetation series." Using a case study of vegetation mosaics from several types of managed oak forests in central-southern Portugal, Capelo verifies the concept and shows that one can deduce the "climax" type, despite its local absence or removal. Such information can be useful for typological systems of plant landscapes and is relevant for vegetation science, for management of natural areas, and for vegetation mapping.

Part II involves the global aspect of plant function and has two chapters. Both deal with plant functional types or characters (traits), a field that Elgene pioneered with his first book (Box 1981).

Andrew N. Gillison (Chap. 2) presents original data on plant functional traits that he collected from polar to tropical sites. This is the first time that a uniform methodology was employed in consideration of plant adaptation to such a large range of environmental conditions.

Javier Loidi (Chap. 3) looks at different biomes around the world and examines the question whether ecophysiological traits are truly related to climate or if they are dependent upon phylogenetic relationships that preserve traits despite variation in environmental factors.

Part III involves effects of prehistoric landscapes and regional vegetation types on species organization and distributions. This section has nine chapters, five from Europe, one from the United States, and two from East Asia; one compares Mediterranean-type vegetation in Europe and California.

Richard Pott (Chap. 4) focuses on the historic impact of human agriculture and animal husbandry practices, which changed original landscapes to present-day cultivated landscapes, concluding that anthropogenic changes have increased landscape diversity considerably.

Rosario G. Gavilan, Beatriz Vilches, and colleagues (Chap. 5) present a thorough review of the mainly oak-dominated evergreen and semi-evergreen forests in the Mediterranean regions of Spain. The ecological requirements of deciduous forests in those regions are discussed in conjunction with the evergreens.

Mark A. Blumler (Chap. 6) shows that evergreen sclerophylls are not always well adapted to, and do not flourish in, strongly winter-wet summer-dry climates. What vegetation is then best adapted to Mediterranean climates? Forests are not discussed but rather some other vegetation types that dominate extensive areas in the Mediterranean area, comparing them with climate data and looking at the percent of therophytes on Mediterranean islands. Studies of vegetation dynamics indicate that annuals can be highly competitive and perhaps even "climax" vegetation where winters are wet (but not too wet) and summers are hot, long, and dry. Tall, laterally expanding clonal plants of the sort that Grime classifies as competitors do not thrive under such conditions. Biome modeling and vegetation mapping would improve if the assumption of one biome per climate type were relaxed and instead different life forms or functional types were mapped as overlays in a GIS.

Irina N. Safronova and T. K. Yourkovskaya (Chap. 7) consider the major latitudinal zones of vegetation in western Russia, recognizing tundra, taiga, broad-leaved forest, steppe, and desert. Within this zonal framework, they distinguish subzones and major edaphic types.

Franco Pedrotti (Chap. 8) describes the processes by which human activity through the years has caused the degeneration and regression of a locally rare forest, one that is dominated by *Carpinus betulus,* in the hills of Gocciadoro, Val d'Adige.

Yakiw Didukh, Irina Kontar, and Adam Boratynski (Chap. 9) bring to the international community of vegetation ecologists a new technique for classifying and ordinating syntaxa, using semiquantitative environmental data. This technique, originally developed for local, Ukrainian vegetation, is given the name synphytoindication (Didukh 2011). In this chapter, the synphytoindication method is used to analyze relevés of syntaxa typical of the Polish Tatra Mountains, the Ukrainian Carpathians, and the Crimean Mountains. The technique was applied to determine the place of the syntaxa in relation to the main ecofactors, and the degree of correlation between the change of values of the latter, the estimation of vegetation differentiation, and the place of corresponding regions in the general ecospace. The authors determined the grades for certain types of syntaxa and mountain ecosystems, as well as the nature of correlation between ecofactor values. They determined further which of the syntaxa are most vulnerable to climate change, by estimating the place of syntaxa in relation to the values of the ecofactors.

Andrew M. Greller (Chap. 10) reviews the distribution of *Fagus grandifolia* (American beech) in woody communities on Long Island, in the Eastern Deciduous

Forest region of North America. He identifies seven forest types in which beech is present, describes the ecological requirements of beech in those situations, and then examines the phytosociological role of *Fagus grandifolia* in all the forest regions of eastern North America and Mexico.

You Hai-Mei, Kazue Fujiwara, and Tang Qian (Chap. 11) present original data on the composition of forests dominated by this endemic tree genus of China. They use Braun-Blanquet phytosociological analysis to prepare a synoptic table of the taxa present and then subject the data to multivariate statistical analysis (DCCA) to yield a new Alliance with two associations.

Wang Zheng-Xiang, Li Ting-Ting, and Lei Yun (Chap. 12) use Braun-Blanquet phytosociological techniques to provide a detailed analysis of taxonomically based assemblages (associations). The variety of communities is unexpectedly large.

Part IV involves vegetation and plant ecology, and environmental mapping, with four chapters. These chapters describe dieback of Ohi'a [*Metrosideros polymorpha*] in Hawaiian Rainforest; plant assemblages on abandoned ore-mining heaps in Romania; autoecological and synecological resilience of *Angelica heterocarpa* M.J. Lloyd in France; and environmental mapping trial and proposal.

Linda Mertelmeyer, James D. Jacobi, Hans Juergen Boehmer, and Dieter Mueller-Dombois (Chap. 13) use very high-resolution aerial imagery (POL) of present-day Ohia forests to examine the consequences of Ohia forest dieback, which was documented originally in 1976 by Mueller-Dombois. The authors show the value of using POL and conclude that the forests are recovering.

Anamaria Roman, Dan Gafta, Tudor-Mihai Ursu, and Vasile Cristea (Chap. 14) studied a chronosequence of mining dumps to see how age, slope, slope position, terrain curvature, and potential solar radiation affect the composition of spontaneous plant communities. Age of the site and substrate acidity were determined to be the most important factors in differentiating the sites. The two successional series, differing in the degree of acidity, both progress to natural regeneration of native woodland.

Kevin Cianfaglione and Frédéric Bioret (Chap. 15) include bibliographic data analysis to examine the response of this local endemic, hydrophytic, *Angelica* species to anthropogenic changes in its habitat. An extensive bibliography follows the discussion.

Marcello Martinelli and Franco Pedrotti (Chap. 16) proposed and tried to make a synthesis map (environmental map) from analysis-type maps of geology, potential vegetation, and actual vegetation.

This book is unique and each chapter has original ideas, reflecting the originality of Prof. Box's work, using mathematics in his research.

Flushing, NY	Andrew M. Greller
Yokohama, Japan	Kazue Fujiwara
Camerino, Italy	Franco Pedrotti

Contents

Dedication v

Preface/Overview xvii

Part I General Problems of Vegetation Interpretation

1 Evidence of a Unique Association Between Single Forest Vegetation-Types and Seral Sequences: Praise for the Concept of 'Vegetation Series' 3
Jorge Capelo

Part II Global Aspects of Plant Function

2 Latitudinal Variation in Plant Functional Types 21
Andrew N. Gillison

3 Plant Eco-Morphological Traits as Adaptations to Environmental Conditions: Some Comparisons Between Different Biomes Across the World 59
Javier Loidi

Part III Regional Vegetation Types and Distribution, Prehistoric Landscapes

4 Changes in the Landscape and Vegetation Under the Influence of Prehistoric and Historic Man in Central Europe 75
Richard Pott

5 Sclerophyllous Versus Deciduous Forests in the Iberian Peninsula: A Standard Case of Mediterranean Climatic Vegetation Distribution 101
Rosario G. Gavilán, Beatriz Vilches, Alba Gutiérrez-Girón, José Manuel Blanquer, and Adrián Escudero

6 What Is the 'True' Mediterranean-Type Vegetation? 117
Mark A. Blumler

7	Characterization of Vegetation on the Plains of European Russia.. 141
	Irina N. Safronova and Tatiana K. Yurkovskaya
8	Degeneration and Regression Processes in *Carpinus betulus* Forests of Trentino (North Italy)........................... 169
	Franco Pedrotti
9	Phytoindicating Comparison of Vegetation of the Polish Tatras, the Ukrainian Carpathians, and the Mountain Crimea.......... 185
	Ya. Didukh, I. Kontar, and A. Boratynski
10	Beech (*Fagus grandifolia*) in the Plant Communities of Long Island, New York, and Adjacent Locations, and Some Comparisons Across Eastern North America.................. 211
	Andrew M. Greller
11	Phytosociological Study of *Pteroceltis tatarinowii* Forest in the Deciduous-Forest Zone of Eastern China..................... 239
	Hai-Mei You, Kazue Fujiwara, and Qian Tang
12	Vegetation Ecology of *Sphagnum* Wetlands in Subtropical Subalpine Regions: A Case Study in Qi Zimei Mountains........ 281
	Ting-Ting Li, Zheng-Xiang Wang, and Yun Lei

Part IV Vegetation and Plant Ecology

13	High-Resolution Aerial Imagery for Assessing Changes in Canopy Status in Hawai'i's 'Ōhi'a (*Metrosideros polymorpha*) Rainforest.. 291
	Linda Mertelmeyer, James D. Jacobi, Hans Juergen Boehmer, and Dieter Mueller-Dombois
14	Plant Assemblages of Abandoned Ore Mining Heaps: A Case Study from Roşia Montană Mining Area, Romania............. 303
	Anamaria Roman, Dan Gafta, Tudor-Mihai Ursu, and Vasile Cristea
15	Autoecological and Synecological Resilience of *Angelica heterocarpa* M.J. Lloyd, Observed in the Loire Estuary (France)... 333
	Kevin Cianfaglione and Frédéric Bioret
16	Environmental Mapping: From Analysis Maps to the Synthesis Map... 347
	Marcello Martinelli and Franco Pedrotti

Part I
General Problems of Vegetation Interpretation

Evidence of a Unique Association Between Single Forest Vegetation-Types and Seral Sequences: Praise for the Concept of 'Vegetation Series'

Jorge Capelo

Abstract
Succession has always been a central issue in ecological theory. Ever since first historical formulations, mosaics of vegetation-types taken to be the result of successional processes were envisaged as the pieces composing the *Plant Landscape*. Following the seminal ideas of F.E. Clements and A. Tansley, first European vegetation scientists, namely G.E. Du Rietz and R. Tüxen, laid the foundations of a system of describing successional vegetation mosaics as a higher complexity level of the Plant Landscape. Such repetitive mosaic units would correspond to 'elementary' successional sequences. These included both metastable seral stages and an ecologically mature stable state: the *climax* stage. Such a successional unit could then be used, as a reference, to systematize Plant Landscape. Early views on succession, expressed by a linear chronological sequence leading to a single territorial stable climatic climax, have been nowadays greatly challenged. Models encompassing sets of non-equilibrium *alternative vegetation states,* as a function of environmental disturbance, have been developed throughout. Moreover, phytocoenotical diversity also arises from *zonation* processes, i.e the spatial differentiation of vegetation mosaics following spatial environmental gradients. We do not consider that subject in this paper. In spite of criticism, continental Europe's phytosociology schools have been elaborating on contemporary concepts of 'elementary successional units' or *vegetation series* as relevant to Vegetation Science. A widely-followed

---

This chapter is dedicated to Elgene.
Electronic supplementary material: The online version of this article (https://doi.org/10.1007/978-3-319-68738-4_1) contains supplementary material, which is available to authorized users.

J. Capelo (✉)
Herbarium of the National Institute of Agrarian and Veterinary Research, Oeiras, Portugal
e-mail: jorge.capelo@iniav.pt

contemporary formulation of *vegetation series* is that of S. Rivas-Martínez (An Inst Bot Cavanilles 33:179–188, 1976): *for a given biogeographical context, in spatially contiguous units possessing uniform environmental conditions, a single successional sequence develops and includes a sole ecologically mature vegetation type—the climax stage* (our paraphrase). The real-world existence of coherent vegetation mosaics as the result of succession in homogeneous habitats, possessing a single 'climax' stage, is still challenged by several authors. On the contrary, by testing this very hypothesis by means of a case-study on vegetation mosaics from several types of managed oak forests in central-southern Portugal, we found a verification of S. Rivas-Martínez model. Also, accounting for bibliographic sources of evidence supporting verification of the vegetation series hypothesis, we found that putative falsifications issued mostly from distinct semantic or methodological settings. Thus, we find that the concept is, in spite of this, scientifically sound as a basis for a typological system of Plant Landscape which is relevant for Vegetation Science and useful for Nature Management and Vegetation Mapping.

Keywords

Succession • Vegetation mosaics • Landscape phytosociology • Synrelevé • Vegetation series • Classification

1.1 Introduction

Early formulations of succession as being the main process of making up the *Plant Landscape* of a territory are found back in the late Eighteenth century, notably in A. von Humboldt (Humboldt 1814–1825): the 'global character' of a given region of the Earth, i.e. it's Landscape, could be perceived as the result of concurrent geological forces and climate that, in turn, shaped the specific character of its plant formations. The formulation of Landscape as the spatial arrangement of characteristic combinations of ecosystems is well-known from Carl Troll (Troll 1968). Incidentally, Bolós (1963, also 1984) stated earlier that Plant Landscape could be systematized by the study of vegetation-type mosaics. In parallel, in Phytosociology, the idea is older; Braun-Blanquet and Furrer (1913) and Braun-Blanquet and Pavillard (1922) recognized that given mosaics of vegetation-types are found repetitively in sites of analogous geomorphology and in general, of similar homogeneous environmental settings. Nevertheless, they do not state clearly from which ecological processes mosaics derive, merely referring to succession and zonation along spatial gradients as determinants along with environmental similarity of sites. Later, Braun-Blanquet (1951) explicitly states that many mosaics of plant communities (*gesellschaftmosaik*) issue from differential human action on a previous climax vegetation. He then cites examples of mosaics of woodland, hedges, scrub and meadows as the result of several events and degrees of disturbance on a former dominant forest stage (Braun-Blanquet 1964). It was Schithüsen (1968) who proposed that vegetation complexes could be systematically studied by a methodological analogue of the phytosociological relevé method, but *mutatis mutandi*, applied to mosaic-level. This is would be a method of describing the β-diversity of the plant landscape, here conceived of as the diversity of vegetation-types in a territory. An alternative formulation of β-diversity is possible via the accounting of spatial species turnover, but both definitions are ontologically equivalent (Whittaker 1960; Anderson et al. 2011). A formal method describing mosaics was then formally set by Reinhold Tüxen (1973). He described a synrelevé method of listing all plant communities in an environmentally homogeneous territory. Minimal-area method was also implied (i.e. an extinction curve for vegetation-types was accounted for). He proposed to use an abundance-dominance-scale similar to that of Braun-Blanquet for *taxa* in vegetation plots, but now used instead, to evaluate the cover of each vegetation-type on a much larger plot, in the order of thousands of square meters for forest mosaics, for instance. Géhu (1974, 1976) and Béguin and Hegg (1975, 1976) presented the first tables made by this method. Tables could then be ordinated or classified as floristic vegetation tables are. As most mosaics were recognized as the result of disturbance (mostly on forests), human-induced secondary succession was the main ecological feature looked for in the creation of mosaics. The later scientific activity was named *Symphytosociology* or *Landscape Phytosociology* and a system to classify vegetation mosaics, as describing a higher ecological complexity level—that of Plant Landscape—was then presumed to be feasible (Béguin and Hegg 1975, 1976; Géhu 1974, 1976, 1977, 1979, 1986; Rivas-Martínez 1976). For the later authors, for typology's sake, the elementary unit of such system would be the *synassociation* to stand as the model of the landscape-repetitive mosaic. The question 'what is in Symphytosociology, a homogeneous area?' pervades the new method (Vigo 1998; Biondi et al. 2011). Of course, environmental homogeneity is presupposed, as it already was for the vegetation-type level. The answer seems to be that homogeneity of 'climax' was implicit and being sought for when choosing *synrelevé* areas (Tüxen 1973; Rivas-Martínez 1976). What would be the reasoning behind this to escape the hint of tautology?

Continental European vegetation scientists did embrace the early notions of climax, namely that of Clements (1916), but did it with some criticism. The well-known concept of F.E. Clements is that of a 'mono-climax'. Specifically, he assumed that, for a given territory and climate, irrespective of substrate type (rock type, soil depth) and hydrological regime of the soil, *all* successional processes would lead to a single climax community. This would stand for the state of maximum biomass and physiognomic complexity (i.e. number of strata) and would be the end-result of progressive succession. In absence of disturbance, this stage would stay in a steady-state condition in terms of composition and species dominance. (Eventually, he conceived a senescence stage, whence a regression to an earlier stage could occur). In present day, the Clementesian model is taken as overwhelmingly simplistic. First, successional pathways are seldom arranged in time in single linear sequence, being more of an intricate network of stages intertwined by succession events and depending on initial frontier conditions, disturbance regime and stochastic fluctuation (Peterson and Reich 2001; Habeck and Much 1973). Second, secondary sequences (succession following a disturbance) don't usually follow the same linear sequence observed in primary succession. Third, disturbance regimes favor meta-stable stages that are secondary and act as climaxes (e.g. recurrent wildfire: e.g. Dovčak et al. 2005). Fourth, such a model does not account for variation of mosaics along spatial gradients; e.g. from crest, midslope to valley bottom). Fifth and most important, and this was very early falsified (Du Rietz 1936; Tüxen and Preising 1942), even in homogeneous climate, lithologically-, physiographically- and hydrologically differentiated biotopes (sites) lead to distinct seral sequences (both primary and secondary) and the climax stage, whenever it is found, is not the same vegetation type for the whole set of sequences. Thus, continental European vegetation scientists, although assuming the general idea of Clementesian succession assumed a *polyclimax* model (Du Rietz, *op. cit.*; Tüxen and Preising, *op. cit.*; Géhu and Rivas-Martínez 1980). Loosely, this means that each combination of physiographic positions expressing general hydrology and lithology/soil-types, even in the same bioclimate could lead to: (i) seral sequences composed of distinct vegetation-types; and (ii) distinct climax vegetation-types.

It was Rivas-Martínez (1976), Géhu (1979) and Géhu and Rivas-Martínez (1980) who, in short, stabilized the, up to then, fragmentary or still uncertain concepts of Symphytosociology. They recognized that coherent mosaics of vegetation-types in a landscape issued mainly from succession and that mosaic-types could be typified by *synassociations*. And they recognized that mosaics found in analogous biotopes, that are ecologically homogeneous, correspond to a single successional 'sequence' and most importantly, to a single 'climax' type. This elementary successional unit is called a *vegetation series*. Thus, the vegetation series is primarily characterized by a characteristic unique combination of plant types in a single habitat strongly related by succession corresponding to a single climax stage. It may share some vegetation-types with neighboring series, either those that are in fact spatially adjacent or in abstract terms of syn-composition. The shared vegetation-types will be called syn-companions. Some vegetation-types will not be diagnostic of climax (indifferent): in general, those promoted by human disturbance, namely crops and nitrogen-prone vegetation. Moreover, the term 'sequence' is, in the author's opinion, somewhat misleading. In fact, what is factual is that a "set" of vegetation-types with strong successional relationships among them is present. The pathways might not be chronologically linear or even clear. One could assume that ranking stages in a linear sequence by biomass is useful, but does not imply a chronological or deterministic sequence. Ranking could be done just for the sake of simplifying the vegetation series depiction. In fact, complex network representations of successional pathways are common in symphytosociological studies (Schwabe 1997). Also, we propose that the occurrence of a maximum biomass (i.e. climax) should not be taken as a forceful terminal stage in time. As discussed earlier, metastable disturbance-driven stages might be dominant in space and permanent in time and even the only ones present.

The subject of vegetation series relates strongly to that of Potential Natural Vegetation (PNV). As defined by Rivas-Martínez (2005): PNV would correspond to the vegetation-type of maximum complexity and biomass, as the result of the process of succession, and would develop: (i) if succession was instantaneous; (ii) in complete absence of disturbance; (iii) in the actual environmental conditions (our paraphrase). Thus, in practice, the "climax" vegetation-type in vegetation series could be taken as the PNV. Therefore, including PNV in the vegetation series concept ensures that PNV can even be diagnosed in the absence of a climax stage if characteristic stages of the vegetation series are present. Note that PNV is an operational concept that does not correspond strongly to *primitive vegetation* nor to *future vegetation* as ecological conditions might meanwhile have changed or be likely to change, namely those of climate. Thus, its validity should not be extrapolated either further back in time beyond recent Holocenic phases, or beyond the near future due to climate change prospects. The concept of PNV *sensu* Rivas-Martínez has been recently criticized based on this very same misunderstanding between PNV and primitive vegetation by Carrión and Fernández (2009) and by Chiarucci et al. (2010) as noted by Loidi and Fernández-González (2012) and Loidi et al. (2010). Apart from theoretical considerations, the *praxis* of determination of the PNV is made by side-by-side comparison of similar successional sequences in similar ecological field conditions where, *ipso facto*, the climax is actually found. As to the model of vegetation series including a single PNV, there are variable degrees of compliance to its full extent, even among practitioners of Symphytosociology. For instance, Theurillat (1992) suggests that mosaic types are recognizable; but that each is associated in a one-to-one relationship to a single forest is debatable. This was also the cautionary position of Tüxen (1973).

As mentioned before, we now come to propose that even if such one-to-one correspondence may be presupposed, no special status should be given to the climax stage other than it represents the maximum potentiality of vegetation succession in a given homogeneous biotope. We mean that among all the alternative vegetation states to be found in a vegetation series, none is to be taken as the end-product of the succession process in the sense that once reached it stays as such indefinitely. Disturbance regimes act as selective pressure on what stages dominate in a landscape in a given time. Nevertheless, PNV reference value should be preserved for practical reasons. This is because in most areas of the Earth where rainfall is sufficient and in soils of normal depth, the PNV stage is a usually a forest and therefore the prediction of its most adapted potential species composition might be paramount to Forest, Nature Planning and Conservation. Moreover, by convention, the PNV stage might be also taken as a reference to the whole vegetation series, i.e. to stand for the name of it. If we might embrace nomenclature for series, the continental European vegetation scientist usually does this by adding a Latin designation—*sigmetum*—to the formal name of the PNV vegetation type (i.e. that according to the International Code of Phytosociological Nomenclature, Weber et al. 2000). E.g. *Arisaro simorrhini-Querco broteroi sigmetum* stands for the whole set of vegetation-types that might be included in the usual set of successional stages directly associated with the forest community *Arisaro simorrhini-Quercetum broteroi* (portuguese oak mesophytic mediterranean thermophile subhumid forests) in central Portugal on limestone dominated by *Quercus faginea* subsp. *broteroi*; for details see Costa et al. (2012).

Finally, in this paper we do not address other levels of ecological complexity, that in contemporary Landscape Phytosociology are also taken into account; namely, sequences of permanent communities along gradients, sequences of vegetation series along gradients, vegetation series not fully developing successionally due to habitat constraints or because of cyclic disturbance (as an example of the first, series in rock outcrops; as an example of the second, vegetation in avalanche pathways or under regular catastrophic floods where succession is regularly taken 'downwards'). For definitions of these additional subjects, please refer to Rivas-Martínez (2005); for application of Landscape Phytosociology units to vegetation mapping, please see Biondi et al. (2011). Symphytosociology has developed a great deal in some European countries and elsewhere (Biondi et al. 2011; Blasi et al. 2005; Bioret 2010; Pedrotti 2013; Delbosc et al. 2015).

In this paper, by means of a case-study, we aim to test the hypothesis of S. Rivas-Martínez by addressing the following questions:

- Can distinct sets of coherent vegetation mosaics be obtained by classification of synrelevé tables excluding the putative PNV stage? Coherent mosaics would be clearly different from neighbors in terms of possession of a set of uniquely characteristic combinations of plant types. The later should have high constancy and fidelity to each group.
- Is there, then, in a *post hoc* analysis, a single PNV forest type associated to each group of synrelevés?

The case-study uses a set of synrelevés from traditionally low-intensity managed oak-forest mosaics of central-southern Portugal. The case-study may either suggest falsification of the above hypothesis or be one-in-many verifications of it.

1.2 Material and Methods

1.2.1 The Research Area

Woodland vegetation mosaics were sampled in the area of continental Portugal south of the Tagus River. The formal biogeographic units included are the Portuguese parts of the Luso-Extremadurese Subprovince of the West-Iberian Mediterranean Province (Rivas-Martínez 2014). Macrobioclimate is mediterranean, including both thermo- and mesomediterranean belts and ombrotypes dry to subhumid (Rivas-Martínez et al. 2011). Lithology is very diverse and includes hard crystalline silicates (schist, granite, sandstone, gabbro and basalt, metavulcanite), loose sand siliceous metasediments and paleodunes with podzols, and several types of limestone. Most mosaics are agro-forest parklands including extensive-grazing grasslands with scattered trees or small woodland nuclei, along with rotation crops and scrub in older fallow land. Small mountain ranges, rocky outcrops, steep river canyons or protected areas might bear mature or even close-to-pristine vegetation. Dominant trees are oaks: *Quercus suber*, *Q. rotundifolia*, *Quercus faginea* subsp. *broteroi*. PNV forest communities and sclerophyllous high scrub/forest hedges are all included in the evergreen mediterranean forest vegetation class *Quercetea ilicis* (order *Quercetalia ilicis* for forests and *Pistacio-Rhamnetalia alaterni* for high scrub; in turn, the former subdivides in two forest alliances: *Quercion broteroi*—thermo-mesomediterranean) and *Querco-Oleion sylvestris*—strictly thermomediterranean). Several low scrub classes are present: *Calluno-Ulicetea* (Western Europe's heathlands) and *Cisto-Lavanduletea* (*Cistus* scrub.). Perennial grasslands are included in *Festuco-Brometea*; non-nitrogen prone annual grasslands are included in *Helianthemetea guttati*; and high-nitrogen annual communities (weed communities) are classified in *Stellarietea mediae* vegetation class (Mucina et al. 2016).

1.2.2 Sampling Design and Data Collection

One hundred synrelevés were distributed in the study area in proportion to sampling strata defined by thermotype, ombrotype, and lithology. Only biotopes that include mostly zonal vegetation were used. Therefore, riverine, active dune, salt marsh, rocky outcrops and wetlands mosaic complexes were not accounted for in this study. Plots were set in uniform physiographies corresponding to homogeneous topographical features and lithology (Biondi et al. 2011). Main topographical types distinguished were: hilltops, mid-slopes and foothills. Valley bottoms, because they usually bear azonal vegetation influenced by water bodies or water-tables were, as explained, not considered. Plots were, in principle, of rectangular shape i.e. those obtained by doubling in sequence an initial 10×10 m quadrat, sometimes excluding parts of borders if overlapping neighboring physiographic units. According to minimal-area procedure, areas ranged from *ca*. 1000 m^2 to 4000 m^2. Each synrelevé consisted of the list of communities associated to its abundance-cover in the Braun-Blanquet scale (Géhu and Rivas-Martínez 1980). The survey of each plot, focused mainly on distinct physiognomical vegetation units: woodland, high scrub or hedge, tall grassland, scrub, annual grassland; but also included other vegetation-types found in the plot, namely: rock vegetation, nitrogen-prone vegetation, synanthropic vegetation and crops. Furthermore, each community in the plot was itself sampled by the standard floristic relevé Braun-Blanquet method (Braun-Blanquet 1964), for *syntaxa* assignment and also for forest communities to be related to corresponding synrelevés in the ordinated table, *post hoc* (see Sect. 1.2.3). Names of *taxa* follow the checklist of Sequeira et al. (2011). The relevés can also be found in the SIVIM public database (SIVIM 2016). The syntaxonomical designation of communities listed as attributes in the synrelevés, to association level when possible, follows the syntaxonomical checklist of Portugal by Costa et al. (2012). Floristic combinations that are characteristic of associations and used to identify them were obtained elsewhere by Capelo (2007).

1.2.3 Data Analysis

The first step was to obtain an ordinated table of synrelevés using a modification of the procedure by Wildi (1989) and Wildi and Orloci (1990)—Table in Electronic Annex (Table S1). Braun-Blanquet indexes (Braun-Blanquet 1964) were first transformed to percent cover using the middle point of the interval scale. The *log (x + 1)* transformation was then applied to cover values. Both the synrelevés (columns) and vegetation-types (lines) were first classified by *minimum-variance clustering* (Ward's method), using for the first the *similarity ratio* (Campbell 1978; Wildi and Orloci 1990; Jongman et al. 1995) and Euclidian distance for vegetation-types (Wildi 1989). After both synrelevé and vegetation-type groups were obtained by cutting of the correspondent dendrograms, their respective classification was used in *Analysis of Concentration*—AOC (Feoli and Orlóci 1979). AOC was used to set the relative position of synrelevé and vegetation-type groups in the table so as to obtain a diagonal structure. Detrended Correspondence Analysis—DCA (Hill and Gauch 1980) was performed and the rank of its scores used to set the order of synrelevés inside each group. Ordering of vegetation-types (lines) inside each group was made by ranking the *phi-coeficient mean square contingency* (Chytrý et al. 2002; Tichý and Jason 2006). Lower *phi* vegetation-types, i.e. those non-discriminant of mosaic types, were moved to the bottom of the table as syn-characteristics of higher units. Also, lower *phi*-value ubiquitous annual synanthropic vegetation associated with nitrogen was also moved to the bottom of the table as non-discriminant of mosaic types. Note that a few annual communities may be nonetheless differential between mosaic types. Numerical operations and table editing were performed with the JUICE© program by L. Tichý vers. 7.0 (Tichy and Holt 2006); classifications, AOC and DCA were performed by means of program MULVA 4.0 by O. Wildi (1989).

The formal elementary mosaic unit used in the table (Table in Electronic Annex—Table S1), following Rivas-Martínez (2005), is the *synassociation* or *sigmetum*. It may be subdivided in *faciations*: variations of a vegetation series having the same PNV but differing in some seral stages according to distinct biogeography. Each of the later might even be subdivided in regional *synvariants* differing in a single seral stage. Taxonomy of series is not yet set among the practitioners of Symphytosociology. Therefore, the parsimonious solution is to follow the PNV communities syntaxonomical affiliation. One can express the higher complexity level by adding 'syn-' to the alliance, order and class names. A group of similar series (*sigmetum*) is joined in a synalliance (*sigmion*). Similar synalliances are joined in synorders (*sigmetalia*) and synorders are joined in synclasses (*sigmetea*). For the case-study, the *sigmeta* (plural of *sigmetum*) are joined among two synalliances: *Querco broteroi sigmion* and *Querco-Oleo sylvestris sigmion*. Both are included in *Querco ilicis sigmetalia* which, in turn are included in the synclass *Querco ilicis sigmetea*. Faciations of *sigmetum* are denominated by an unambiguous dominant species in the seral stage that is differential of it (e.g. *Aro neglecti-Querco suberis sigmetum* faciation of *Stauracanthus genistoides*, because the vegetation-type *Thymo capitellati-Stauracanthetum genistoidis* is differential to the other faciation with *Stauracanthus vicentinus*). One of them is taken to be the typical one.

The second step was to associate in the table, and in a passive fashion, the floristic relevés of woodland communities found in each synrelevé (a single floristic relevé to each corresponding synrelevé). The aim was to see if floristic relevés can be re-organized coherently in the original forest floristic associations, or else be dispersed throughout mosaic types. Re-composing of floristic combinations that are characteristic of those forest types and also coherence with previous relevé classification to a single forest association (Capelo 2007) were the criteria used for matching. No consensus measure was used; rather matching was evaluated by visual inspection of the table. In the case that 'one-forest-to-one-mosaic' is the emergent pattern, it will mean that a single forest type associates to a single mosaic type, for the universe studied. This would verify Rivas-Martínez (1976) definition, i.e. a vegetation series being a coherent set of alternative vegetation states, related by succession, corresponding to a single PNV vegetation type.

1.3   Results and Discussion

By analyzing the syn-phytosociological table (Table 1.1) and inspecting also the full table (Table S1: including also the floristic relevés, which is in the electronic annex), it can be seen that:

- a clear coherent 'packet' diagonal structure is found, i.e. clear-cut characteristic combinations of vegetation-types correspond almost one-to-one to synrelevé groups (mosaic types or syn-associations).
- by *post hoc* matching, the classification of forest associations that results from respective floristic relevés' re-arrangement produces a very good one-to-one correspondence: *each mosaic type (synassociation) corresponds to a single forest association (PNV)*.

Briefly, the vegetation series that are found in the study-area are characterized, according to Tables 1.1 and S1, as follows:

- *Aro neglecti-Querco suberis sigmetum*
  Thermomediterranean dry to subhumid hyperoceanic series of cork-oak (*Quercus suber*) forests on Pleistocene sandy substrata: paleodunes with podzol horizons and some influence of a variable water table. Characteristic seral stages are: (1) high scrub/hedge: *Asparago aphylli-Myrtetum communis* (tall myrtle scrub), *Junipero-Quercetum lusitanicae* (shrubby oak tall scrub) and *Daphno gnidi-Juniperetum navicularis* (Tagus juniper tall scrub); (2) tall grassland: *Chamamelo mixti-Vulpietum alopecuroris*; (3) low scrub: *Thymo capitellati-Stauracanthetum genistoidis* (Sado River basin), *Stipo-Stauracanthetum vicentini* (coastal SW); (4) annual grasslands: *Anthyllido-Malcolmion lacerae*

Table 1.1 Synthetic table of vegetation series including faciations

Vegetation series and faciations	1a	1b	2a	2b	3a	3b	4a	4b	4c	4d	4e	5	6	7
Syn-caracteristics of series and macroseries *Querco broteroi sigmion*, *Querco-Oleo sylvestris sigmion*														
Com. of *Anthyllido-Malcolmion lacerae*	■													
Asparago-Myrtetum communis	■													
Junipero-Quercetum lusitanicae	■													
Daphno-Juniperetum navicularis	■													
Thymo-Sauracanthetum genistoidis	■													
Stipo-Stauracanthetum viventini		■												
Erico-Quercetum lusitanicae	■													
Cisto-Ulicetum argentei typicum			■											
Querco-Sauracanthetum boivinii				■										
Bupleuro fruticosi-Arbutetum unedonis				■										
Senecio-Cheirolophetum sempervirentis					■									
Phillyreo-Arbutetum unedonis typicum					■									
Erico-Cistetum ladaniferi							■							
Polygalo-Cistetum populifolii						■								
Cisto-Ulicetum minoris						■								
Hyacinthoido-Quercetum cocciferae								■		■				
Genisto-Cistetum ladaniferi typicum								■						
Clematido-Pistacietum lentisci									■					
Ulici eriocladi-Cistetum ladaniferi										■				
Asparago-Calicotemetum villosae										■				
Asparago-Rhamnetum quercetosum cocciferae											■			
Asparago-Rhamnetum oleetosum sylvestris											■			
Phlomido-Juniperetum turbinatae											■			
Genistetum polianthi											■			
Genisto-Cistetum cistetosum monspeliensis										■				
Asparago-Quercetum cocciferae												■		
Lavandulo-Cistetum albidi												■		
Velezio-Anteriscetum aquaticae								■						
Com. of *Brachypodion phoenicoidis*												■		
Melico-Quercetum cocciferae													■	
Phlomido-Cistetum albidi														■
Syn-characteristics of *Querco ilicis sigmetea* & *Querco ilicis sigmetalia*														
Trifolio-Plantaginetum bellardii	■	■	■		■		■							
Clinopodio-Origanetum virentis	■		■		■									
Retamo-Cytisetum bourgaei								■						
Pimpinello-Origanetum virentis			■											
Phillyreo-Arbutetum quercetosum cocciferae	■							■						
Genisto-Cistetum variant of *Genista triacanthos*					■									
Community of *Trachynion dystachiae*														■
Halimio ocymoidis-Cistetum psilosepali	■		■											

(continued)

Table 1.1 (continued)

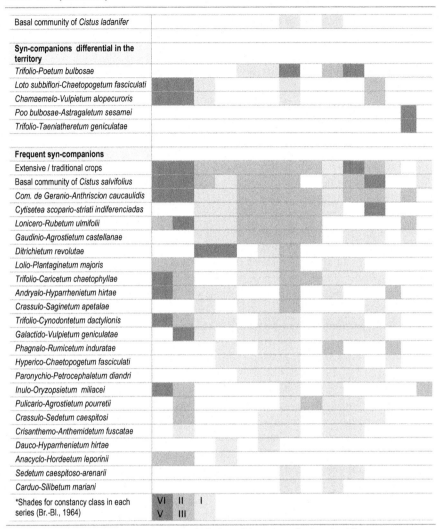

Basal community of *Cistus ladanifer*
Syn-companions differential in the territory
Trifolio-Poetum bulbosae
Loto subbiflori-Chaetopogetum fasciculati
Chamaemelo-Vulpietum alopecuroris
Poo bulbosae-Astragaletum sesamei
Trifolio-Taeniatheretum geniculatae
Frequent syn-companions
Extensive / traditional crops
Basal community of *Cistus salvifolius*
Com. de Geranio-Anthriscion caucaulidis
Cytisetea scopario-striati indiferenciadas
Lonicero-Rubetum ulmifolii
Gaudinio-Agrostietum castellanae
Ditrichietum revolutae
Lolio-Plantaginetum majoris
Trifolio-Caricetum chaetophyllae
Andryalo-Hyparrhenietum hirtae
Crassulo-Saginetum apetalae
Trifolio-Cynodontetum dactylionis
Galactido-Vulpietum geniculatae
Phagnalo-Rumicetum induratae
Hyperico-Chaetopogetum fasciculati
Paronychio-Petrocephaletum diandri
Inulo-Oryzopsietum miliacei
Pulicario-Agrostietum pourretii
Crassulo-Sedetum caespitosi
Crisanthemo-Anthemidetum fuscatae
Dauco-Hyparrhenietum hirtae
Anacyclo-Hordeetum leporinii
Sedetum caespitoso-arenarii
Carduo-Silibetum mariani
*Shades for constancy class in each series (Br.-Bl., 1964)   VI II I   V III

1a *Aro neglecti-Querco suberis sigmetum* faciation of *Stauracanthus genistoides*
1b *Aro neglecti-Querco suberis sigmetum* faciation of *Stauracanthus vicentinus*
2a *Lavandulo viridis-Querco suberis sigmetum* faciation of *Stauracanthus boivinii*
2b *Lavandulo viridis-Querco suberis sigmetum* faciation of *Ulex argenteus*
3a *Sanguisorbo agrimonioidis-Quercetum suberis sigmetum* typical faciation
3b *Sanguisorbo agrimonioidis-Quercetum suberis sigmetum* fac. of *Quercus marianica*
4a *Pyro bourgaeanae-Querco rotundifoliae sigmetum* typical faciation
4b *Pyro bourgaeanae-Querco rotundifoliae sigmetum* faciation with *Erica australis*
4c *Pyro bourgaeanae-Querco rotundifoliae sigmetum* faciation with *Pistacia lentiscus*
4d *Pyro bourgaeanae-Querco rotundifoliae sigmetum* faciation with *Ulex eriocladus*
4e *Pyro bourgaeanae-Querco rotundifoliae sigmetum* faciation with *Calicotome villosa*
5 *Myrto communis-Querco rotundifoliae sigmetum*
6 *Rhamno laderoi-Querco rotundifoliae sigmetum*
7 *Arisaro simorrhinii-Querco broteroi sigmetum*

(dry soil surface) and *Loto subbiflori-Chaetopogetum fasciculati* (temporary wet soil).

- **Lavandulo viridis-Querco suberis sigmetum**
  Thermomediterranean subhumid to humid hyperoceanic mixed cork-oak (*Quercus suber*, *Q. marianica*, *Q. faginea* ssp. *broteroi*) forests series in hard silicate rock cambisols (schist, sienite, sandstone). Characteristic seral stages are: (1) high scrub/hedge: *Bupleuro fruticosi-Arbutetum unedonis* (strawberry-tree tall scrub); (2) tall forb/grassland: *Senecio lopezii-Cheirolophetum sempervirentis*; (3) low scrub: *Querco-Stauracanthetum boivinii* (heathland, humid) or *Cisto-Ulicetum argentei* (gorse scrub, subhumid); (4) annual grassland: *Trifolio cherleri-Plantaginetum bellardii* (shared with other series).

- **Sanguisorbo agrimonioidis-Querco suberis sigmetum**
  Mesomediterranean subhumid to humid oceanic to semi-continental cork-oak (*Quercus suber*) forests series on hard silicate (schist, granite, greywacke). Characteristic seral stages are: (1) high scrub/hedge: *Phillyreo angustifoliae-Arbutetum unedonis* (strawberry tree tall scrub); (2) tall grassland: *Trifolio subterranei-Poetum bulbosae* (usually grazed, shared with series #4); (3) low scrub (heath/cistus): *Erico australis-Cistetum ladaniferi* (low altitude), *Polygalo microphylli-Cistetum populifoli* (mid altitude), locally other heathland associations appear; (4) annual grassland: *Trifolio cherleri-Plantaginetum bellardii*.

- **Pyro bourgaeanae-Querco rotundifoliae sigmetum**
  Mesomediterranean dry to subhumid oceanic to semi-continental live-oak (*Quercus rotundifolia*) forests' series on hard silicate (schist, granite, greywacke).
  Characteristic seral stages are: (1) high scrub/hedge: *Hyacinthoido hispanicae-Quercetum cocciferae* (kermes-oak tall scrub), locally in sienite *Asparago-Calicotemetum villosae*; (2) tall grassland: *Trifolio subterranei-Poetum bulbosae* (usually grazed, shared with series #3); (3) low scrub: *Genisto hirsutae-Cistetum ladaniferi* (Baixo Alentejo cistus scrub) and *Ulici eriocladi-Cistetum ladaniferi* (Alto Alentejo cistus scrub); (4) annual grassland: *Trifolio cherleri-Plantaginetum bellardii*.

- **Myrto communis-Querco rotundifoliae sigmetum**
  Thermomediterranean dry to lower subhumid hyperoceanic mixed live-oak forests (*Quercus rotundifolia*, *Olea europaea* ssp. *sylvestris*) series on crystalline hard silicates (schist, sienite, sandstone). Characteristic seral stages are: (1) high scrub/hedge: *Asparago albi-Rhamnetum oleoidis* with oleaster, *Phlomido purpureae-Juniperetum turbinatae*; (2) tall grassland: absent; (3): low scrub: *Genistetum polyanthi* (rocky leptosols), *Genisto hirsutae-Cistetum ladaniferi cistetosum monspeliensis* (soft leptosols); (4) annual grassland: *Trifolio cherleri-Plantaginetum bellardii*.

- **Rhamno laderoi-Querco rotundifoliae sigmetum**
  Mesomediterranean subhumid to humid oceanic live oak (*Quercus rotundifolia*) forests series on hard metamorphic paleozoic limestone. Characteristic seral stages are: (1) high scrub/hedge: *Asparago albi-Quercetum cocciferae* (kermes

oak high scrub); (2) tall grassland: community of *Brachypodium phoenicoides/ Poo-Astragaletum sesamei* (if grazed); (3) low scrub: *Lavandulo pedunculatae-Cistetum albidi* (cistus scrub); (4) annual grassland: *Velezio rigidae-Asteriscetum aquaticum* (shared with #7).

- **Arisaro simorrhini-Querco broteroi sigmetum**
  Thermomediterranean subhumid hyperoceanic portuguese oak (*Quercus faginea* ss. *broteroi*) forests series on mesozoic dolomitic limestone. Characteristic seral stages: (1) high scrub/hedge: *Viburno tini-Quercetum rivasmartinezii* (tall hedge), *Melico arrectae-Quercetum cocciferae* (high scrub); (2) tall grassland: *Phlomido lychnitidis-Brachypodietum phoenicoidis*; (3) low scrub: *Phlomido purpureae-Cistetum albidi*; (4) annual grassland: *Velezio rigidae-Asteriscetum aquaticum* (shared with #6).

Performing a generalist interpretation of the landscapes' ecological context in relation to the formal syn-phytosociological tables (Tables 1.1 and S1), it might be noted that 'core' syn-characteristic vegetation-types are those associated with progressive successions following moderate disturbance events of physical nature (a.k.a 'degradation stages' in phytosociological jargon), e.g.: individual tree cutting facilitates tall hedges/scrubs; severe cutting down combined with fire and extensive grazing promotes tall grassland followed by phases of shrub encroachment. The syntaxonomical identity of the total set of characteristic vegetation-types, for each series, seem to be determined by biogeographical and natural environmental contexts (regional species pool, bioclimate-lithology combinations, respectively); whereas human-induced succession seems to determine the concrete mosaic configurations by means of selective disturbance intensity. On the other hand, under the more intensive agriculture (more nitrogen, intense tillage, herbicides and irrigation), the species pool becomes more generalist, thus less liable to be organized in discriminant vegetation-types. Thus, synanthropic generalist weed communities or, on the other hand, widespread annuals, tend to be shared by more and more series. Therefore, in addressing classifications of the Landscape that are focused in natural drivers (vs. human-induced), mosaic classifications should focus, for reliance, on forest, pre-forest, at the most adding low-nutrient input induced phases: scrub, non-nitrophile grasslands.

1.4 Conclusions

We may assume that the thesis of S. Rivas-Martínez is verified, i.e. his model of vegetation series is not falsified and therefore is a good model for the universe studied. The evidence obtained is regional but biogeographically representative for the SW part of the Iberian Peninsula. Moreover, taking into account information from literature (see Ichter et al. 2014 for an overview in Europe; for North America see Rivas-Martínez 1997 and Rivas-Martínez et al. 1999), we suggest that further evidence needs to be collected which extends to other Eurosiberian territories that might also verify similar correspondences and therefore tests the goodness of the

Vegetation Series model. Functional traits involved in the assembly of each stage of vegetation series are not addressed here, but this also might be an important future line of research to consider. Despite criticism, mostly based on inductive nature of the sampling method and also in its base assumptions seeming putatively too simplified at first glance (linear succession, finalism), the concept of vegetation series *sensu* Rivas-Martínez and encompassing that of PNV is, on the contrary, undoubtedly a far-reaching one when put into test against real-world field conditions.

References

Anderson MJ et al (2011) Navigating the multiple meanings of β diversity: a roadmap for the practicing ecologist. Ecol Lett 14:19–28. https://doi.org/10.1111/j.1461-0248.2010.01552.x
Béguin C, Hegg O (1975) Quelques associations d'associations (sigmassociations) sur les anticlinaux jurassiens recouvertes d'une vegetation naturelle potentielle (essai de l'analyse scientifique du paysage). Doc Phytosoc 9–14:9–18
Béguin C, Hegg O (1976) Une sigmassociation remarquable au pied du premier anticlinal jurassin. Doc Phytosoc 15–18:15–24
Biondi E, Casavecchia S, Pesaresi S (2011) Phytosociological synreleves and vegetation plant mapping: from theory to practice. Plant Biosys 145(2):261–273
Bioret F (2010) Un siècle de phytosociologie sigmatiste en France: du temps des pionniers aux applications modernes. Braun-Blanquetia 46:27–40
Blasi C, Caporti F, R. (2005) Defining and mapping typological models at the landscape scale. Plant Biosyst 139(2):155–163
Bolòs O (1963) Botánica y geografía. Mem Real Acad Cienc Artes Barc 34:443–480
Bolòs O (1984) Plant landscape (phytotopography). In: Kuhbier H, Alcover JA, Guerau T (eds) Biogeography and ecology of the Pityusic Islands. Dr. W. Junk Publisher, The Hague, pp 185–221
Braun-Blanquet J (1951) Pflanzensoziologie, 2nd edn. Springer, Wien, p 631
Braun-Blanquet J (1964) Pflanzensoziologie, 3rd edn. Springer, Wien, p 865
Braun-Blanquet J, Furrer E (1913) Remarques sur l'étude des groupements de plantes. Bull Soc Languedocienne de géographie 1913:20–41
Braun-Blanquet J, Pavillard J (1922) Vocabulaire de sociologie végétale, 1st edn. Roumégous & Déhan, Montpelier, p 16
Campbell B (1978) Similarity coefficients for classifying relevés. Vegetatio 37:101–109
Capelo J (2007) Nemorum Transtaganae Descriptio. Syntaxonomia numérica das comunidades florestais e pré-florestais do Baixo Alentejo. PhD Thesis, Technical University of Lisbon, 529 pp
Carrión JS, Fernández S (2009) The survival of the 'Natural Potential Vegetation' concept (or the power of tradition). J Biogeogr 36:2202–2203
Chiaruccii A et al (2010) The concept of potential natural vegetation: an epitaph? J Veg Sci 21:1178–1128
Chytrý M, Tichý L, Holt J, Botta-Dukát Z (2002) Determination of diagnostic species with statistical fidelity measures. J Veg Science 13:79–90
Clements FE (1916) Plant sucession: an analysis of the development of vegetation. Carnegie Institution, Washington DC
Costa JC et al (2012) Vascular plant communities in Portugal (continental, Azores and Madeira). Glob Geobot 2:1–180
Delbosc P, Bioret F, Panaïotis C (2015) Les séries de végétation de la vallée D'Asco (typologie et cartographie au 1 :25.000). Ecol Mediter 41(1):5–87

Dovčak M, Frelich LE, Reich PB (2005) Pathways in old-field sucession to white-pine: seed rain, shade and climate effects. Ecol Monogr 75(3):363–378

Du Rietz GE (1936) Classification and nomenclature of vegetation units 1930–1935. Sven Bot Tidskr 30:580–589

Feoli E, Orlóci L (1979) Analysis of concentration and detection of underlying factors in structured tables. Vegetatio 40:49–54

Géhu J-M (1974) Sur l'emploi de la méthode phytosociologique sigmatiste dans l'analyse, la définition et cartographie des paysages. CR Acad Sci Paris 379:1167–1170

Géhu J-M (1976) Sur les paysages végétaux ou sigmassociations des pariries salées du Nord-Ouest de la France. Doc Phytosociol 15–18:57–62

Géhu J-M (1977) Le concept de sigmassociation et son application a l'étude du paysage vegetal des falaises atlantiques françaises. Vegetatio 34:117–125

Géhu J-M (1979) Pour une approche nouvelle des paysages végétaux: la symphytosociologie. Bull Soc Bot France 126:213–223

Géhu J-M (1986) Des complexes des groupements végétaux à la phytosociologie paysagère contenporaine. Inf Bot Ital 18:53–83

Géhu J-M, Rivas-Martínez S (1980) Notions fondamentales de phytosociologie. Ber Int Symp Int Vereinigung Vegetationsk 1980:5–33

Habeck JR, Much RW (1973) Fire dependent forests in the northern Rocky Mountains. Quat Res 3 (3):408–424

Hill MO, Gauch HG (1980) Detrended correspondence analysis, an improved ordination technique. Vegetatio 42:47–58

Humboldt A (1814–1825) Personal narrative of a journey to the equinoctial regions of the new continent [Trad. Jason Wilson 1995. Penguin Classics], 310 p

Ichter J, Evans D, Richard D (2014) Terrestrial habitat mapping in Europe: an overview. European Environment Comission, 154 p

Jongman RHG, Ter Braak CFJ, Van Tongeren OFR (1995) Data analysis in community and landscape ecology. Cambridge University Press, Cambridge, p 299

Loidi J, Fernández-González F (2012) Potential natural vegetation: reburying or reboring? J Veg Sci 23(3):596–604

Loidi J et al (2010) Understanding properly the 'potential natural vegetation' concept. J Biogeogr 37(11):2209–2211

Mucina L et al (2016) Vegetation of Europe: hierarchical floristic classification system of vascular plant, bryophyte, lichen, and algal communities. Appl Veg Sci 19(S1):3–264

Pedrotti F (2013) Plant and vegetation mapping. Springer, Berlin, p 294

Peterson WD, Reich PP (2001) Prescribed fire in oak savanna: fire effects on stand structure and dynamics. Ecol Appl 11(3):913–917

Rivas-Martínez S (1976) Sinfitosociología, una nueva metodología para el estudio del paisage vegetal. An Inst Bot Cavanilles 33:179–188

Rivas-Martínez S (1997) Syntaxonomical synopsis of the potential natural plant communities of North America, I. Itinera Geobot 10:5–148

Rivas-Martínez S (2005) Notions on dynamic-catenal phytosociology as a basis of landscape science. Plant Biosyst 139(2):135–114

Rivas-Martínez S (2014) Biogeography of Spain and Portugal. Preliminary typological synopsis. Int J Geobot Res 4:1–64

Rivas-Martínez S, Sánchez-Mata D, Costa M (1999) North american boreal and western temperate forest vegetation (syntaxonomical synopsis of the potential natural plant communities of North America II). Itinera Geobot 12:5–316

Rivas-Martínez S, Rivas Sáenz S, Penas Merino A (2011) World bioclimatic classification system. Global Geobotany 1:1–634. + 4 maps

Schithüsen J (1968) Allgemeine Vegetationsgeographie. Lerbuch der Allgemeinen Geographie 4, Berlin

Schwabe A (1997) Sigmachorology as a subject of phytosociological research: a review. Phytocoenologia 27(4):463–507

Sequeira M et al. (2011) Checklist da Flora de Portugal (continental, Açores e Madeira). ALFA, The Portuguese Association of Phytosociology, 74 p. https://bibliotecadigital.ipb.pt/bitstream/10198/6971/4/2011%20Checklist%20da%20Flora%20de%20Portugal.pdf. Accessed 23 Jun 2016

Sivim (2016) Sistema de Información de la Vegetación Ibérica y Macaronésica [database]. http://www.sivim.info/sivi/. Accessed 23 Jun 2016

Theurillat JP (1992) L'analyse du paisage végétale en symphytocoenologie: ses niveaux et leurs domaines spatiaux. Bull Ecol 23:83–92

Tichy L, Holt J (2006) JUICE Program for management, analysis and classification of ecological data. Masaryk University, Brno Czech Republic, p 103

Troll C (1968) Landschaftsökologie. Ber Int Symp Int Vereinigung Vegetationsk 1963:1–21

Tüxen R (1973) Vorschlag zur Aufnahme von Gesellschaftkomplexen in potentiell natürlichen Vegetationsgebieten. Acta Bot Acad Sci Hung 19:379–384

Tüxen R, Preising E (1942) Grundbegriffe und Methoden zum studium der Wasser und Sumpfpflanzen-Gesellschaften. Dtsch Wasserw 37:10–69

Vigo J (1998) Some reflections on geobotany and vegetaion mapping. Acta Bot Barc 45:535–556

Weber HE, Moravec J, Theurillat JP (2000) International Code of Phytosociological Nomenclature 3rd ed. J Veg Sci 11(5):739–768

Whittaker RH (1960) Vegetation of the Siskiyou Mountains, Oregon and California. Ecol Monogr 30:279–338

Wildi O (1989) New numerical solution to traditional phytosociological tabular classification. Vegetatio 81:95–106

Wildi O, Orlóci L (1990) Numerical exploration of community patterns. SPB Academic Publishing, The Hague, p 124

Part II

Global Aspects of Plant Function

Latitudinal Variation in Plant Functional Types

Andrew N. Gillison

Abstract
Relationships between species richness, diversity and latitudinal gradients present "the oldest problem in ecology and biogeography" (Hawkins 2008). Overall biotic richness increases toward equatorial regions, but underlying explanations for this trend are inconclusive. One reason is that comparative studies of latitudinal variation at global scale are limited by a pervasive lack of uniformity in purpose and scale of data collection and analysis. This chapter explores patterns of functional diversity along latitudinal gradients across all major biomes using a global database in which all data were collected using a uniform survey protocol (VegClass). Plant functional attributes are considered at two levels: individual traits (plant functional elements or PFEs) and whole-plant syndromes or Plant Functional Types (PFTs). Together with species richness, the data reveal departures from a commonly assumed latitudinal trend that are manifested by as yet unexplained mid-latitudinal peaks. Spatial patterning between PFTs and functional traits and certain environmental factors (mainly climate and substrate) across multiple scales exhibit non-linear relationships that are generally consistent with known responses of plant functional characteristics. Other key factors that may contribute to latitudinal variation in PFTs are briefly reviewed including nutrient stoichiometry, differences in continental gene pools and land use history.

Keywords
Latitudinal gradients · PFT global database · VegClass · Stoichiometry · Metabolic theory · Functional modus · Cortical photosynthesis · Functional leaf · Plant strategies

---

A.N. Gillison (✉)
Center for Biodiversity Management, Yungaburra, QLD, Australia

© Springer International Publishing AG, part of Springer Nature 2018
A.M. Greller et al. (eds.), *Geographical Changes in Vegetation and Plant Functional Types*, Geobotany Studies,
https://doi.org/10.1007/978-3-319-68738-4_2

2.1 Introduction

2.1.1 Historical Evidence

Some of the earliest attempts to record changes in species diversity along biophysical gradients at differing biogeographic scales (Forster 1778; von Humboldt 1808) were followed by suggestions from Darwin (1859) and Wallace (1878) that adaptations in the temperate zone were due mainly to a limiting climate, whereas those in tropical regions were caused by biotic interactions. Raunkiær and Gilbert-Carter (1937), on the other hand, contended that because a tropical climate was optimal for plant growth, the tropics could be regarded as the starting point from which adaptation of species to temperate climates originated between its center of origin and dispersal at tropical latitudes, that, according to Axelrod (1959) ranged between 45°N and 45°S in pre-Cretaceous time. Overall there is a general acceptance that species richness increases towards the equator, with this pattern more evident at regional than at local scale. Evidence to date also suggests that this trend is consistent in both northern and southern hemispheres although lacking symmetry around the equator (Hillebrand 2004; Mittelbach et al. 2007). To date the global data are extremely patchy, being collected by different workers for diverse purposes using different methods and represent varying environmental and geospatial scales. It is therefore hardly surprising that a universal theory that explains the reasons underlying a latitudinal biodiversity gradient remains elusive. The aim of this chapter is, first, to review findings that contribute to a better understanding of factors that determine observed patterns of biodiversity along latitudinal gradients; and, second to provide recent evidence from a unique global database of plant functional types that both supports and contradicts some leading assumptions about the drivers of biodiversity gradients at regional and global scale.

2.1.2 Underlying Hypotheses for a Latitudinal Gradient in Biodiversity

Within ecology no search for a generalizable theory has attracted greater attention or more ongoing debate than the search for key determinants of the distribution of taxa along latitudinal gradients. Despite current advances, relationships between species richness, diversity and latitudinal gradients still present "*the oldest problem in ecology and biogeography*" (Hawkins 2008). A latitudinal gradient in biodiversity has persisted since the evolution of the world's biotas, yet how and why this gradient arose remains unresolved. Theories and hypotheses proposed for this phenomenon have been proposed by plant and animal ecologists for more than two centuries. Many evolutionary hypotheses for higher diversification rates in tropical than temperate regions have been extensively reviewed by Mittelbach et al. (2007) and focus on two key elements:

Higher speciation rates due to:

1. Genetic drift in small populations accelerating evolutionary rates,
2. Climatic variation resulting in higher speciation at lower latitudes,
3. Higher likelihood of parapatric and sympatric speciation in the tropics
4. Larger area of the tropics providing more opportunities for isolation,
5. Narrower physiological tolerances in tropical organisms which reduce dispersal across unfavourable environments
6. Higher temperatures which result in increased evolutionary speed,
7. Stronger biotic interactions which lead to greater specialization and faster speciation;

Lower extinction rates due to:

1. Stability of tropical climates reducing the chance of extinction
2. Larger tropical area resulting in larger populations,
3. Larger species ranges, and thus lower chance of extinction

2.1.3 Biophysical Drivers of a Latitudinal Diversity Gradient

According to Allen and Gillooly (2006), two general classes of mechanisms account for variation in biodiversity: rates of speciation, extinction and dispersal, referred to as dispersal–assembly mechanisms; and species differences, species (biotic) interactions and environmental heterogeneity, namely niche–assembly mechanisms. Their analysis suggested that latitudinal biodiversity gradients and speciation rates are highly correlated even when controlled for variation in sampling effort and increased equatorial habitat area. Although habitat area is a significant factor in taxon distribution with latitude, the precise effect is difficult to quantify due to other confounding factors (Rosenzweig and Sandlin 1997; Chown and Gaston 2000; Losos and Schluter 2000; Gilman 2007; Marini et al. 2012; Nogué et al. 2012). Current evidence suggests key drivers of ecological niche assembly along biodiversity gradients are inevitably connected with modes of speciation especially with respect to climate, and by association, latitude and elevation. So far, it has been established that, whereas niche conservatism (where many closely related taxa occupy the same or similar habitat) predominates in the temperate zone, niche divergence (many taxa occupying different habitats) is higher in the tropics where latitudinal differences in elevational climatic zonation may increase opportunities for geographical isolation, speciation, and the associated build-up of species diversity compared to the temperate zone (Kozak and Wiens 2007). While the interactive effects at different levels of elevational gradients within latitude continue to be debated, there is evident consensus that the relative strength of biotic and abiotic drivers on community assembly varies in a more or less predictable way with increasing complexity towards the equator (Stevens 1992; Nogué et al. 2012; Hulshof et al. 2013; Randin et al. 2013; Hsu et al. 2014). It is nonetheless, increasingly apparent that the widely accepted global spectrum of equator-centric biodiversity may be significantly mediated by environmental variability at regional scale. Latitudinal patterns and regionalization of plant diversity along a 4270-km gradient in continental Chile (Bannister et al. 2012) for example, revealed a unimodal pattern of plant richness at the family, genus and species level that contrasts with the global diversity gradient observed in the northern hemisphere.

2.1.4 Temperature, Elevation and Precipitation

When considered in isolation from other environmental factors, it is difficult to assess the value of elevation as a primary driver of biotic richness along environmental gradients. Rather, elevation serves as a general surrogate for physiologically more meaningful temperature that, in turn, is a reflection of both latitude and elevation (Nakamura et al. 2007; Randin et al. 2013; de Oliveira et al. 2015). Thermal-latitudinal relationships are widely assumed to be operating in predictive modelling of vegetation (Box 1996; Harrison and Prentice 2003), but disentangling temperature interaction from other biophysical traits requires more critical assessment than is usually the case. The translation of latitudinal clines of many plant life history traits into responses to temperature is therefore a crucial step where adaptive differentiation of populations and confounding environmental factors other than temperature need to be addressed (Graae et al. 2012; de Frenne et al. 2013). Another environmental factor often considered in association with temperature is precipitation. Although an important agent in supporting ecosystem performance at a range of scales, meaningful precipitation measurements can be problematic in the absence of other known sources of atmospheric (fog and mist) and groundwater supply. Nonetheless, on a global scale, Moles et al. (2009) found that one of the best predictors of plant height was precipitation in the wettest month. In many cases, underlying hydrology can be more significant than precipitation in influencing community dynamics (Lou et al. 2015), whereas precipitation can be closely associated with species distribution in harsh desertic environments when combined with other factors such as topographic heterogeneity (Zhang et al. 2015). A global analysis of trait variation in climbing plants (Gallagher and Leishman 2012) revealed significant relationships between annual temperature and precipitation and leaf size, seed mass and specific leaf area (SLA). The response pattern of plant functional types with respect to thermal, precipitation and latitudinal gradients is examined further in this chapter.

2.1.5 Substrate

The scope of this chapter mitigates against an adequate review of the highly significant role of substrate within the multiplicity of factors that together contribute to a latitudinal biodiversity gradient. In almost every situation, environmental drivers of biodiversity are to varying degrees affected by the nature of the substrate, especially with respect to soil nutrient-moisture availability (Bardgett 2005). A broad-ranging study of woodlands and shrublands (Naveh and Whittaker 1979) showed that, unlike recent Mediterranean areas, Gondwanan heath-like communities are adapted to very old, nutrient-poor soils and that these communities have some of the highest plant alpha diversities in the world because their floristic richness (especially in annual plants) results from relatively rapid evolution under stress by drought, fire, grazing, and cutting. Convergence in vegetation structure in the Mediterranean type communities of California, Chile and South Africa was found to be strongly influenced by soil nutrient availability by Cowling and Campbell (1980) who also maintained that much of the divergence between the South African fynbos and the vegetation of the other continents can be attributed to the nutrient-poor soils on which fynbos has evolved. Long-standing vegetation and soil chronosequences commonly show strong correlates among nutrient stocks, species and functional type diversity. Within a 120,000 year chronosequence on the Franz-Josef glacier of New Zealand, Mason et al. (2012a, b) noted that leaf N and P declined and leaf thickness and density increased directly with declining total soil P along the sequence. In central African moist forests, Gourlet-Fleury et al. (2011) concluded that soil physical conditions (notably texture) constrained the amount of biomass stored in tropical moist forests and that biomass is likely to be similar on resource-poor and resource-rich soils. For north-western Europe, Wasof et al. (2013) found that, with increasing latitude, most species of understorey plant species shifted their realized-niche[1] position for soil nutrients and pH towards nutrient-poorer and more acidic soils, suggesting local adaptation and/or plasticity. At global scale, an investigation of relationships between leaf traits, climate and soil measures of nutrient fertility by Ordoñez et al. (2009) showed that in a mainly forested environment, relationships of leaf traits to soil nutrients were stronger than those of growth form *versus* soil nutrients. In that study, climate determined distribution of growth form more strongly than it did leaf traits more or less confirming that climatic influence operates across differing temporal response scales of vegetation. From the above it is clear that much more needs to be done with respect to environmental and geospatial scale in formulating and testing generalizeable hypotheses about the role of substrate as a key driver in latitudinal biodiversity gradients.

---

[1]Factors influencing plant ecological niche space have long concerned ecologists. Hutchinson (1957) conceived the niche as a hyper-volume in multi-dimensional environmental space delimiting where stable populations can be maintained. When biotic interactions such as predation and competition are included in the calculation of niche space, one obtains the 'realized niche', as opposed to the 'fundamental' or 'physiological niche' that ignores such interactions (see also Kearney et al. 2010). Use of the term is widespread in plant ecology (Beck et al. 2005; Broennimann et al. 2006; McGill et al. 2006; Pellissier et al. 2013; Wasof et al. 2013; Serrano et al. 2015).

2.1.6 Productivity Gradients

Vegetation-based productivity is yet another trait that is known to vary with latitude, but here the picture also tends to be confounded with substrate effects. Temperature related plant-physiological stoichiometric processes and biogeographical gradients in soil substrate age are known to exhibit a related increase in leaf N and P from the tropics to the cooler and drier mid-latitudes while the N:P ratio increases towards the equator due to limiting P in older tropical soils and because N is a limiting nutrient in younger and high-latitude soils (Reich and Oleksyn 2004). Apparent paradoxical patterns across ecosystems are discussed by Enquist et al. (2007) who point out that, compared with warmer low latitudes, ecosystems in cold northerly latitudes are described firstly by a greater temperature normalized flux of CO_2 and energy; and secondly by similar annual values of gross primary production (GPP), and possibly net primary production (NPP). Reich and Oleksyn (2004) maintain that directional and adaptive changes in metabolic and stoichiometric traits of autotrophs may mediate patterns of plant growth across broad temperature gradients where, on average, mass-corrected whole-plant growth rates are not related to differences in growing season temperature or latitude. According to Elser et al. (2010), plant stoichiometry exhibits large-scale macroecological patterns, including stronger latitudinal trends and environmental correlations for phosphorus concentration. On the other hand, Gillman et al. (2014) also found that forest NPP is increasingly correlated with species richness and vascular plant richness towards the equator. According to these authors, this phenomenon is consistent with ecological theories that predict a positive relationship between species richness and productivity and thus obviate the need to explain peaked richness–productivity relationships over broad spatial extents "since they do not appear to exist".

By focusing on plant functional traits, and based on empirical data, Falster et al. (2011) found that the response of NPP to changes in site productivity in single species forests was larger than the response of NPP to trait variation when considering the influence of four key traits (leaf economic strategy, height at maturation, wood density, and seed size). In a parallel study involving models of functional trait performance at the scale of an earth system model, Verheijen et al. (2015) argue that the inclusion of ecologically-based trait variation reduced the projected level of a land carbon sink. In addition, trait-induced differences in productivity and relative respirational costs led to an stronger carbon sink in the mid- and high latitudes and to a stronger carbon source in the tropics.

2.1.7 Metabolic Theory

Whereas biological stoichiometry theory considers the balance of multiple chemical elements in living systems, metabolic scaling theory considers how size affects metabolic properties from cells to ecosystems (Elser et al. 2010). A temperature-based 'metabolic theory of ecology' (MTE) was proposed by Allen et al. (2002) as an explanation of a range of macroecological patterns, including latitudinal diversity gradients, by linking ecological and evolutionary processes to plant and animal metabolic rates. The theory has some attraction as, in principle, it makes precise predictions about the temperature-driven metabolic rates of biota. However, as with many other global data, the level of precision is critical and in reality may be open to question. Initial formulations of metabolic theory assumed a precise energy of activation (Allen et al. 2002), whereas, as indicated by Hawkins et al. (2007), other studies (Brown et al. 2003; Enquist et al. 2003) suggested that levels of precision may be highly variable. A comprehensive investigation by Hawkins et al. (2007) noted that, when richness decreases with increasing temperature, as occurs in many parts of the world, notably in mesic environments, (Rosenzweig and Sandlin 1997; Gaston 2000) the theory can be rejected unless energies of activation are allowed to assume biologically impossible values. Hawkins et al. (2007) concluded that "...*metabolic theory, as currently formulated, is a poor predictor of observed diversity gradients in most terrestrial systems*". From the foregoing it may be concluded that, while there are some clear lines of theoretical support for MTE, adequate testing along latitudinal gradients would be very difficult in the short term due to practical limitations and a continuing critical shortage of meaningful data and a lack of harmonization of data collection methods.

2.1.8 Rapoport's Rule and the Mid-Domain Effect

Rapoport's rule is an ecological hypothesis which states that *'latitudinal ranges of plants and animals are generally smaller at lower latitudes than at higher latitudes'* (see also Stevens 1989, 1992). Under this rule, species can have narrower tolerances in more stable (e.g. tropical) climates, leading to smaller ranges and allowing coexistence of more species. The rule has been generally dismissed as being over-simplistic and misleading by many authors including Sizling et al. (2009) who used a simple geometric model to show that the postulated decrease of species' potential range sizes toward the tropics would itself lead to a latitudinal gradient contrary to that observed. In contrast, an increase in extent of *potential* ranges toward the tropics would lead to the observed diversity gradient. Using simulated sampling from a parametric distribution of ranges that incorporates a richness gradient but no Rapoport effect, Colwell and Hurtt (1994) showed that a spurious Rapoport effect can be caused by sampling bias alone (e.g. with respect to sample area and unequal sampling effort), even when total sampling effort is equal at all latitudes or elevations. The effect has been compared with a null model—the Mid-Domain Effect (MDE) described by Colwell and Lees (2000) as a statistical pattern that is generated if species' latitudinal ranges are randomized within the geometric constraints of a bounded biogeographical domain. Under these conditions, species' ranges would tend to overlap more toward the center of the domain than towards its limits, forcing a mid-domain peak in species richness. At the time it was felt that this hypothesis, might lead to better insights about the latitudinal gradient in species richness. However, positive applications of the MDE are rare—one exception being a detailed study in the Guyana highlands by Nogué et al. (2012) who found that a combination of elevation, area and MDE contributed to a basic explanation for the diversity of vascular plants. Although biophysical constraints of the real world mitigate against the use of MDE, Gotelli and McGill (2006) argue that as a null model, MDE could be important in the evaluation of the more process-based, neutral model but stated that in most cases data are limiting—a view supported by Zapata et al. (2005). More strident critiques of the MDE include those of Hawkins et al. (2005) who argue the MDE misrepresents the nature of species ranges and that an internal logical inconsistency underpins the MDE because the range size frequency distribution cannot exist in the absence of environmental gradients. One of the few statistical analyses of the distribution of plant functional types and traits (Swenson et al. 2012) showed that the overall distribution of functional traits in tropical regions often exceeds the expectations of random sampling given the species richness, and that conversely, temperate regions often had narrower functional trait distributions. Swenson et al. (2012) concluded that while the overall distribution of function does increase towards the equator, the functional diversity within regional-scale tropical assemblages is higher than that expected given their species richness. Further, this outcome supports the hypothesis that physical limitations (abiotic filtering) constrain the overall distribution of function more in temperate than in tropical assemblages.

2.1.9 Differences Between Continents

Numerous studies show that any generalizeable theory which aims to explain underlying causality within the biodiversity-latitude gradient must consider potential intercontinental differences in biodiversity and biogeography and their evolutionary origins. Key intercontinental disparities in ecosystem performance have been demonstrated by many authors including Pausas et al. (2004) who showed that the pattern of correlation between two basic postfire persistent traits and other plant traits varies between continents and ecosystems. In their study of land use patterns along a series of disturbance gradients Rusch et al. (2003) found that repeatable patterns in the changes of plant functional traits along gradients are evident among continents and conclude that life-history classifications can be used for comparisons and predictions between regions and at global scales. In this context, a comparison of changes in plant and animal species diversity along land use intensity gradients in two biogeographically separate areas (Indonesia and Brazil) Gillison et al. (2013) revealed significant similarities and differences in both biotic and abiotic factors among correlations between plant functional types and species. The data from Gillison et al. (2013) further suggest that variation in intercontinental biodiversity gradients can be attributed to separate evolutionary lineages and gene pools with resulting differences in phylogeny and niche conservatism. Evidence for differential responses by species and plant functional types and traits to environments in different countries has been demonstrated by many others (Beadle 1951; Naveh and

Whittaker 1979; Cowling and Campbell 1980; Nakamura et al. 2007; Ordoñez et al. 2009; Gallagher and Leishman 2012; Banin et al. 2012; Tomlinson et al. 2013; Moncrieff et al. 2014; Rossetto et al. 2015; Yang et al. 2015). In summary, it would seem that significant differences in evolutionary pathways and in the continental gene pools, combined with subsequent and diverse impacts through global change, present clear challenges to studies that seek common threads in causal patterns of biodiversity response along latitudinal gradients.

2.1.10 Vegetation Structure and Latitude

2.1.10.1 Plant Height

Plant height (usually expressed as mean canopy height) is one of the most widely recorded and readily observable plant functional traits. Height exerts a major influence on functional processes in ecological succession especially in the competition for light and other biophysical resources. Although recorded across ecoregions and, more recently included in developing global databases such as TRY (Kattge et al. 2011) and the LEDA-Traitbase of Northwestern European flora (Kleyer et al. 2008), the global data are surprisingly patchy, being dominated by trees rather than a broader range of vegetation including herbs (*cf.* Klimešová et al. 2015). For this reason, studies that rely on data of diverse origins across latitudinal gradients raise questions about the validity of research outcomes. Along gradients of regional land use intensity, plant height is a powerful indicator of plant and animal diversity within and between continents (Gillison et al. 2013). Plant height also exerts significant influence on microclimate, hydrology, substrate and thus ecosystem performance. Regional studies including plant height are numerous although few specifically consider the functional role of height. For example, there are few plant height data in landscape mosics involving shifting agriculture, an exception being that of Lebrija-Trejos et al. (2008), who found that along a successional sequence in a dry tropical forested landscape, canopy height together with canopy cover recovered faster than most other traits. The first published comparative latitudinal study of plant height conducted by Moles et al. (2009) used maximum height data for 7084 plant species and found a remarkably direct relationship between latitude and height thus indicating a major difference in plant strategy between high and low latitude systems. In that study there was no evidence that the latitudinal gradient in plant height differed in either hemisphere, although a 2.4-fold drop in plant height at the edge of the tropics may indicate a switch in plant strategy between temperate and tropical zones.

2.1.10.2 Basal Area

As with plant height, basal area (BA) is widely recorded throughout ecological literature. As a functional trait BA is closely correlated with biomass density, above ground plant carbon, and thus constitutes an important element in allometric estimates of plant based carbon sinks and sources (Gillison et al. 2013; Verheijen et al. 2015). BA is commonly measured as diameter at breast height (DBH) with diameter thresholds varying from <5 cm to >30 cm; and then only applied to trees and not to other woody plants such as undershrubs or lianas. For the most part, BA tends to be used as a standard mensurational unit in forestry (Vanclay et al. 1997), although there are many instances of its measurement as an ecological factor in local, and to a lesser extent, regional studies. Apart from newly developing global databases derived from multiple sources, there are few independent investigations involving BA that can be applied at global latitudinal scale (but see: Banin et al. 2012). Comparative studies at intercontinental scale (Gillison et al. 2013) reveal that BA is a potentially powerful indicator of biodiversity. Also, as with plant height, the devil is clearly in the detail where estimates of BA rely on widely varying techniques (Gillison 2002; Lebrija-Trejos et al. 2008; Falster et al. 2011; Mason et al. 2012a, b; Hernández-Calderón et al. 2014; Barry et al. 2015). More particularly, comparative studies at world scale are frequently restricted to trees to the exclusion of other plant life forms. Given the background of inconsistency in BA measurement, its application in studies at latitudinal scale is, at the very least, questionable when estimates rely on multiple data sources.

2.1.11 Plant Life Forms and Latitudinal Gradients

Within the broad spectrum of plant life forms or growth forms, tree architecture tends to dominate global ecological studies of vegetation structure whereas other life forms and growth forms are rarely included. Yet all such elements are critical when assessing community ecology at almost any scale. Among the best studied non-tree structural forms at latitudinal scale are understorey plants, lianas and cushion plants and their ecological or realized niche. Despite long-standing observations by ecologists that the realized niche of understorey shrubs frequently extends beyond that of their overstorey counterparts, few studies have addressed this phenomenon in detail at latitudinal scale. In north-western Europe, Wasof et al. (2013) conducted just such a study of understorey plants along a latitudinal gradient. Among their findings were that while few species shifted their realized-niche width, all changed their realized-niche position and that, with increasing latitude, most species changed their realized-niche position according to the type of soil nutrients. Wasof et al. (2013) therefore argued that this macroecological pattern casts doubt on a key assumption of species distribution models, namely that the realized niche is stable in space and time.

The distribution of climbing plants or lianas has attracted increasing attention at both regional and global scale with implications for their occurrence along latitudinal gradients. At regional scale, in tropical environments, especially those that are subject to hurricane wind damage, lianas can exert significant influence along forest successional gradients (Gentry 1988; Murphy et al. 2014; Barry et al. 2015). A global analysis of climbing plants by Gallagher and Leishman (2012) found highly significant relationships between latitude and a range of plant functional traits.

These results were largely consistent with phylogenetic analyses across species where functional traits and phylogenetic patterns of climbers differ between biogeographical regions, and from other better studied plant growth forms. Species-level trait differences may hold the key to understanding why climbers are increasing in abundance in some regions of the world, but not in others. Environmental range limits to lianas are widely considered to be constrained by freeze and thaw effects on internal vascular structure. An examination of this phenomenon by Jiménez-Castillo and Lusk (2013) revealed that hydraulic performance of temperate lianas exceeds that of coexisting trees, in similar magnitude to tropical counterparts. The same authors point out that, under forecast scenarios of global warming, liana range could be predicted to increase along latitudinal gradients. Similar consequences of global warming on plants at environmental extremes such as cushion plants have been predicted by Cranston et al. (2015).

2.1.12 Leaf Life Span and Latitude

Plant leaf life span is most commonly reflected in the seasonal response of plants according to either a short-term (deciduous) or long-term (evergreen) life strategy. In one of the first classifications of plants in the Western world by Theophrastus (370–285 B.C.) (see Hort 1916), while categories of tree, shrub, under-shrub and herb were recognized, Theophrastus emphasized the difficulties of achieving precision in such definitions especially between tree and shrub, deciduous and non-deciduous. While the literature is replete with studies of this essentially seasonal phenomenon at local and regional scales, there are few sufficiently comprehensive studies that adequately explain how these two strategies relate to latitudinal gradients. Most literature sources focus on the northern hemisphere (Meher-Homji 1978; Niinemets and Kull 1998; Jankowska-Blaszczuk and Grubb 2006; Kleiman and Aarssen 2007; Markesteijn et al. 2007; Lebrija-Trejos et al. 2008; Lusk et al. 2008; Gillison 2012; Kikuzawa et al. 2013) with very few in the southern hemisphere (Prior et al. 2004). Here again, results will be influenced by data quality and precision, as many indeterminate stages can occur between these two extremes in addition to which vegetational studies frequently include only subsections of plant assemblages (e.g. trees) rather than complete communities, thus limiting adequate analyses. Further, as noted by Meher-Homji (1978), there are some functional differences between plants that are winter-deciduous and those that are drought deciduous. One of the few studies of deciduousness relative to latitude was undertaken by Kikuzawa (1991) who undertook a cost-benefit analysis of leaf habit and leaf longevity of trees and their geographical pattern and found a bimodal distributional pattern with two peaks, one at lower and the other at higher latitudes for the percentages of evergreenness while percentages of deciduousness showed a unimodal distribution pattern with a peak at midlatitude. New evidence that may counter such modality is discussed further in this chapter.

2.1.13 Plant Functional Types and Traits

In the evolution of ecological literature, plant functional types (PFTs) are a relatively recent innovation with wide-ranging studies rapidly increasing in the past two decades. An initial approach using plant functional attributes (Gillison 1981) paralleled the first development of mechanistic models of global classifications of structural-functional plant functional types by Box (1981, 1996), who constructed a set of pheno-physiognomically defined plant types associated with ecophysiological functional traits that could be related directly to climate variables such as water balance and evapotranspiration (see also Box and Fujiwara 2005). The model proposed by Box provided a framework for subsequent investigations of functional typology (Gillison and Carpenter 1997; Lavorel et al. 1997, 2007, 1999; Díaz et al. 2002). More recent applications focus on functional traits as they apply to plant functional strategies that have attracted a diverse array of proposed models. These models are described in more detail by Gillison (2013, 2016).

Definitions of PFTs are as numerous as they are diverse. In its simplest form PFTs can be regarded as *'functionally similar plant types.'* (Box 1996). Other PFT definitions can vary according to whether the response to an environment or the effects on an ecosystem, singly or both, are intended. Smith et al. (1992) defined PFTs as *'sets of species showing similar responses to the environment and similar effects on ecosystem functioning'*. This approach is supported by others (Díaz and Cabido 1997, 2001; Lavorel and Garnier 2002). PFTs represent specific assemblages of individual functional traits or plant functional elements (PFEs, Gillison 2002) and as such may be regarded as trait assemblages or trait syndromes. Elsewhere PFTs are commonly equated with life forms or growth forms (tree, shrub etc.) representing distinct vegetation units devoid of specific combinations of traits. A 'functional trait' has been described as *'Any measurable feature at the individual level affecting its fitness directly or indirectly'* (Albert et al. 2010). Further background as to trait characteristics is described in Gillison (2013) who defined a **functional trait** as *'any measureable plant trait with potential to influence whole-plant fitness'*—a definition retained in this chapter.

Ecologically important, correlated plant traits may be said to constitute an ecological 'strategy' when matched against trade-offs in investment (Westoby et al. 2002; Wright et al. 2007). In this context, **'plant strategy'** usually indicates a combination of plant characteristics that optimize trade-off in resource allocation patterns in order to achieve maximum growth rate, maximum size and maximum age along with the plant's growth response to different combinations of light and water availability (*cf.* Smith and Huston 1989; Fortunel et al. 2012). Different assemblages of plant functional traits, especially those described as orthogonal or independent, have been shown to contribute to a global leaf economic spectrum (LES, Wright et al. 2004) and a leaf-height-seed (LHS) strategy (Westoby 1998). These LES core traits are known to exhibit an invariant response to key climate factors (e.g. temperature and precipitation) at biome scale (Wright et al. 2004, 2007). However, their positioning along a latitudinal gradient, although inferred at biome level (Wright et al. 2004), has not been shown directly.

These and other strategies relating to functional traits including Grime's (2001) CSR strategy are described in more detail by Gillison (2013, 2016). A different strategy involving whole trait syndromes (PFTs) is briefly described here Gillison (2016) and is the basis for the the data analyses used in this chapter.

2.1.14 The Leaf-Life-Form-Root (LLR) Strategy and Its Relationship to Latitude

The LHS and LES strategies contribute useful information about the individual values of parsimonious, (independently functioning) traits that, although exhibiting a clear case for a biome-invariant pattern, ignore additional functional information at a 'whole-plant' level or as a functional 'Gestalt' that represents more than just the apparent sum of the functional parts. To this extent the LLR approach considers ways in which multiple traits can be used to construct plant functional types (PFTs) or trait syndromes via an assembly system that addresses whole-plant performance rather than economically-acquisitive, single traits such as leaf mass per area (LMA) or SLA. This is achieved in part by coupling photosynthetic traits with life form and above-ground rooting structures, and is consistent with observed stem-root interaction (*cf*. Fortunel et al. 2012). When coupled with additional information that describes vegetation structure, the methodology facilitates comparative analysis across a range of environmental scales (Gillison 1981, 2002, 2013). The LLR strategy complements significant gaps in the CSR, LHS and LES systems that otherwise exclude critical photosynthetic traits such as leaf inclination (Falster and Westoby 2003; Posada et al. 2009), leaf phyllotaxis or insertion pattern such as rosettes (Withrow 1932; Lavorel et al. 1998, 1999; Díaz et al. 2007; Ansquer et al. 2009; Bernhardt-Römermann et al. 2011) and green-stem photosynthesis, all of which constitute significant plant adaptations to irradiance, nutrition and water availability.

One of the best plant functional classifications that has stood the test of time is that of the life-form system of Raunkiær (1934). That system is built on a fundamental survival adaptation to cyclic environmental and edaphic (nutritional) extremes and remains in wide use partly because of its elegant simplicity and limited set of readily observable traits. In its basic form, however, and despite external reference to a table of leaf size classes, the Raunkiær model excludes photosynthetic traits. To resolve this issue but to retain the essential Raunkiær format, Gillison (1981) devised a whole-plant classification system based on plant functional attributes in which an individual plant is classified as a 'functionally coherent' unit composed of a photosynthetic 'envelope' supported by a modified Raunkiaer life form and an above-ground rooting system presented as the Leaf-Life form-Root' or LLR spectrum (Gillison 2013).

A fundamental principle of the LLR is that a combination of plant functional traits such as leaf size class assumes additional functional significance when combined with leaf-inclination and other morphological (e.g. leaf stomatal distribution) and temporal (e.g. deciduous) descriptors of photosynthetic tissue. In this case the photosynthetic attributes describe a 'functional leaf' that includes any part of the plant (including the primary stem cortex but excluding fruit) capable of photosynthesis. Unlike the LES and LHS systems, the LLR system also includes succulent vegetation (*ca*. 10,000 species, Oldfield 1997) that involves significant elements of world flora. For convenience, and to indicate the unique type of PFT, specific LLR combinations are termed functional *modi* (from *modus* Latin SM II, meaning mode or manner of behaviour) (compare also the "modality" of Violle et al. 2007). This initial model (Gillison 1981) was the first coordinated use of plant functional attributes (PFAs) or functional traits to relate whole-plant PFTs to environmental conditions. The method was subsequently formalized (Gillison and Carpenter 1997) using an assembly-rule set and syntactical grammar to construct *modal* PFTs based on 36 generic plant functional elements (PFEs) (Gillison and Carpenter 1997; Gillison 2002) (Table 2.1). In this method for example, a typical PFT *modus* for an individual of *Acer palmatum* might be a mesophyll (*me*) leaf size class with pendulous (*pe*) inclination, dorsiventral (*do*) (hypostomatous), deciduous (*de*) leaves with green-stem (cortex) (*ct*) photosynthesis supported by a phanerophyte (*ph*), the resulting *modal* PFT combination being *me-pe-do-de-ct-ph*.

Within the same species on the same or other site, variation in any one functional element, such as leaf size class, results in a new *modus*, thereby facilitating comparison of intra- as well as inter-specific variability at a described location. This can be especially useful where phenotypic expression within a species may be expressed in different modal combinations along an environmental gradient ranging, for example, from a dry ridge (e.g. small vertically inclined leaves) with skeletal soil to a river margin on alluvium (larger, laterally inclined leaves). With the public-domain VegClass software package (Gillison 2002), quantitative and statistical comparisons within and between species and plots are facilitated via a predetermined 'costing' of lexical distances between different PFTs (Gillison and Carpenter 1997). The system comprises a many-to-many mapping whereby more than one modal PFT can be represented within a species and *vice versa*. While ~7.2 million combinations are theoretically possible, a data set compiled using a standard recording proforma (Table 2.2) from 1138 field sites worldwide covering all major latitudes and climates indicates the 'real' number of unique modal PFTs approximates 3500 or 1% of the world's approximately 350,000 vascular plant species.

Functional diversity measures based on the abundance of species per PFT (Shannon H', Simpson's D and Fisher's Alpha) can also be generated on demand in VegClass, with the inverse of the PFT Simpson measure equating to Rao's quadratic entropy (Botta-Dukat 2005; Lepš et al. 2006; Mason et al. 2012a, b; Vandewalle et al. 2014). A separate measure of plant functional complexity (PFC), based on a minimum spanning-tree distance of PFT values within a transect (Gillison and Carpenter 1997), also provides a useful comparator between sites

Table 2.1 Plant functional attributes and elements

Attribute	Element	Description	
Photosynthetic envelope			
Leaf size	nr	No repeating leaf units	
	pi	Picophyll	<2 mm^2
	le	Leptophyll	2–25
	na	Nanophyll	25–225
	mi	Microphyll	225–2025
	no	Notophyll	2025–4500
	me	Mesophyll	4500–18,200
	pl	Platyphyll	18,200–36,400
	ma	Macrophyll	36,400–18 × 10^4
	mg	Megaphyll	>18 × 10^4
Leaf inclination	ve	Vertical	>30° above horizontal
	la	Lateral	±30° to horizontal
	pe	Pendulous	>30° below horizontal
	co	Composite	
Leaf chlorotype	do	Dorsiventral	
	is	Isobilateral or isocentric	
	de	Deciduous	
	ct	Cortic	(photosynthetic stem)
	ac	Achlorophyllous	(without chlorophyll)
Lf. morphotype	ro	Rosulate or rosette	
	so	Solid 3-D	
	su	Succulent	
	pv	Parallel-veined	
	fi	Filicoid (fern)	(Pteridophytes)
	ca	Carnivorous	(e.g. *Nepenthes*)
Supporting vascular structure			
Life form	ph	Phanerophyte	
	ch	Chamaephyte	
	hc	Hemicryptophyte	
	cr	Cryptophyte	
	th	Therophyte	
	li	Liane	
Root type	ad	Adventitious	
	ae	Aerating	(e.g. pneumatophore)
	ep	Epiphytic	
	hy	Hydrophytic	
	pa	Parasitic	

where the number of PFTs is the same but where their species binomials differ. Across biodiversity gradients, the PFC measure has been found useful as a bioindicator, as it is also independent of species—an advantage where species identification is problematic.

Table 2.2 VegClass data variables recorded for each 40 × 5 m transect

Site feature	Descriptor	Data type
Location reference	Location	Alpha-numeric
	Date (dd-mm-year)	Alpha-numeric
	Plot number (unique)	Alpha-numeric
	Country	Text
Observer/s	Observer/s by name	Text
Physical	Latitude deg.min.sec. or decimal deg. (GPS)	Alpha-numeric
	Longitude deg.min.sec. or decimal deg. (GPS)	Alpha-numeric
	Elevation (m.a.s.l.) (aneroid or GPS)	Numeric
	Aspect (compass. deg.) (perpendicular to plot)	Numeric
	Slope percent (perpendicular to plot)	Numeric
	Soil depth (cm)	Numeric
	Soil type (US Soil taxonomy)	Text
	Parent rock type	Text
	Litter depth (cm)	Numeric
	Terrain position	Text
Site history	General description and land-use/landscape context	Text
Vegetation structure	Vegetation type	Text
	Mean canopy height (m)	Numeric
	Canopy cover percent (total)	Numeric
	Canopy cover percent (woody)	Numeric
	Canopy cover percent (non-woody)	Numeric
	Cover-abundance (Domin)—bryophytes	Numeric
	Cover-abundance woody plants <2 m tall	Numeric
	Cover-abund. lichens (crustose, fruticose, foliose)	Numeric
	Basal area (mean of 3) ($m^2 ha^{-1}$);	Numeric
	Furcation index (mean and cv % of 20)	Numeric
	Profile sketch of 40 × 5 m plot (scannable)	Digital
Plant taxa	Family	Text[a]
	Genus	Text[a]
	Species	Text[a]
	Botanical authority	Text[a]
	If exotic (binary, presence-absence)	Numeric
Plant functional type	Plant functional elements (36) combined according to published rule set[b]	Text
Quadrat listing	Unique taxa and PFTs per quadrat (for each of 8 (5 × 5 m) or more quadrats)	Numeric
Photograph	Hard copy and digital image	JPEG

[a]Where identified, usually with voucher specimens, used directly in numerical analysis
[b]See Gillison and Carpenter (1997). Table From Gillison (2002)

2.1.15 Pattern Analyses of Plant Functional Types, Vascular Plant Richness and Vegetation Structure Along a Latitudinal Gradient

The methodology applied to the following section is described fully in the VegClass manual obtainable by download via the internet at: http://cbmglobe.org/pdf/VegMan2006.pdf. It is important to point out (see Table 2.2) that to a large extent the VegClass method differs from most other field recording procedures in that *all* vascular plant species are recorded in a 40 × 5 m transect together with all PFTs. Basal area is recorded for *all* woody plants that are detectable using a radial plotless system employing a Bitterlich prism with readings positioned at the centre and ends of each transect. This includes understorey plants as well as lianas. The data held by the author for all 1138 transects (Fig. 2.1) were recorded by the author on site, in most cases assisted by other colleagues or technical assistants. Climate data were extracted for each transect location from the CGIAR Climate Database. Where possible, in cases where species could not be identified on site, voucher specimens were taken or else arbitrary 'morphospecies' names attached for convenience. In every case however, this did not prevent the recording of all PFTs. The resulting database is therefore unique in many respects as, unlike many other global databases, all the data were recorded using a standard protocol thus facilitating comparative analyses.

In undertaking a visual appraisal of the following patterns it is necessary to state that, while >95% of climate variability for the globe is covered by the transects shown in (Fig. 2.1), 67% of sites were located in the southern hemisphere.

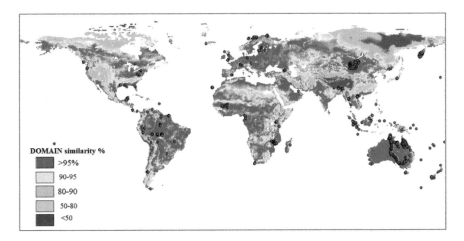

Fig. 2.1 Climatic representation of 1138 VegClass transect locations (black dots) used in this chapter illustrated in a DOMAIN similarity map based on Elevation (m), Minimum temperature coldest month (°C), Mean annual precipitation (mm year^{-1}) and Mean annual actual evapotranspiration (W m^2 year^{-1}) (Modified from Gillison 2016)

2.1.16 Species Richness and PFT Richness at Global Scale

Figure 2.2 Provides evidence of a close match expressed as a $\log_{10}$ scale. In analysing this graph it is necessary to point out that the relationship between species and PFTs is a many-to-many mapping, (i.e. where more than one species can occur in a PFT and *vice versa*). This system allows for the detection of infraspecific variation where, for example, a species may exhibit different functional morphologies within a transect depending on its position and pheno-morphological response along an environmental gradient.

2.1.17 Variation in Species and PFT Richness with Latitude

Figures 2.3 and 2.4 illustrate how richness of both species and PFT richness exhibit a more or less similar response with an evident peak in richness at or near the equator with secondary peaks at approximately 35°N and S of the equator but with overall higher richness in the northern hemisphere. Note also the drop in richness around 10°N that is consistent with a similar pattern reported by Moles et al. (2009) for plant height.

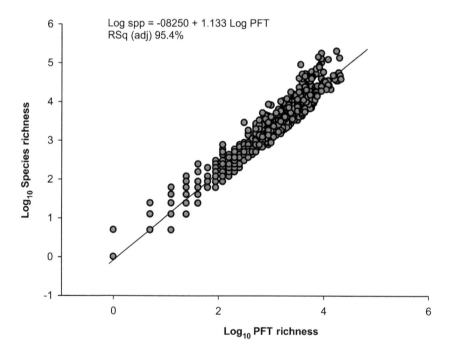

Fig. 2.2 Species richness and PFT richness (counts) expressed on a $\log_{10}$ scale. Despite the close correlation, species and PFT can and often do vary independently of each other. Each point represents a 40 × 5 m VegClass transect (From Gillison 2016)

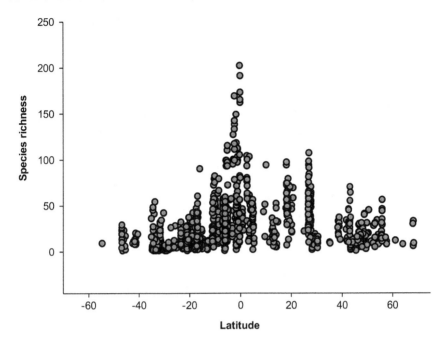

Fig. 2.3 Species richness and latitude. Each point represents a 40 × 4 m VegClass transect. A secondary peak is evident at 20–30°N with decrease at approximately 10°N and S of the equator (From Gillison 2016)

2.1.18 PFT Richness Matched Against Minimum Temperature of the Coldest Month

This environmental variable was chosen as the distribution of land based biota appears to be most sensitive to this thermal extreme where, for example, internal embolisms in vascular tissue can occur. Such values may not be so evident if average annual temperatures are used. In this graph (Fig. 2.5) there is a clear evidence for three obvious peaks at 0 °C (freezing point?) and again at about −18 °C and centred around 20 °C. Lower frequencies appear at around −5 and 5 °C.

2.1.19 Species and PFT Richness Combined Matching Against Latitude

The pattern in Fig. 2.6 shows a generally similar response with latitude but, beyond the equator with overall highest richness in the southern hemisphere. An unexplained peak is apparent at ~38°S. with a marked dip at ~30°N.

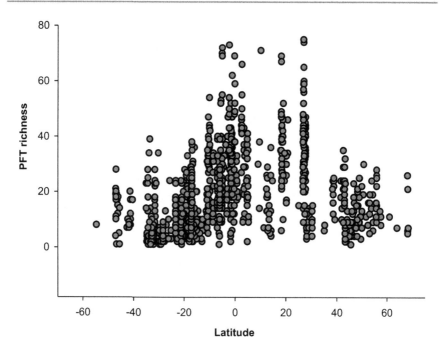

Fig. 2.4 PFT richness and latitude. Apart from equatorial region, richness peaks between approximately 20 and 30°N, with decrease at about 10°N and S of the equator (From Gillison 2016)

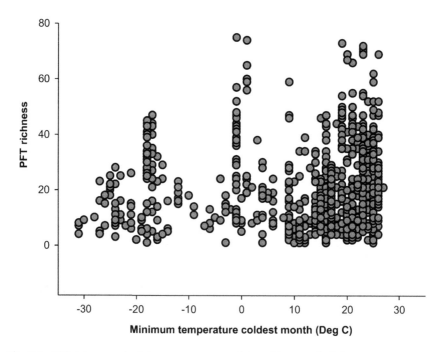

Fig. 2.5 PFT richness and minimum temperature of the coldest month (°C)

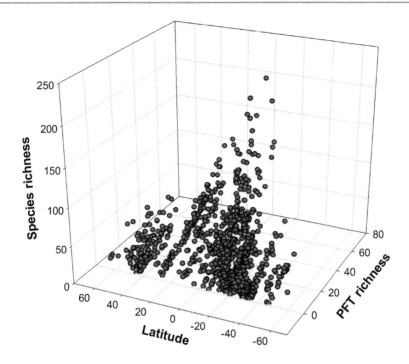

Fig. 2.6 Species and PFT richness together matched against latitude

2.1.20 An Example of Latitudinal Distribution of One PFT or Functional *Modus* (*me-la-do-ct-ph*)

This particular PFT is one of the more widespread of all PFTs recorded in this exercise. It is included here as it contains a specific PFE (*ct*—for green stem or *cortical* photosynthesis). As shown in Fig. 2.7, there is a clear peak towards the equator and with a secondary hump at ~20°N. While green stem photosynthesis is evident in holarctic and boreal regions (Gillison 2012) for example in several *Betula* species, it is also evident in *Nothofagus pumilio* in Tierra del Fuego in the extreme south of South America. The majority by far occur in the tropics however, in particular among the gum-barked eucalypts of Australia.

2.1.21 Mesophyll Leaf Size Distribution with Latitude

Mesophyll leaf size class (4500–18,200 mm^2) (the '*me*' of *me-la-do-ct-ph*) is one of the more widely distributed plant functional traits on the planet. In this example (Fig. 2.8), a conspicuous equatorial spike is offset by two secondary peaks in northern latitudes. This size class is characteristic of plants in mesic environments and optimal growth conditions usually with higher rates of resource acquisition and SLA.

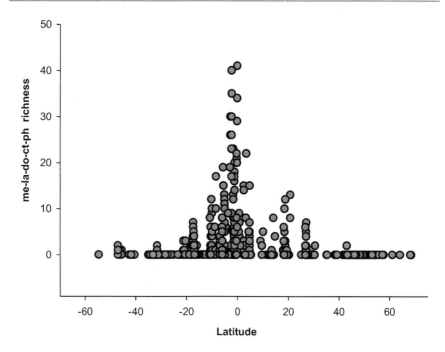

Fig. 2.7 Latitudinal distribution of one PFT (functional *modus*) **me-la-do-ct-ph** (plant individual syndrome comprising mesophyll leaf size class (**me**), lateral leaf inclination (**la**), dorsiventral (hypostomatous) leaves (**do**), green stem (cortical) photosynthesis (**ct**), and perennial woody plant >2 m tall (phanerophyte) (**ph**). Classification follows VegClass system

2.1.22 Leptophyll Leaf Size Distribution with Latitude

Leptophyll leaf size class (2–25 mm^2) (Fig. 2.9) shows a very different latitudinal pattern to that of mesophyll leaf size with a generally broad distribution across all latitudes but with noticeable spike at ~36°S. and much higher peaks (one at ~ 30°N, increasing towards the boreal-arctic regions). This PFE tends to be consistent with measures of LMA or SLA that typify a slow rate of resource acquisition.

2.1.23 Latitudinal Distribution of PFTs with Lateral Leaf Inclination

Unlike most vegetation recording protocols the VegClass system takes notice of the fact that leaf inclination can play a positive role as an adaptation to a resource acquisition strategy involving light (see the rationale in the LLR strategy above). In Fig. 2.10 The distribution tends to follow that of plants with a mesophyll trait (Figs. 2.7 and 2.8) that is consistent with an adaptation for light-seeking plants in relatively mesic environments (*cf.* high SLA).

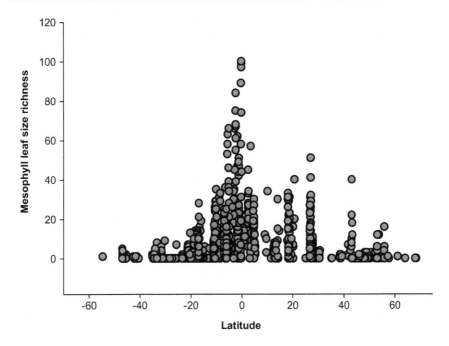

Fig. 2.8 Mesophyll leaf size distribution shows secondary peaks of richness in northern latitudes

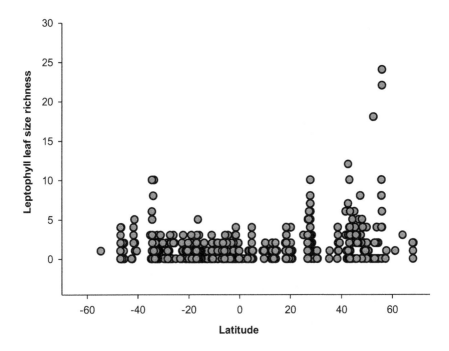

Fig. 2.9 Leptophyll leaf size distribution shows highest values in north temperate boreal and arctic regions with an additional spike at approximately 35°S

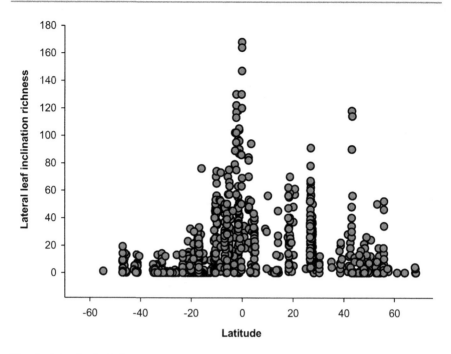

Fig. 2.10 Latitudinal distribution of PFTs with lateral leaf inclination show an equatorial maximum and otherwise highest values in northern latitudes

2.1.24 Latitudinal Distribution of PFTs with a Vertical Leaf Inclination

Compared to the distribution of leaves in mesic environments (Figs. 2.7 and 2.8), leaves with a vertical inclination tend to be more frequent in increasingly harsh environments (notably hot or cold desertic). In this case (Fig. 2.11), vertical inclination tends to be associated with leptophyll leaves, again, with visible peaks at ~35°S. and again at 20–30°N. but with highest peaks above 55°N. (*cf.* low SLA).

2.1.25 Latitudinal Distribution of PFTs with a Deciduous Trait

From the foregoing, it is very clear that the great majority of ecological studies are derived from the northern hemisphere. This would appear to be consistent from the pattern in Fig. 2.12 that shows a pronounced swing to toward the highest northern latitudes. Possible reasons are discussed below.

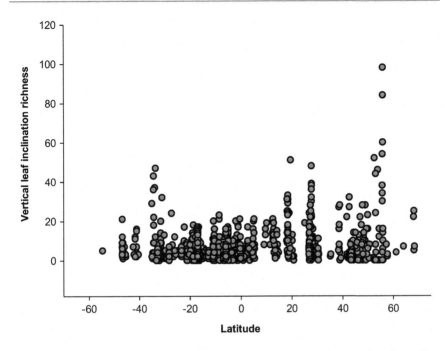

Fig. 2.11 Distribution of PFTs with vertical leaf inclination with highest values in northern latitudes and at approximately 35°S (see also Fig. 2.8)

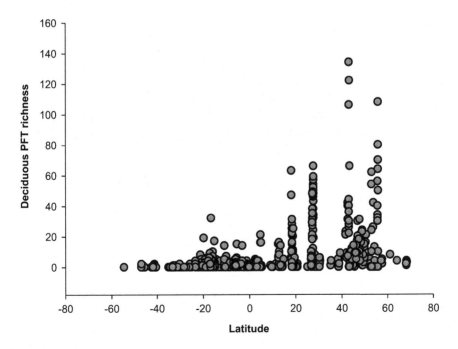

Fig. 2.12 PFTs with a deciduous trait show a marked increase toward high northern latitudes but also with gaps at ~10°N and ~35°N

2.1.26 Latitudinal Distribution of Plants with Hypostomatous Leaves

The physiological dynamics of leaf stomates play a significant role in gas exchange where feed-forward systems can come into play to anticipate and thus quickly adapt to sudden changes in atmospheric moisture levels, notably vapour pressure deficit. The VegClass system provides for the recording of two basic types of leaf stomate distribution: one being hypostomatous (dorsiventral) leaves with stomates mainly on the abaxial surface or underside; the other being leaves that are amphistomatous or isostomatous, with stomates mainly on both surfaces (e.g. most *Eucalyptus* spp.). The latter can occur in plants with flattened leaf laminas or in cylindrical or isocentric leaves (e.g. many *Euphorbia* or *Hakea* spp.). In the present example (Fig. 2.13), hypostomatous plants show an equatorial spike with secondary patterns in the N. hemisphere, but with an unexplained decrease at ~35°N.

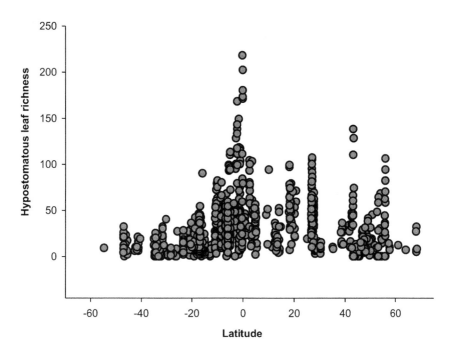

Fig. 2.13 Hypostomatous leaves (stomates mainly on the abaxial or underside) are predominantly equatorial with secondary richness patterns highest in northern latitudes

2.1.27 Latitudinal Distribution of Plants with Amphistomatous Leaves

In a reverse pattern to plants with hypostomatous leaves, Fig. 2.14. Displays a clear swing towards peaks of richness in the southern hemisphere. This may be a reflection of the higher frequency of phyllodineous plants such as *Acacia* spp. and Casuarinaceae as well as *Eucalyptus* spp., Cactaceae and Euphorbiaceae. While there is a perceptible dip at the equator, note the secondary hump towards areas above 40°N.

2.1.28 Latitudinal Distribution of Mean Canopy Height and Basal Area

In every example displayed so far there are evident patterns of increase or decrease along latitudinal gradients. Such is not the case in Figs. 2.15 and 2.16 where mean canopy height and basal area reveal a far more diffuse patterning. It is noteworthy that Fig. 2.15 indicates a sudden dip at ~ 10°N. that is consistent with the finding by Moles et al. (2009). The same phenomenon tends to be repeated in Fig. 2.16.

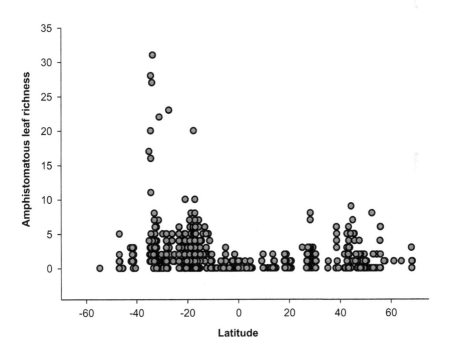

Fig. 2.14 Amphistomatous leaves (stomates on all leaf lamina surfaces) are found predominantly in southern latitudes

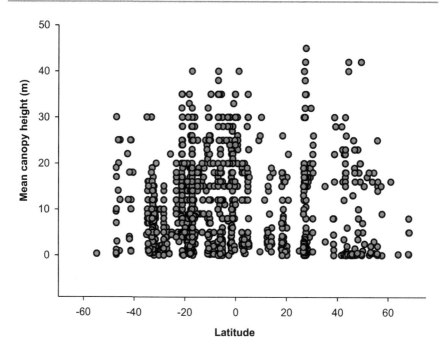

Fig. 2.15 Mean canopy height is largely independent of latitude

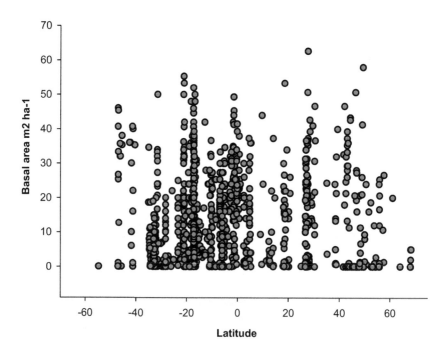

Fig. 2.16 Basal area of all woody plants is largely independent of latitude

2.2 Discussion

This chapter represents a far from complete review of available evidence for and against elements of causality that can explain many more or less obvious patterns of latitude-related change along biodiversity gradients. Nonetheless the visual evidence apparent in the foregoing graphs does raise questions about many common assumptions, among them claims for a monotonic trend in biodiversity richness towards low latitudes. The present treatment is a departure from the norm for two reasons: the first is that it represents the most comprehensive, albeit visual, analysis of plant functional types in a latitudinal context using whole-plant trait syndromes (PFTs) in addition to specific individual traits or PFEs contained within each PFT. The second is that, unlike nearly all other databases at global scale, the data have been recorded using a common protocol. Databases such as LEDA and TRY are improving in quality and access but, inevitably, will continue to be sourced from multiple origins of varying accuracy that, in turn, will also continue to influence the quality and validity of outcomes of investigations such as those related to biophysical and related latitudinal gradients. Recent technological advances however, (Schrodt et al. 2015) now make possible more 'intelligent' data searches that employ a hierarchical Bayesian approach to filling in gaps in knowledge and trait prediction for macroecology and plant functional biogeography.

Although the findings are tentative, several new outcomes are evident. Distributional patterns of vascular plant species and PFTs do, in fact, show richness trends toward the equator; however, there are sufficient departures from the trend that generate questions about other environmental factors that might be responsible for this pattern. At this point it is necessary to depart from an analysis of holistic richness of species and PFTs alone, and examine more specific patterns that show highly divergent departures from equator-centric peaks. These peaks are represented by different functional components such as leaf size classes and leaf inclination that, for example, reflect varying ecophysiological strategies and adaptations to resource acquisition and survival. These suggest some consistency with related specific leaf area (SLA) that is known to increase with resource rich environments and to decrease correspondingly under resource poor conditions. Another unusual and novel example of latitude-related trait response is the evident increase in PFT richness towards regions that experience coolest minimum temperature of the coldest month at ~20 °C (e.g. the Greater Caucasus and upland Eastern Himalaya) but, at the same time exhibit a spike at 0 °C or near freezing point (e.g. in some mid-elevation Himalayan sites). From this one may speculate that the conspicuous spike at 0 °C could represent a confluence of plants with PFTs that are able to employ strategies with facultative control of the molecular composition of vascular tissues in order to avoid or tolerate embolisms caused by freezing.

A further unexpected trend is the increase in PFTs containing a deciduous trait toward the high northern latitudes. The reasons for this are not entirely clear but it is likely that the northern clines are dominated by winter-deciduous rather than drought-deciduous plants, which tend to be more numerous in southern latitudes, e.g. southern Africa and Australia. The clearly visible trend in Fig. 2.12 is contrary to the distribution described by Kikuzawa (1991) that supported evidence for a unimodal, mid-latitude peak. The assumption that deciduousness arises through either temperature or drought-induced conditions (*cf.* Meher-Homji 1978) may now require re-examination given that in tropical regions especially, drought effects may be confounded by seasonal variation in solar insolation that can also influence synchronous leaf and flower flushing (Borchert et al. 2015). While the finding by Borchert et al. (2015) may help explain why many tropical trees (e.g. Bignoniaceae, Combretaceae, Fabaceae, Moraceae) become deciduous even where water supply is non-limiting as in lacustrine and riparian environments, it is clearly a factor to be considered when addressing leaf longevity along biodiversity gradients.

Other than deciduousness, among the most striking outcomes is the contrasting distribution of PFTs with hypostomatous vs. amphistomatous leaves that suggest significantly different evolutionary and likely phylogenetic pathways that distinguish the southern hemisphere plants from the north. The data acquired for basal area and mean canopy height are more comprehensive than any other existing database located to date as they are more inclusive of community attributes. As indicated above, basal area (BA) estimates are for *all* woody plants and thus differ from estimates that are solely tree based. In this respect they cannot be compared directly with research outcomes from other, solely tree-based, studies. However, the data for both BA and mean canopy height have one thing in common: they show little if any consistency with widespread assumptions about relative increases from high to low latitudes. For plant height, the patterns displayed in this treatment clearly differ from the results described by Moles et al. (2009). On the other hand the patterning for BA tends to be consistent with metabolic theory and the stoichiometric modelling of trees (Enquist and Niklas 2001; Enquist et al. 2003; Reich and Oleksyn 2004). Contrasting peaks in richness in PFEs and PFTs peaks at various points along the latitudinal gradient, especially those in equatorial and mid-northern latitudes, may be consistent with underlying gradients in productivity and distributional patterns of carbon sinks and sources proposed by Verheijen et al. (2015). The results outlined in this chapter suggest that the search for a generalizeable theory of causality underlying the relationship between biodiversity, especially with respect to taxa, plant functional types and traits and latitude, has a long way to go. With this perspective in mind it is hoped that the analyses presented here may point the way to more productive investigation.

References

Albert CH, Thuiller W, Yoccoz NJ, Douzet R, Aubert S, Lavorel S (2010) A multi-trait approach reveals the structure and the relative importance of intra- vs. interspecific variability in plant traits. Funct Ecol 24:1192–1201

Allen AP, Gillooly JF (2006) Assessing latitudinal gradients in speciation rates and biodiversity at the global scale. Ecol Lett 9:947–954

Allen AP, Brown JH, Gillooly JF (2002) Global biodiversity, biochemical kinetics, and the energy-equivalence rule. Science 297:1545–1548

Ansquer P, Duru M, Theau JP, Cruz P (2009) Convergence in plant traits between species within grassland communities simplifies their monitoring. Ecol Indic 9:1020–1029

Axelrod DI (1959) Poleward migration of early angiosperm flora: angiosperms only displaced the relict Jurassic-type flora at high latitudes in late cretaceous time. Science 130:203–207

Banin L, Feldpausch TR, Phillips OL, Baker TR, Lloyd J, Affum-Baffoe K, Arets EJJM, Berry NJ, Bradford M, Brienen RJW, Davies S, Drescher M, Higuchi N, Hilbert DW, Hladik A, Iida Y, Salim KA, Kassim AR, King DA, Lopez-Gonzalez G, Metcalfe D, Nilus R, Peh KS-H, Reitsma KJM, Sonké B, Taedoumg H, Tan S, White L, Wöll H, Lewis SL (2012) What controls tropical forest architecture? Testing environmental, structural and floristic drivers. Glob Ecol Biogeogr 21:1179–1190

Bannister JR, Vidal OJ, Teneb E, Sandoval V (2012) Latitudinal patterns and regionalization of plant diversity along a 4270-km gradient in continental Chile. Austral Ecol. 37:500–509

Bardgett RD (2005) The biology of soil: a community and ecosystem approach. Oxford University press, Oxford

Barry KE, Schnitzer SA, van Breugel M, Hall JS (2015) Rapid liana colonization along a secondary forest chronosequence. Biotropica 47:672–680

Beadle NCW (1951) The misuse of climate as an indicator of vegetation and soils. Ecology 32:343–345

Beck PSA, Kalmbach E, Stien A, Joly D, Nilsen L (2005) Modelling local distribution of an arctic dwarf shrub indicates an important role for remote sensing of snow cover. Remote Sens Environ 98:110–121

Bernhardt-Römermann M, Gray A, Vanbergen AJ, Bergès L, Bohner A, Brooker RW, De Bruyn L, De Cinti B, Dirnböck T, Grandin U, Hester AJ, Kanka R, Klotz S, Loucougaray G, Lundin L, Matteucci G, Mészároz I, Oláh V, Preda E, Prévosto B, Pykälä J, Schmidt W, Taylor ME, Vadineanu A, Waldmann T, Stadler J (2011) Functional traits and local environment predict vegetation responses to disturbance: a pan-European multi-site experiment. J Ecol 99:777–787

Borchert R, Calle Z, Strahler AH, Baertschi A, Broadhead RE, Kamau JS, Njoroge E, Muthuri C (2015) Insolation and photoperiodic control of tree development near the equator. New Phytol 205:7–13

Botta-Dukat Z (2005) Rao's quadratic entropy as a measure of functional diversity based on multiple traits. J Veg Sci 16:533–540

Box EO (1981) Macroclimate and plant forms: an introduction to predictive modeling in phytogeography. Dr. W. Junk, The Hague, 258 p

Box EO (1996) Plant functional types and climate at the global scale. J Veg Sci 7:309–320

Box EO, Fujiwara K (2005) Vegetation types and their broadscale distribution. In: van der Maarel E (ed) Vegetation ecology. Blackwell, Oxford, pp 106–128

Broennimann O, Thuiller W, Hughes G, Midgley G, Alkemade JMR, Guisan A (2006) Do geographic distribution, niche property and life form explain plants' vulnerability to global change? Glob Chang Biol 12:1079–1093

Brown JH, Gillooly JF, West GB, Savage VM (2003) The next step in macroecology: from general empirical patterns to universal ecological laws. In: Blackburn TM, Gaston KJ (eds) Macroecology: concepts and consequences. Blackwell, Oxford, pp 408–442

Chown SL, Gaston KJ (2000) Areas, cradles and museums: the latitudinal gradient in species richness. Trends Ecol Evol 15:311–315

Colwell RK, Hurtt GC (1994) Nonbiological gradients in species richness and a spurious Rapoport's rule. Am Nat 144:570–595

Colwell RK, Lees DC (2000) The mid-domain effect: geometric constraints on the geography of species richness. Trends Ecol Evol 15:70–76

Cowling RM, Campbell BM (1980) Convergence in vegetation structure in the Mediterranean communities of California, Chile and South Africa. Vegetatio 43:191–197

Cranston BH, Monks A, Whigham PA, Dickinson KJM (2015) Variation and response to experimental warming in a New Zealand cushion plant species. Austral Ecol 40:642–650. https://doi.org/10.1111/aec.12231

Darwin CR (1859) The origin of species by means of natural selection, or the preservation of favoured races in the struggle for life. John Murray, London

De Frenne P, Graae BJ, Rodríguez-Sánchez F, Kolb A, Chabrerie O, Decocq G, De Kort H, De Schrijver A, Diekmann M, Eriksson O, Gruwez R, Hermy M, Lenoir J, Plue J, Coomes DA, Verheyen K (2013) Latitudinal gradients as natural laboratories to infer species' responses to temperature. J Ecol 101:784–795. https://doi.org/10.1111/1365-2745.12074

de Oliveira CC, Zandavalli RB, de Lima ALA, Rodal MJN (2015) Functional groups of woody species in semi-arid regions at low latitudes. Austral Ecol 40:40–49

Díaz S, Cabido M (1997) Plant functional types and ecosystem function in relation to global change. J Veg Sci 8:463–474

Díaz S, Cabido M (2001) Vive la différence: plant functional diversity matters to ecosystem processes. Trends Ecol Evol 16:646–655

Díaz S, McIntyre S, Lavorel S, Pausas JG (2002) Does hairiness matter in Harare? Resolving controversy in global comparisons of plant trait responses to ecosystem disturbance. New Phytol 154:1–14

Díaz S, Lavorel S, McIntyre S, Falsczuk V, Casanoves F, Milchunas DG, Skarpe C, Rusch G, Sternberg M, Noy-Meir I, Landsberg J, Zhang W, Clark H, Campbell B (2007) Plant trait responses to grazing – a global synthesis. Glob Chang Biol 13:313–341

Elser JJ, Fagan WF, Kerkhoff AJ, Swenson NG, Enquist BJ (2010) Biological stoichiometry of plant production: metabolism, scaling and ecological response to global change. New Phytol 186:593–608

Enquist BJ, Niklas KJ (2001) Invariant scaling relations across tree-dominated communities. Nature 410:655–660

Enquist BJ, Economo EP, Huxman TE, Allen AP, Ignace DD, Gillooly JF (2003) Scaling metabolism from organisms to ecosystems. Nature 423:639–942

Enquist BJ, Kerkhoff AJ, Huxman TE, Economow EP (2007) Adaptive differences in plant physiology and ecosystem paradoxes: insights from metabolic scaling theory. Glob Chang Biol 13:591–609

Falster DS, Westoby M (2003) Leaf size and angle vary widely across species: what consequences for light interception? New Phytol 158:509–525

Falster DS, Bränstrom Å, Dieckmann U, Westoby M (2011) Influence of four major plant traits on average height, leaf-area cover, net primary productivity, and biomass density in single-species forests: a theoretical investigation. J Ecol 99:148–164

Forster JR (1778) Observations made during a voyage round the world, on physical geography, natural history, and ethnic philosophy. G. Robinson, London (quoted by Hawkins et al. 2007)

Fortunel C, Fine PVA, Baralot C (2012) Leaf, stem and root tissue strategies across 758 Neotropical tree species. Funct Ecol 26:1153–1161

Gallagher RV, Leishman MR (2012) A global analysis of trait variation and evolution in climbing plants. J Biogeogr 39:1757–1771

Gaston KJ (2000) Global patterns in biodiversity. Nature 405:220–227

Gentry AH (1988) Changes in plant community diversity and floristic composition on environmental and geographical gradients. Ann MO Bot Gard 75:1–34

Gillison AN (1981) Towards a functional vegetation classification. In: Gillison AN, Anderson DJ (eds) Vegetation classification in Australia. Commonwealth Scientific and Industrial Research Organization and the Australian National University Press, Canberra, pp 30–41

Gillison AN (2002) A generic, computer-assisted method for rapid vegetation classification and survey: tropical and temperate case studies. Conserv Ecol 6:3. http://www.consecol.org/vol6/iss2/art3

Gillison AN (2012) Circumboreal gradients in plant species and functional types. Bot Pac 1:97–107

Gillison AN (2013) Plant functional types and traits at the community, ecosystem and world level. In: van der Maarel E, Franklin J (eds) Vegetation ecology, 2nd edn. Wiley, Oxford, pp 347–386. https://doi.org/10.1002/9781118452592.ch12

Gillison AN (2016) Vegetation functional types and traits at multiple scales. In: Box EO (ed) Vegetation structure and function at multiple spatial, temporal and conceptual scales, Geobotany studies, pp 53–97. https://doi.org/10.1007/978-3-319-21452-8_2

Gillison AN, Carpenter G (1997) A plant functional attribute set and grammar for dynamic vegetation description and analysis. Funct Ecol 11:775–783

Gillison AN, Bignell DE, Brewer KRW, Fernandes ECM, Jones DT, Sheil D, May PH, Watt AD, Constantino R, Couto EG, Hairiah K, Jepson P, Kartono AP, Maryanto I, Neto GG, Neto RJV, van Noordwijk M, Silveira EA, Susilo F-X, Vosti SA, Nunes PC (2013) Plant functional types and traits as biodiversity indicators for tropical forests: two biogeographically separated case studies including birds, mammals and termites. Biodivers Conserv 22:1909–1930

Gillman LN, Wright SD, Cusens J, McBride PD, Malhi Y, Whittaker RJ (2014) Latitude, productivity and species richness. Glob Ecol Biogeogr 24:107–117

Gilman AC (2007) Biodiversity patterns in tropical montane rainforest flora of Costa Rica. Dissertation, University of California, Los Angeles, p 123; AAT 3295742

Gotelli NJ, McGill BJ (2006) Null versus neutral models: what's the difference? Ecography 29:793–800

Gourlet-Fleury S, Rossi V, Rejou-Mechain M, Freycon V, Fayolle A, Saint-André L, Cornu G, Gérard J, Sarrailh J-M, Flores O, Baya F, Billand A, Fauvet N, Gally M, Henry M, Hubert D, Pasquier A, Picard N (2011) Environmental filtering of dense-wooded species controls above-ground biomass stored in African moist forests. J Ecol 99:981–990

Graae BJ, De Frenne P, Kolb A, Brunet J, Chabrerie O, Verheyen K, Pepin N, Heinken T, Zobel M, Shevtsova A, Nijs I, Milbau A (2012) On the use of weather data in ecological studies along altitudinal and latitudinal gradients. Oikos 121:3–19

Grime JP (2001) Plant strategies, vegetation processes, and ecosystem properties. Wiley, Chichester

Harrison SP, Prentice IC (2003) Climate and CO_2 controls on global vegetation distribution at the last glacial maximum: analysis based on palaeovegetation data, biome modelling and palaeoclimate simulations. Glob Chang Biol 9:983–1004

Hawkins BA (2008) Recent progress toward understanding the global diversity gradient. Int Biogeogr Soc Blog Newsl 6:5–7

Hawkins BA, Diniz-Filho JAF, Weis AE (2005) The mid-domain effect and diversity gradients: is there anything to learn. Am Nat 166:E140–E143

Hawkins BA, Albuquerque FS, Araújo MB, Beck J, Bini LM, Cabrero-Sañudo FJ, Castro Parga I, Diniz- Filho JAF, Ferrer-Castán D, Field R, Gómez JF, Hortal J, Kerr JT, Kitching IJ, León-Cortés JL, Lobo JM, Montoya D, Moreno JC, Olalla-Tárraga MÁ, Pausas JG, Qian H, Rahbek C, Rodríguez MÁ, Sanders NJ, Williams P (2007) A global evaluation of metabolic theory as an explanation for terrestrial species richness gradients. Ecology 88:1877–1888

Hernández-Calderón E, Méndez-Alonzo R, Martínez-Cruz J, González-Rodríguez A, Oyama K (2014) Altitudinal changes in tree leaf and stem functional diversity in a semi-tropical mountain. J Veg Sci 24:921–931

Hillebrand H (2004) On the generality of the latitudinal diversity gradient. Am Nat 163:192–211

Hort A (1916) Theophrastus. An enquiry into plants. Book 1 (English trans: Hort A). Harvard University Press, Cambridge, MA

Hsu RC-C, Wolf JHD, Tamis WLM (2014) Regional and elevational patterns in vascular epiphyte richness on an East Asian Island. Biotropica 46:549–555

Hulshof CM, Violle C, Spasojevic MJ, McGill B, Damschen E, Harrison S, Enquist BJ (2013) Intra-specific and inter-specific variation in specific leaf area reveal the importance of abiotic and biotic drivers of species diversity across elevation and latitude. J Veg Sci. https://doi.org/10.1111/jvs.12041

Hutchinson GE (1957) Concluding remarks. Cold Spring Harb Symp Quant Biol 22:415–427. https://doi.org/10.1101/SQB.1957.022.01.039

Jankowska-Blaszczuk M, Grubb PJ (2006) Changing perspectives on the role of the soil seed bank in northern temperate deciduous forests and in tropical lowland rain forests: parallels and contrasts. Perspect Plant Ecol Evol Syst 8:3–21

Jiménez-Castillo M, Lusk CH (2013) Vascular performance of woody plants in a temperate rain forest: lianas suffer higher levels of freeze–thaw embolism than associated trees. Funct Ecol 27:403–412

Kattge J, Díaz S, Lavorel S et al (2011) TRY – a global database of plant traits. Glob Chang Biol 17:2905–2935

Kearney M, Simpson SJ, Raubenheimer D, Helmuth B (2010) Modelling the ecological niche from functional traits. Proc R Soc Lond B 365:3469–3483

Kikuzawa K (1991) A cost-benefit analysis of leaf habit and leaf longevity of trees and their geographical pattern. Am Nat 138:1250–1263

Kikuzawa K, Onoda Y, Wright IJ, Reich PB (2013) Mechanisms underlying global temperature-related patterns in leaf longevity. Glob Ecol Biogeogr 22:982–993

Kleiman D, Aarssen LW (2007) The leaf size/number trade-off in trees. J Ecol 95:376–382

Kleyer M, Bekker RM, Knevel IC, Bakker JP, Thompson K, Sonnenschein M, Poschlod P, van Groenendael JM, Klime L, Klimesová J, Klotz S, Rusch GM, Hermy M, Adriaens D, Boedeltje G, Bossuyt B, Dannemann A, Endels P, Götzenberger L, Hodgson JG, Jackel A-K, Kühn I, Kunzmann D, Ozinga WA, Römermann C, Stadler M, Schlegelmilch J, Steendam HJ, Tackenberg O, Wilmann B, Cornelissen JHC, Eriksson O, Garnier E, Peco B (2008) The LEDA traitbase: a database of life-history traits of the Northwest European flora. J Ecol 96:1266–1274

Klimešová J, Tackenberg O, Herben T (2015) Herbs are different: clonal and bud bank traits can matter more than leaf–height–seed traits. New Phytol 210:13–17. https://doi.org/10.1111/nph.13788

Kozak KHJ, Wiens J (2007) Climatic zonation drives latitudinal variation in speciation mechanisms. Proc R Soc Lond B 274:2995–3003

Lavorel S, Garnier E (2002) Predicting changes in community composition and ecosystem functioning from plant traits: revisiting the Holy Grail. Funct Ecol 16:545–556

Lavorel S, McIntyre S, Landsberg J, Forbes TDA (1997) Plant functional classifications: from general groups to specific groups based on response to disturbance. Trends Ecol Evol 12:474–478

Lavorel S, Touzard B, Lebreto J-D, Clément B (1998) Identifying functional groups for response to disturbance in an abandoned pasture. Acta Oecol 19:227–240

Lavorel S, McIntyre S, Grigulis K (1999) Plant response to disturbance in a Mediterranean grassland: How many functional groups? J Veg Sci 10:661–672

Lavorel S, Díaz S, Cornelissen JHC, Garnier E, Harrison SP, McIntyre S, Pausas JG, Pérez-Harguindeguy N, Roumet C, Urcelay C (2007) Plant functional types: are we getting any closer to the Holy Grail? In: Canadell JG et al (eds) Terrestrial ecosystems in a changing world, IGBP Series. Springer, Berlin, pp 149–160

Lebrija-Trejos E, Bongers F, Pérez-García EA, Meave JA (2008) Successional change and resilience of a very dry tropical deciduous forest following shifting agriculture. Biotropica 40:422–431

Lepš J, de Bello F, Lavorel S, Berman S (2006) Quantifying and interpreting functional diversity of natural communities: practical considerations matter. Preslia 78:481–501

Losos JB, Schluter D (2000) Analysis of an evolutionary species-area relationship. Nature 408:847–850

Lou Y, Zhao K, Wang G, Jiang M, Lu X, Rydin H (2015) Long-term changes in marsh vegetation in Sanjiang Plain, northeast China. J Veg Sci 26:643–650

Lusk CH, Reich PB, Montgomery R, Ackerly D, Cavender-Bares J (2008) Why are evergreen leaves so contrary about shade? Trends Ecol Evol 23:299–303

Marini L, Bruun HH, Heikkinen RK, Helm A, Honnay O, Krauss J, Kühn I, Lindborg IR, Pärtel M, Bommarco R (2012) Traits related to species persistence and dispersal explain changes in plant communities subjected to habitat loss. Divers Distrib 18:898–908

Markesteijn L, Poorter L, Bongers F (2007) Light-dependent leaf trait variation in 43 tropical dry forest tree species. Am J Bot 94:515–525

Mason NWH, de Bello F, Mouillot D, Pavoine S, Dray S (2012a) A guide for using functional diversity indices to reveal changes in assembly processes along ecological gradients. J Veg Sci 24:794–806

Mason NWH, Richardson SJ, Peltzer DA, de Bello F, Wardle DA, Allen RB (2012b) Changes in coexistence mechanisms along a long-term soil chronosequence revealed by functional trait diversity. J Ecol 100:678–689

McGill BJ, Enquist BJ, Weiher E, Westoby M (2006) Rebuilding community ecology from functional traits. Trends Ecol Evol 21:178–185

Meher-Homji VM (1978) Vegetation classification: need we dissociate environmental terminologies from the physiognomic nomenclature? Indian For 10:653–660

Mittelbach GG, Schemske DW, Cornell HV, Allen AP, Brown JM, Bush MB, Harrison SP, Hurlbert AH, Knowlton N, Lessios HA, McCain CM, McCune AR, McDade LA, McPeek MA, Near TJ, Price TD, Ricklefs RE, Roy K, Sax DF, Schluter D, Sobel JM, Turelli M (2007) Evolution and the latitudinal diversity gradient: speciation, extinction and biogeography. Ecol Lett 10:315–331

Moles AT, Warton DI, Warman L, Swenson NG, Laffan SW, Zanne AE, Pitman A, Hemmings FA, Leishman MR (2009) Global patterns in plant height. J Ecol 97:923–932

Moncrieff GR, Lehmann CER, Schnitzler J, Gambiza J, Hiernaux P, Ryan CM, Shackleton CM, Williams RJ, Higgins SI (2014) Contrasting architecture of key African and Australian savanna tree taxa drives intercontinental structural divergence. Glob Ecol Biogeogr 23:1235–1244

Murphy HT, Metcalfe DJ, Bradford MG, Ford AJ (2014) Community divergence in a tropical forest following a severe cyclone. Austral Ecol. 39:696–709

Nakamura Y, Krestov PV, Omelko AM (2007) Bioclimate and zonal vegetation in Northeast Asia: first approximation to an integrated study. Phytocoenologia 37:443–470

Naveh Z, Whittaker RH (1979) Structural and floristic diversity of shrublands and woodlands in northern Israel and other mediterranean areas. Vegetatio 41:171–190

Niinemets Ü, Kull O (1998) Stoichiometry of foliar carbon constituents varies along light gradients in temperate woody canopies: implications for foliage morphological plasticity. Tree Physiol 18:467–479

Nogué S, Rull V, Vegas-Vilarrúbia T (2012) Elevational gradients in the neotropical table mountains: patterns of endemism and implications for conservation. Divers Distrib 19:676–687

Oldfield S (comp) (1997) Cactus and succulent plants – status survey and conservation action plan. IUCN/SSC Cactus and Succulent Specialist Group. IUCN, Gland Switzerland and Cambridge UK, 212 p

Ordoñez JC, Bodegom PM, van Witte J-PM, Wright IJ, Reich PB, Aerts R (2009) A global study of relationships between leaf traits, climate and soil measures of nutrient fertility. Glob Ecol Biogeogr 18:137–149

Pausas JG, Bradstock RA, Keith DA, Keeley JE, Network tGF (2004) Plant functional traits in relation to fire in crown-fire ecosystems. Ecology 85:1085–1100

Pellissier L, Bråthen KA, Vittoz P, Yoccoz NG, Dubuis A, Meier ES, Zimmermann NE, Randin CF, Thuiller W, Garraud L, Van Es J, Guisan A (2013) Thermal niches are more conserved at cold than warm limits in arctic-alpine plant species. Glob Ecol Biogeogr 22:933–941

Posada JM, Lechowicz MJ, Kitajima K (2009) Optimal photosynthetic use of light by tropical tree crowns achieved by adjustment of individual leaf angles and nitrogen content. Ann Bot Lond 103:795–805

Prior LD, Bowman DJMS, Eamus D (2004) Seasonal differences in leaf attributes in Australian tropical tree species: family and habitat comparisons. Funct Ecol 18:707–718

Randin CF, Paulsen J, Vitasse Y, Kollas C, Wohlgemuth T, Zimmermann NE, Körner C (2013) Do the elevational limits of deciduous tree species match their thermal latitudinal limits? Glob Ecol Biogeogr 22:913–923

Raunkiær C (1934) The life forms of plants and statistical plant geography. Clarendon Press, Oxford

Raunkiær C, Gilbert-Carter H (1937) Plant life forms. Clarendon Press, Oxford. 104 pp

Reich PB, Oleksyn J (2004) Global patterns of plant leaf N and P in relation to temperature and latitude. Proc Natl Acad Sci USA 101:11001–11006

Rosenzweig ML, Sandlin EA (1997) Species diversity and latitudes: listening to area's signal. Oikos 80:172–176

Rossetto M, McPherson H, Siow J, Kooyman R, van der Merwe M, Wilson PD (2015) Where did all the trees come from? A novel multispecies approach reveals the impacts of biogeographical history and functional diversity on rain forest assembly. J Biogeogr 42:2172–2186. https://doi.org/10.1111/jbi.12571

Rusch GM, Pausas JG, Lepš J (2003) Plant functional types in relation to disturbance and land use: introduction. J Veg Sci 14:307–310

Schrodt F, Kattge J, Shan H, Fazayeli F, Joswig J, Banerjee A, Reichstein M, Bönisch G, Díaz S, Dickie J, Gillison AN, Karpatne A, Lavorel S, Leadley P, Wirth CB, Wright IJ, Wright SJ, Reich PB (2015) BHPMF – a hierarchical Bayesian approach to gap-filling and trait prediction for macroecology and functional biogeography. Glob Ecol Biogeogr 24:1510–1521

Serrano HC, Antunes C, Pinto MJ, Máguas C, Martins-Louçāo MA, Branquinho C (2015) The ecological performance of metallophyte plants thriving in geochemical islands is explained by the inclusive niche hypothesis. J Plant Ecol 8:41–50

Sizling AL, Storch D, Keil P (2009) Rapoport's rule, species tolerances, and the latitudinal diversity gradient: geometric considerations. Ecology 90:3575–3586

Smith T, Huston M (1989) A theory of spatial and temporal dynamics of plant communities. Vegetatio 83:49–69

Smith TM, Shugart HH, Woodward FI, Burton PJ (1992) Plant functional types. In: Solomon AM, Shugart HH (eds) Vegetation dynamics and global change. Chapman & Hall, New York, pp 272–292

Stevens GC (1989) The latitudinal gradient in geographical range: how so many species coexist in the tropics. Am Nat 133:240–256

Stevens GC (1992) The elevational gradient in altitudinal range: an extension of Rapoport's latitudinal rule to altitude. Am Nat 140:893–911

Swenson NG, Enquist BJ, Pither J, Kerkhoff AJ, Boyle B, Weiser MD, Elser JJ, Fagan WF, Forero-Montaña J, Fyllas N, Kraft NJB, Lake JK, Moles AT, Patiño S, Phillips OL, Price CA, Reich PB, Quesada CA, Stegen JC, Valencia R, Wright IJ, Wright SJ, Andelman S, Jørgensen PM, Lacher TE Jr, Monteagudo A, Núñez-Vargas MP, Vasquez-Martínez R, Nolting KM (2012) The biogeography and filtering of woody plant functional diversity in North and South America. Glob Ecol Biogeogr 21:798–808

Tomlinson KW, Poorter L, Sterck FJ, Borghetti F, Ward D, de Bie S, van Langevelde F (2013) Leaf adaptations of evergreen and deciduous trees of semi-arid and humid savannas on three continents. J Ecol 101:430–440

Vanclay J, Gillison AN, Keenan RJ (1997) Using plant functional attributes to quantify site productivity and growth patterns in mixed forests. For Ecol Manag 94:149–163

Vandewalle M, Purschke O, de Bello F, Reitalu T, Prentice HC, Lavorel S, Johansson LJ, Sykes MT (2014) Functional responses of plant communities to management, landscape and historical factors in semi-natural grasslands. J Veg Sci 25:750–759

Verheijen LM, Aerts R, Brovkin V, Cavender-Bares J, Cornelissen JHC, Kattge J, van Bodegom PM (2015) Inclusion of ecologically based trait variation in plant functional types reduces the projected land carbon sink in an earth system model. Glob Chang Biol 21:3074–3086

Violle C, Navas M-L, Vile D, Kazakou E, Fortunel C, Hummel I, Garnier E (2007) Let the concept of trait be functional. Oikos 116:882–892

von Humboldt A (1808) Ansichten der Natur mit wissenschaftlichen Erlaüterungen. J. G. Cotta, Germany

Wallace AR (1878) Tropical nature and other essays. Macmillan, New York

Wasof S, Lenoir J, Gallet-Moron E, Jamoneau A, Brunet J, Cousins SAO, De Frenne P, Diekmann M, Hermy M, Kolb A, Liira J, Verheyen K, Wulf M, Decocq G (2013) Ecological niche shifts of understorey plants along a latitudinal gradient of temperate forests in north-western Europe. Glob Ecol Biogeogr 22:1130–1140

Westoby M (1998) A leaf–height–seed (LHS) plant ecology strategy scheme. Plant Soil 199:213–227

Westoby M, Falster DS, Moles AT, Vesk PA, Wright IJ (2002) Plant ecological strategies: some leading dimensions of variation between species. Annu Rev Ecol Syst 33:125–159

Withrow AP (1932) Life forms and leaf size classes of certain plant communities of the Cincinnati region. Ecology 13:12–35

Wright IJ, Reich PB, Westoby M, Ackerly DD, Baruch Z, Bongers F, Cavender-Bares J, Chapin FS, Cornelissen JHC, Diemer M, Flexas J, Garnier E, Groom PK, Gulias J, Hikosaka K, Lamont BB, Lee T, Lee W, Lusk C, Midgley JJ, Navas M-L, Niinemets Ü, Oleksyn J, Osad N, Poorter H, Poot P, Prio L, Pyankov VI, Roumet C, Thomas SC, Tjoelker MG, Veneklaas E, Villar R (2004) The worldwide leaf economics spectrum. Nature 428:821–827

Wright IJ, Ackerly DD, Bongers F, Harms KE, Ibarra-Manriquez G, Martinez-Ramos M, Mazer SJ, Muller-Landau HC, Paz H, Pitman N, Poorter L, Silman MR, Vriesendorp CF, Webb CO, Westoby M, Wright SJ (2007) Relationships among ecologically important dimensions of plant trait variation in seven neotropical forests. Ann Bot 99:1003–1015

Yang J, Spicer RA, Spicer TEV, Arens NC, Jacques FMB, Su T, Kennedy EM, Herman AB, Steart DC, Srivastava G, Mehrotra RC, Valdes PJ, Mehrotra NC, Zhou Z–K, Lai J-S (2015) Leaf form–climate relationships on the global stage: an ensemble of characters. Glob Ecol Biogeogr 24:1113–1125. https://doi.org/10.1111/geb.12334

Zapata FA, Gaston KJ, Chown SL (2005) The mid-domain effect revisited. Am Nat 166:E144–E148

Zhang R, Liu T, Zhang J-L, Sun Q-M (2015) Spatial and environmental determinants of plant species diversity in a temperate desert. J Plant Ecol. https://doi.org/10.1093/jpe/rtv053

Plant Eco-Morphological Traits as Adaptations to Environmental Conditions: Some Comparisons Between Different Biomes Across the World

Javier Loidi

Dedication: I want to dedicate this work to my friend Elgene O. Box, comrade of many meetings and excursions over many years and in various parts of the world in that we always had a good understanding and sympathy forged through stimulating conversations and experiences together.

Abstract

Vegetation has been studied by means of eco-morphological trait syndromes of dominant plants (physiognomy) since the beginning of vegetation science. Different approaches have tried to match eco-morphological traits with environmental conditions in the basic belief that physiognomically similar vegetation types occur in different geographical areas where similar environmental conditions occur, in a physiognomic convergence phenomenon. This has been considered a fundamental phenomenon in plant geography and a validation for this approach. Comparison of physiognomic types from the Southern Hemisphere vegetation with those from the Northern Hemisphere reveals that this statement is of limited value. Winter deciduousness is a common feature in woody vegetation in the temperate biomes of the Northern Hemisphere, but it is rare in the Southern Hemisphere. Sclerophylly is a unifying trait in all Australian vegetation regardless of the different climatic areas in which they live. This can be related to the common origin of its flora. Traits may have developed as an adaptation to environmental conditions but show no particular adaptive advantages under current conditions, giving rise to the phenomenon of

---

J. Loidi (✉)
Department of Plant Biology and Ecology, University of the Basque Country (UPV/EHU), Bilbao, Spain
e-mail: javier.loidi@ehu.es

exaptation. The fact that ecological and phenotypical similarity of related taxa is greater than one would expect on a random basis, can be explained by phylogenetic niche conservatism and phylogenetic signal. A part of the observed traits are inherited from ancestors and not all of them conserve an adaptive functionality nowadays, in spite of having had it when they differentiated a long time ago. Phylogeny matters, also for eco-morphological trait syndromes.

Keywords

Physiognomy • Plant functional traits • Exaptation • Niche conservatism • Phylogenetic niche • Sclerophylly • Evergreen plants • Deciduous plants

3.1 Introduction

There is a long tradition of studying the vegetation and plant ecology in the context of morphological traits that appear in relation to environmental conditions, and the assumption is made that such traits are a response to environmental changes. The first contributions can be traced to the dawn of botany, but we will consider as the initial stage of this approach what has been called the "physiognomic tradition" (Whittaker 1962) which developed in Europe along the nineteenth and twentieth centuries, with relevant contributions from von Humboldt (1805), Grisebach (1872), Warming (1909), Raunkjaer (1905), Brockmann-Jerosch and Rübel (1912) and, more recently, from Ellenberg and Mueller-Dombois (1967) and Box (1981) among others. In the physiognomic tradition, individual species as well as vegetation types are taken into account, usually in a worldwide perspective. More recently, the term *eco-morphological traits* has been used to name the traits of organisms that have a clear adaptive role relative to environmental conditions, a term that is more frequent in zoology than in botany, although a few modern authors have applied it to plants (Latorre and Cabezudo 2012). Nevertheless, the term "functional traits" is more frequently used to include these and other features of plants, as explained below.

The basic assumption in this approach is that a vegetation type with a defined set of observable traits, also called *physiognomy*, usually occurs in a defined range, or ranges, of environmental conditions, to which the dominant growth forms are adapted. This thus underlies the idea that physiognomically similar vegetation inhabit different continents where similar environmental conditions coincide, resulting in physiognomic convergence of vegetation from widely separated regions. This convergence has been considered one of the major phenomena in plant geography and a validation to this approach. A typical example of this is succulence in vascular plants as an adaptation to drought which is exhibited by cacti in the Americas and by *Mesembryanthaceae* and many *Euphorbia* species in Africa, groups that are dominant plants in their respective deserts. Succulence could be considered the perfect example of convergence of eco-morphological traits in taxonomically unrelated plant taxa occurring in different continents with similar climatic zones. This is the basis for several world-wide classifications of structural or eco-physiognomic types (Brockmann-Jerosch and Rübel 1912; Ellenberg and Mueller-Dombois 1967; Box 1981).

More recently it has been assumed that the visible structural attributes of plants are surrogates for functional patterns (Grime 1979; Box 1996), and there has been a steady increase in the use of the approach known as Plant Functional Traits (PFTs) to study plant communities. The difference is that PFT encompasses not only morphological features of plant species, but also a diverse range of traits (physiological, structural, etc.) which respond to the ecosystem conditions, including disturbances and environmental constraints (Noble and Gitay 1996; Gitay and Noble 1997; Pillar 1999; Díaz and Cabido 1997; Gillison 2013), playing a major role in ecosystem processes (Díaz and Cabido 2001). Accordingly, they can be used in the assessment of ecosystem diversity, ecosystem services and estimation of global change. Initially, this approach was conceived for the simplification of floristic complexity on a broad scale (Lavorel et al. 2007) and it has led some authors to consider it as an illusory escape from taxonomy (Goodall 2014). Research in this field has shown explosive growth in the recent past and we have to recognize that it has yielded significant achievements in the understanding of functional aspects of plant communities.

This paper will focus on the morphological features of plants, which have been considered as an adaptive consequence to environment, particularly to climatic conditions, resembling the *ecophysiognomic growth forms* of Box (1981), and thus we will call them **eco-morphological traits.** We do not know if they have a real adaptive function or are surrogates of physiological processes, at least under current conditions.

The aim of this paper is to point out some of the mismatches revealed in a number of physiognomic types in their relation to a particular climatic type when considered worldwide.

3.2 Sclerophylly: Australia and the Northern Hemisphere Mediterranean Climatic Areas

Sclerophylly has been defined as a type of foliage that is reduced in size, thickened, and hardened by sclerenchma tissue. Sclerophylly is a trait present in the vegetation of very different climatic areas, such as Mediterranean type climates, but also in tropical pluviseasonal areas (e.g. Brazilian cerrado, northern Australian savannas, etc.), or in conditions of low temperatures (boreal and tundra heathland) and in territories of soils with extreme nutrient poverty (Beadle 1967; Loveless 1961).

To illustrate how accurately some eco-morphological syndromes match environmental conditions, we can make a comparison between different regions in four areas which receive comparable annual rainfall and are subjected to seasonal drought periods, two in the Horthern Hemisphere (hereafter NH) and two in Australia, in the Southern Hemisphere (hereafter SH) in order to test the influence of the common origin of the flora influencing the syndromes of some basic eco-morphologic traits (Table 3.1).

In the similarity table (Table 3.2), a clear pairing California-Mediterranean Region and SW Australia-NE Australia is observed. The Mediterranean climatic region of SW Australia is much more similar to the NE Australian tropical region than to the NH Mediterranean regions. This suggests a unifying effect of the

Table 3.1 Comparison of some traits in two distant areas with Mediterranean climate (Mediterranean basin and SW Australia) and the northern Australian tropical pluviseasonal territory

	SW Australia Mediterranean climate	Medit. region Mediterranean climate	NE Australia (dry) tropical pluviseasonal climate	California Mediterranean climate
Light in forest	High	Low	High	Low
Forest canopy density	Low	Variable but high in broadleaved	Low	Variable but high in broadleaved
Closed shrubland	Yes (kwongan)	Yes (maquis)	No	Yes (chaparral)
Sclerophylly	High	High	High	High
Malacophylly	Low	High (wet places)	Low	High (wet places)
Hairiness of vegetative parts	Low	High	Low	High
Gladular hairs and glands	Low	High	Low	High
Aromatic substances	High	High	High	High
Spinescence	Low	High	Low	High
Fire adaptations	High	Medium-low	High	High
Fleshy fruits	Low	High	Low	High
Ornithophylly	High	Absent	High	Absent
Mirmecocory (ants and termites)	High	Medium	High	Medium

Table 3.2 Similarity matrix resulting from comparing the traits in common between the different areas of Table 3.1

Territories	SW Australia	Med. Reg	NE Australia	California
SW Australia	–			
Med. Basin	3	–		
NE Australia	12	2	–	
California	4	12	3	–

phylogenetic structure of the respective floras. The widely recognized similarity between California and Mediterranean basin is not strengthened with the inclusion of SW Australia in the "mediterranean physiognomic" type. To separate it from the rest of Australia seems to be unnatural. Only sclerophylly has been invoked as unifying feature for mediterranean-climatic type ecosystems (Di Castri 1981), but this is not at all an exclusive feature for the vegetation of this climatic type. Australian vegetation shows a high homogeneity among most of the different parts of the continent as it has a particular evolutionary history in long isolation (Potts and Behernsmeyer 1992), and one of the typical features of the genuine Australian flora and vegetation in most of its regions is sclerophylly (Beadle 1967; Barlow 1994). This general sclerophylly of Australian vegetation has been related to the low nutrient content of its soils (similarly to South Africa) supporting the idea that this trait is not only related to summer-drought (Fonseca et al. 2000; Givnish 1987). Few exceptions to this generalized sclerophylly are to be found in the Australian continent, one example is the tropical rainforest of northwestern Queensland (Gillison 1987), that differs substantially in its syndrome of traits from the rest of the Australian biomes. This vegetation similar to the Palaeotropical Indo-Malesian (New Guinean) flora and vegetation (Barlow 1994), presumably because the Pleistocene connection of Australia with New Guinea (Voris 2000) was not interrupted until the end of the last ice age.

The Mediterranean ecosystems existing in both N and S hemispheres of the world have been profusely presented as an example of convergent evolution, where the unifying phenomenon in eco-morphological traits can be observed. This is particularly evident when comparing the Mediterranean basin and the Californian province, which represents a well documented convergence in physiognomic types (Cody and Mooney 1978; Mooney and Dunn 1970; Naveh 1967), although in some cases correspondence that is not so easily explained has been reported (Barbour and Minnich 1990). However, even in this case, there are considerable evidence that several of their main common traits, such as sclerophylly, existed prior to the appearance of such vegetation types (Axelrod 1989; Vankat 1989; Herrera 1992; Verdú et al. 2003). Mediterranean climate appeared quite recently (around 5 million years ago) and by that time the ancestors of modern plants adapted to Mediterranean climates, both in North America and in the Mediterranean Basin, had already acquired some of the traits they bear now. This does not imply that such traits did not played a role in the ecological success of pre-Mediterranean plants. When Mediterranean climates appeared such lineages had to survive under the new conditions. Consequently, the common climatic conditions dominating today probably have less influence than the common history shared by the North American and Eurasian biomes. In the commented cases, particularly for Australia, the common origin of the flora plays a major role, distorting the idea that the eco-morphological traits reflect the general climatic conditions currently ruling. The unifying effect of common ancestry is clearly paramount.

3.3 Evergreen and Deciduous Leaves

Some widely accepted biomes or broad vegetation types defined by physiognomy, such as the lauroid evergreen forest, the sclerophyllous ligneous vegetation, and others, are present in several climatic belts of the earth, as we have seen for Australia. Table 3.3 shows some examples of this and sclerophylly is not an exclusive feature of Mediterranean-climate biomes nor are lauroid forests exclusive of subtropical humid conditions or winter deciduous of temperate climates.

The deciduous character in trees and shrubs is usually divided into summer deciduousness, adapted to a tropical pluviseasonal climate in which drought drives these plants to shed their leaves, and winter deciduousness, typically adapted to temperate climates in which the plant sheds the leaves to avoid cold winter temperatures. Concerning the origin of this trend, it has been suggested (Cronquist 1988) that deciduousness originated in the late Cretaceous and early Tertiary as an adaptation to seasonal darkness at high latitudes and that it has later accommodated to other types of seasonalities (thermic or ombric) at lower latitudes. This could be an example of how a trend which was acquired as a response to certain conditions, is further "used" to adapt to other different seasonal stresses (cold or drought), a typical case of exaptation.

The evergreen character of forest foliage can have very different functioning patterns in the different biomes where it occurs. For example, in an evergreen tropical rain forest, the vegetation always has green leaves but each of the leaves has a short life (often less than 1 year). This is different from an evergreen lauroid forest or even to a Mediterranean or coniferous forest, where leaves live longer than 1 year, sometimes up to 3 or 4 years in a functional state, enduring adverse periods, as long as it is worthwhile for the plant to keep the leaves alive for a longer time in order to offset the costs of having to built them up again.

Table 3.3 Distribution of the main broad forest types of the world in the main climatic types in which they can occur

	Tropical humid	Tropical pluviseasonal	Mediterranean	Temperate	Boreal
Evergreen malacophyllous	+				
Evergreen sclerophyllous		+	+	+ (Australia)	
Evergreen lauroid	+ (mt.)		+ (rel.)	+ (term.)	
Winter deciduous			+ (hum.)	+	+(oc.)
Summer deciduous		+			
Conifer forest			+ (mt. NH)	+ (mt. NH)	+ (NH)

Most of them occur in one of the climatic types as zonal and as extrazonal in others. *mt.* mountain, *rel.* relictual, *term.* thermophilous, *hum.* humid zones, *oc.* oceanic, *NH* Northern hemisphere

The distribution of dominance between evergreen and deciduous trees and shrubs in the NH and in the tropics is quite satisfactorily explained by Givnish (1987, 2002) from the eco-physiological point of view. This could be attributed to the fact that these evergreen and deciduous species have been competing for a long time and, as a result of such concurrence, a tight adjustment has been established between both types. Evergreens have a longer photosynthetic season and they need fewer nutrients as the plant does not need to replace the leaves so frequently. Thus evegreeness is favored by climates with long growing seasons and by nutrient poor substrata. On the other hand, the greater the difference between seasons in terms of temperature and rainfall, the greater is the dominance of deciduous plants because they reduce transpiration and respiration during the dry or the cold season. Soil infertility favors evergreeness and sclerophylly because leaf replacement is too expensive for plants in terms of nutrients; leaves have to be functional for a longer period. Sclerophylly is an additional defence against herbivory. The explanation for evergreen dominance in boreal forests, submitted to extreme seasonal differences and sharing evergreens (coniferous) with deciduous plants, is based on the poverty of the soils, but this coexistence of evergreens with deciduous trees is not entirely convincing. Boreal climate should have an entirely deciduous forest in response to its extremely seasonal climate: the puzzle for boreal evergreenness persists.

The reason why the oceanic temperate Atlantic coasts of Europe are dominated by summergreen deciduous (with a few lauroid evergreens) while in temperate western North America evergreens conifers are the dominants also remains unexplained. The scarcity of European lauroid evergreens can be explained by history (Pleistocenic extinctions). Another not sufficiently explained fact is why, in large areas under severely seasonal climate in tropics, evergreens dominate (e.g. cerrado in Brazil) instead of being populated by summergreen deciduous plants.

The pattern for the SH is entirely different than for the NH. Relatively, evergreenness is much more widespread in SH than in NH non-tropical areas, or if we wish to express it in an opposite way, winter deciduous woodlands are much more rare in extratropical SH (Southern Chile, Tasmania). The case of New Zealand is particularly illustrative as in this archipelago evergreeness dominates in forests under clearly temperate climate, quite comparable to areas in the NH (Europe and North America) where deciduous forests dominate; only in seral and forest marginal habitats are some deciduous plants found. This dominance of evergreens has been explained to be favored by its pronounced oceanic climate and the low fertility of the soils (McGlone et al. 2004) although Wardle (1991) supports the idea that in New Zealand deciduous habitat is poorly represented because of its long isolation which prevented deciduous lineages from colonizing the islands. In any case, areas of similar soil poverty and oceanic character are also found in western Europe and are entirely dominated by deciduous forests (Brittany, western British Isles).

3.4 The Conifers

Less represented in tropical biomes, conifers are present in all climatic zones of the world, where they usually play a marginal role in the landscape, occupying stressed biotopes (rocky places, subalpine belt, highly continental or dry areas, etc.) or forming part of seral stages in succession. This pattern is typical of a group that has been historically replaced in its hegemony by a newcomer (angiosperms) and now is relegated to marginal positions. As a result of this interpretation, it is logical that the role of conifers differs profoundly between hemispheres. In the NH, where the families Pinaceae and Cupressaceae are the most important among the conifers, the classical pattern of conifer dominance is in the boreal zone and in the subalpine belts of the temperate and mediterranean latitudes. This is not symmetrical in the opposite part of the world (the SH), where the main families are Araucariaceae and Podocarpaceae. In the SH there is almost no boreal zone with emerged vegetated land, only in the Falkland Islands and in southern Chile and Tierra del Fuego can such climatic type be recognized (Rivas-Martínez et al. 2011). The subalpine belt is well represented in the SH: in New Zealand (Wardle 1991), SE Australia and Tasmania (Beadle et al. 1981), as well as in the southern Andes. These SH subalpine belts are populated by different types of vegetation, but most often without conifers. Exceptionally, only in South America can *Araucaria araucana* and other gymnosperm species can be found in subalpine levels (Amigo and Ramírez 1998; Enright and Hill 1995). SH conifers, with their needle or scaly leaves but belonging to very different lineages, do not play the same role as they play in the NH and are not always dominant in the vegetation of cold climates.

3.5 The History of the Flora in Both Hemispheres

As seen above, a certain number of mismatches arise when we compare the SH and NH ecosystems. In general, we contrast similar climatic areas of both hemispheres and check if they are inhabited by vegetation types having a similar set of eco-morphological traits. Some early studies were carried out on areas in both hemispheres populated by similar plant types, such as that on the genus *Larrea* in North and South America (Barbour and Díaz 1973), but similarities were guaranteed by using two species of a single genus, evidently narrowly related. An important point concerning those comparisons is that the currently existing non-tropical flora and vegetation in the SH is relatively scarce, if compared with that of the NH, parallelling the extent of the extratropical SH areas. Territories of medium or large extent in the SH south of the tropic of Capricorn are to be found in the southern extremes of South America and Africa (Capensis), Australia and New Zealand, as well as Antarctica. The plant taxa and vegetation of non-tropical Gondwanaland origin survive and dominate in those areas survive. In the case of Australia, this flora has had to adapt to changing conditions in severe isolation while the Australian continent has shifted northwards since it separated from Antarctica ca. 37 Ma ago (Willis and McElwain 2002). Throughout this time, Australian biota adapted to changing climates, from the original cool temperate conditions dominating most of that continent to the diverse climatic conditions present nowadays (Willis and McElwain 2002), as well as to the impoverishment of the soils (Barlow 1994). South America (Patagonia) has been strongly influenced by the Neotropical biota as well as the Cape region by the African Palaeotropical one. New Zealand is perhaps a good representation of the original conditions of extratropical cool temperate Gondwanaland, although severe extinction episodes happened during the long period of isolation which affected this territory (Wardle 1991). Antarctica has been completely covered by ice and its vegetation has been totally extinguished. Thus there is much less extratropical vegetation in the SH than in the NH, and it is furthermore split among different land masses and islands, having suffered massive extinctions, isolation and introgression of neighbouring floras. This asymmetry between both hemispheres has biased the generalizations made by many of the authors we have been learning from, particularly those of the physiognomic tradition. Features such as sclerophylly, deciduous or evergreen character of the forests, and others, acquire a new meaning when analyzing the vegetation of those SH territories; they host the surviving representatives of a much richer flora existing in the past, which covered a much larger territory than it occupies nowadays.

3.6 Traits as Adaptations to Current Conditions or Conserved Along Evolution

Here the basic question is whether a trait is the consequence of an adaptation to the environmental conditions after a directional evolution process, or whether the trait previously existed in a certain population and has found a suitable territory with adequate conditions in which that population can survive. What came first, the trait or the environment? (Ackerly 2004). If the trait originated as an adaptation to a different environment and shows no particular advantages for the fitness in the bearing organisms under the present environmental conditions, then we are faced with a phenomenon of *exaptation*. This has been commented on above and would be the case for traits that have previously been shaped by natural selection for a particular function (an adaptation); or cannot even be ascribed to the direct action of natural selection (non-adaptation), which are co-opted for the new current use (Gould and Vrba 1982).

In evolutionary theory, *Phylogenetic Niche Conservatism* (PNC) is defined as the "tendency of species to retain ecological characteristics" (Wiens and Graham 2005). This is applied mostly to ecological affinities and predicts that related species are more ecologically similar than might be expected due to random causes. Nevertheless, this does not entail a morphological transposition, because morphology does not necessarily predict ecology. Thus, a parallel and related concept has been coined: that of *Phylogenetic Signal* (PS), which indicates that the phenotypic similarity of related taxa is greater than expected by random causes. This entails a relationship of relatedness and morphology (Losos 2008). In other words, PS indicates the relationship between relatedness of taxa and their phenotypic similarity. In addition, recently, by means of surveying the phylogenetic distribution of ecological traits, the conservative nature of plant trait evolution has been reported (Silvertown et al. 2006). This conservative tendency should have a clear influence in the conservation of traits throughout the time without being changed by adaptive forces. Such phenomena could explain the mismatches we have found in the explanations of the fitting of eco-morphological traits to environmental conditions.

In considering adaptation vs niche conservatism or phylogenetic signal, we believe both play an important role in the evolution and assembly of communities; thus no vegetation type can be understood totally when one examines its eco-morphological traits exclusively from the adaptive point of view. The age of the trait can be different from the age of the taxon (Ackerly 2009), as the appearance of a trait and its pace of change can be separated from that of the taxon that bears the trait.

3.7 Final Statements

Most of the eco-morphological traits that are currently observed in Australian, North American-Eurasian and other territories were shaped during the Tertiary, and they survive today having adapted later to new and diverse climatic conditions. The Early and Mid-Tertiary seem to have had more stable climatic conditions than later Pleistocene (Willis and McElwain 2002), a period of numerous and dramatic changes. Such abrupt and relatively quick changes in the climate probably did not give sufficient time to tailor the "perfect" eco-morphological adaptive syndromes to the vegetation under them; this effect has been even more intense in recent times (recent Holocene). The pace of climatic change is too fast to allow plants to tailor the appropriate eco-morphological syndromes to adapt to the new conditions. Historically, climate change has driven morphological and physiological changes upon plants that already had adaptations to previous climatic conditions. Present eco-morphological syndromes are probably in a lag with the current climate and most probably this lag will increase with the acceleration of current climate changes. It is illusory to pretend a perfect match between current physiognomic types and climatic types. When viewed at a lower time scale, climate change drives spatial displacements of the range of species and vegetation types and causes a different species sorting within plant communities. However this occurs at a faster pace than vegetation biomes can move following the track of the spatial displacement of the climatic types. This results in a distorting effect of the species composition and structure of plant communities.

Of course any adaptation is valid if it permits plants to endure adverse events or conditions, but such eco-morphological adaptations are in a great part determined by the ancestors' adaptations. New adaptations happen upon older ones, in a sort of accumulative phenomenon. The weight of evolutionary ancestry is very important in shaping the phygsionomy of plants and of vegetation types, as it was already glimpsed by Warming (1909), Whittaker (1962) or Dansereau (1964) and documented by Wilson and Lee (2012) for New Zealand, but often it seems to be quite ignored in the assumptions of modern plant ecologists. The belief that all traits have an adaptive value to current conditions and have a functional role cannot be sustained as a whole, as far as a part of the observed traits are simply inherited from ancestors. Such traits most probably evolved as a response to an adaptive necessity in some moment of evolution, but they do not necessarily conserve the same function nowadays. Phylogeny matters, also for eco-morphological traits.

Acknowledgements The funds of the project T299-10 financed by the Basque Government have supported this research.

References

Ackerly DD (2004) Adaptation, niche conservatism and convergence: comparative studies on leaf evolution in the Californian chaparral. Am Nat 163:654–671

Ackerly DD (2009) Evolution, origin and age of lineages in the Californian and Mediterranean floras. J Biogeogr 36:1221–1233

Amigo J, Ramírez C (1998) A bioclimatic classification of Chile: woodland communities in the temperate zone. Plant Ecol 136:9–26

Axelrod DI (1989) History of the Mediterranean ecosystem in California. In: Keeley SC (ed) The California chaparral: paradigms revisited. Natural History of Los Angeles Museum County, Los Angeles, pp 7–19

Barbour MG, Díaz DV (1973) Larrea plant communities on bajada and moisture gradients in the United States and Argentina. Vegetatio 28:335–352

Barbour MG, Minnich RA (1990) The myth of chaparral convergence. Isr J Bot 39:453–463

Barlow BA (1994) Biogeography of the Australian region. In: Groves RH (ed) Australian vegetation, 2nd edn. Cambridge University Press, Cambridge, pp 3–36

Beadle NCW (1967) Soil phosphate and its role in molding segments of the Australian flora and vegetation, with special reference to xeromorphy and sclerophylly. Ecology 47:992–1007

Beadle NCW, Stocker GC, Hyland BPM, Wallace BJ (1981) The vegetation of Australia. Gustav Fischer Verlag, Stuttgart, p 690

Box EO (1981) Macrocliamate and plant formations. Junk, The Hague

Box EO (1996) Plant functional types and climate at the global scale. J Veg Sci 7:309–320

Brockmann-Jerosch H, Rübel E (1912) Die Einteilung der Pflanzengesellschaften nach ökologisch-physiognomischen Gesichtpunkten. Ver. Wilhelm Engelmann, Leipzig, p 72

Cody ML, Mooney HA (1978) Convergence versus non-convergence in Mediterranean-climate ecosystems. Annu Rev Ecol Syst 9:265–321

Cronquist A (1988) The evolution and classification of flowering plants, 2nd edn. The New York Botanical Garden, Bronx, NY, p 555

Dansereau PM (1964) Six problems in New Zealand vegetation. Bull Torrey Bot Club 91:114–140

Di Castri F (1981) Mediterranean-type shrublands of the world. In: Di Castri F, Goodall DW (eds) Mediterranean-type shrublands. Ecosystems of the world, vol 11. Elsevier, Amsterdam, pp 1–42

Díaz S, Cabido M (1997) Plant functional types and ecosystem function in relation to global change. J Veg Sci 8:463–474

Díaz S, Cabido M (2001) Vive la différence: plant functional diversity matters to ecosystem processes. Trends Ecol Evol 16(11):646–655

Ellenberg H, Mueller-Dombois D (1967) Tentative physiognomic-ecological classification of plant formations on the earth. Ber Geobot Inst ETHS Stiftg Rübel Zürich 37:21–73

Enright NJ, Hill RS (1995) Ecology of the southern conifers. Smithsonian Press, USA

Fonseca CR, Overton JM, Collins B, Westoby M (2000) Shifts in trait combinations along rainfall and phosphorous gradients. J Ecol 88:964–977

Gillison AN (1987) The dry rainforest of *Terra australis*. In: Kershaw AP, Werren G (eds) The rainforest legacy, Special Australian heritage series number 7(1), vol 1, pp 305–321

Gillison AN (2013) Plant functional types and traits at the community, ecosystem and world level. In: Van der Maarel E, Franklin J (eds) Vegetation ecology, 2nd edn. Wiley-Blackwell, Hoboken, NJ, pp 347–386

Gitay H, Noble IR (1997) What are functional types and how should we seek them? In: Smith TM, Shugart HH, Woodward FI (eds) Plant functional types. Their relevance to ecosystem properties and global change: 3–19. International geosphere-biosphere programme book series 1. Cambridge Univ. Press, Cambridge

Givnish TJ (1987) Comparative studies in leaf form: assessing the relative roles of selective pressures and phylogenetic constraints. New Phytol 106(Suppl):131–160

Givnish TJ (2002) Adaptive significance of evergreen vs. deciduous leaves: solving the triple paradox. Silva Fennica 36(3):703–743

Goodall DW (2014) A century of vegetation science. J Veg Sci 25:913–916

Gould SJ, Vrba E (1982) Exaptation – a missing term in the science of form. Paleobiology 8:4–15

Grime JP (1979) Plant strategies and vegetation processes. Wiley, New York

Grisebach A (1872) Die Vegetation der Erde. Wilhelm Engelmann, Leipzig

Herrera CM (1992) Historical effects and sorting processes as explanations for contemporary ecological patterns: character syndromes in Mediterranean woody plants. Am Nat 140:421–446

Humboldt A (1805) Ansichten der Natur, mit wissenschaftlichen Erläuterungen. J.G. Cotta, Stuttgart

Latorre AVP, Cabezudo B (2012) Phenomorphology and ecomorphological traits in *Abies pinsapo*. A comparison to other Mediterranean species. Phytocoenologia 42:15–27

Lavorel S, Díaz S, Cornelissen HC, Garnier E, Harrison SP, McIntre S, Pérez-Harguindeguy N, Roumet C, Urcelay C (2007) Plant functional types: are we getting any closer to the holy grail? In: Canadell JG, Pataki DE, Pitelka LF (eds) Terrestrial ecosystems in a changing world. Springer, Berlin, pp 149–164

Losos JB (2008) Phylogenetic niche conservatism, phylogenetic signal and the relationship between phylogenetic relatedness and ecological similarity among species. Ecol Lett 11:995–1007

Loveless AR (1961) A nutritional interpretation of sclerophylly based on differences in the chemical composition of sclerophyllous and mesophytic leaves. Ann Bot 25(2):168–184

McGlone M, Dungan RJ, Hall GMJ, Allen RB (2004) Winter leaf loss in the New Zealand woody flora. N Z J Bot 42:1–19

Mooney HA, Dunn EL (1970) Convergent evolution of mediterranean-climate evegreen sclerophyll shrubs. Evolution 24(2):292–303

Naveh Z (1967) Mediterranean ecosystems and vegetation types in California and Israel. Ecology 48:445–459

Noble I, Gitay H (1996) A functional classification to predict the dynamics of landscapes. J Veg Sci 7:329–336

Pillar VP (1999) On the identification of the optimal plant functional types. J Veg Sci 10:631–640

Potts R, Behernsmeyer AK (1992) Late Cenozoic terrestrial ecosystems. In: Behernsmeyer AK, Damuth JD, DiMichele WA, Potts R, Sues H-D, Scott SL (eds) Terrestrial ecosystems through time. Evolutionary paleoecology of terrestrial plants and animals. University of Chicago Press, Chicago, pp 419–541

Raunkjaer C (1905) Types biologiques pour la géographie botanique. Ov. K. Danske Vid. Selsk. Forh

Rivas-Martínez S, Rivas Sáenz S, Penas A (2011) Worldwide bioclimatic classification system. Glob Geobot 1:1–643

Silvertown J, Dodd M, Gowing D, Lawson C, McKoonway K (2006) Phylogeny and the hierarchical organization of plant diversity. Ecology 87(7 Suppl):39–49

Vankat JL (1989) Water stress in chaparral shrubs in summer-rain versus summer-drought climates: wither the Mediterranean type climate paradigm? In: Keeley SC (ed) The California chaparral: paradigms revisited. Natural History of Los Angeles Museum County, Los Angeles, pp 117–124

Verdú M, Dávila P, García-Fayos P, Flores-Hernández N, Valiente-Banuet A (2003) "Convergent" traits of mediterranean woody plants belong to pre-Mediterranean lineages. Biol J Linn Soc 78:415–427

Voris HK (2000) Maps of Pleistocene sea levels in Southeast Asia: shorelines, river systems and time durations. J Biogeogr 27:1153–1167

Wardle P (1991) Vegetation of New Zealand. Blackburn Press, Caldwell, p 672

Warming E (1909) Oecology of plants – an introduction to the study of plant-communities. Clarendon Press, Oxford, p 422

Whittaker RH (1962) Classification of natural communities. Bot Rev 28(1):1–239

Wiens JJ, Graham CH (2005) Niche conservatism: integrating evolution, ecology, and conservation biology. Annu Rev Ecol Syst 36:519–539

Willis KJ, McElwain JC (2002) The evolution of plants. Oxford University Press, Oxford. 378 p

Wilson BJ, Lee WG (2012) Is New Zealand vegetation really 'problematic'? Dansereau's puzzles revisited. Biol Rev 87:367–389

Part III

Regional Vegetation Types and Distribution, Prehistoric Landscapes

Changes in the Landscape and Vegetation Under the Influence of Prehistoric and Historic Man in Central Europe

4

Richard Pott

Abstract

This paper is dedicated to Holocene forest history with a special focus on prehistoric and historic human impact. As the original landscape is turned into cultivated land, humankind's influence on the evolution and formation of central Europe's cultivated landscapes is of major importance.

Today's central European woodlands are the result of utilization and forest change over centuries, locally even millennia. The central European climate is conducive to tree growth and all of central Europe would be a more or less monotonous woodland now if human beings had not created cultivated landscapes with their meadows, pastures and fields, continually pushing the forests back over recent centuries. This paper will focus on whether or not there would have been forest-free habitats of any significant size in the areas covered by deciduous and coniferous forests, that were created and cleared by herds of animals as open landscapes, in addition to the naturally forest-free habitats.

Keywords

History of forests · Ancient woodlands · Megaherbivores · Human impact

4.1 Introduction

Central Europe only has a few landscapes left that still appear natural. These are the coastal regions with their tidelands, dune complexes and a number of ungrazed salt marshes. Further inland, they include some riverscapes and a small number of

R. Pott (✉)
Institute of Geobotany, Leibniz Universität Hannover, Hannover, Germany
e-mail: pott@geobotanik.uni-hannover.de

inland waters with their covers of floating leaves, reeds, and swamp forest complexes. And then there are also a number of xerothermic vegetation complexes on natural rock habitats, and the grass, scree and rock formations at the alpine levels of our higher mountains.

Almost all other areas have been cultivated, changed or at least temporarily used or moulded by human beings for quite some time. Even the once vast, nutrient-poor upland moors have been reduced to a small number of near-natural landscape elements in some conservation areas. Extensive moors are still to be found in the Baltic region and the north of Russia.

Today's forests, also, are the result of centuries and sometimes millennia of use and silvicultural transformation. We are living in a highly cultivated landscape. The attributes "virgin", "natural", "cultivated" and "industrial" derive from the various uses specific sections of the landscape are put to today. What is often forgotten in the process is that apparently natural landscapes with their characteristic ecosystems have also been subjected to steady natural change for many thousands of years. The term "wilderness" is frequently used of late to describe biotope types that are unclaimed by man. Mudflats and islands, lakes and brooks, forests and moderately used meadows and heathlands with their hedges and bushes have hence come to epitomize intact landscapes for many because they often feature a great variety of biotopes in the smallest of spaces.

Such elements are invariably part and parcel of central Europe's cultivated landscape, however, having undergone various different developments in time and space, depending on their endowments as natural regions. The central European climate is conducive to tree growth, and all of central Europe would therefore be a more or less monotonous woodland now if human beings had not created today's cultivated landscapes with their meadows, pastures and fields, pushing the forest ever further back over the course of recent millennia. Only the salty marshes, the windswept and salt-affected coastal dunes, the ombrotrophic, rain-fed hill moors, some low moors, a number of xerothermic vegetation complexes at rocky ridges and rock waste screes, the avalanche paths and alpine regions above climatic timber lines are free from forest by nature, along with some relatively short-lived beaver meadows in the riparian zones. This has also been known for many years from a great number of vegetation history studies (Pott 1993, 1997, 2003).

We are therefore living in a highly cultivated landscape and can hardly imagine what it looked like before, or what it would look like today if people had not intervened in it for thousands of years. This impact of human beings on the vegetation and natural environment in the course of the landscape's prehistoric and historic changes is to be examined in greater detail under various aspects by this essay.

4.2 Cultivated Landscape: What Is It?

Everything designed by human beings is culture. This includes artworks as much as farmed land, the landscape. Nature develops in various landscapes: the growth and decline of living things shapes and changes the land. People influence landscapes by their activities, by cutting down forests, creating fields and pastures, by letting their domestic animals graze in the forest or grasslands. All this is connected with the daily work and practices of land users. The notions of landscape in literature, paintings and the landscape-describing arts are meanwhile often idealized: landscapes are perceived as beautiful, as a paradise or Arcadia, they appear Mediterranean or like a Switzerland, or remind us of England. Landscape is constantly informed by nature, mostly by human agency, and always by ideas. Everything is contained in the "total impression of an area" Alexander von Humboldt spoke of (Humboldt 1845). Today, we need to modify this "total impression": landscape can be viewed subjectively, abstracted objectively, and penetrated intellectually. Its spatiotemporal perspective remains an essential feature of this concept—one can only view landscape from a historico-genetic, i.e. landscape-historical or artistic-architectural perspective (Pott 1998, 1999).

Not everyone is aware of how strongly landscape is influenced by culture. To emphasize this cultural imprinting, one speaks of cultivated landscape. One then perceives an opposition between natural landscape and cultivated landscape. Cultivated landscape is understood as an area that is informed by human agency, while natural landscape is meant to be untouched, a wilderness. But not every wilderness perceived as such is really that. The notion that a landscape is a wilderness is frequently only an interpretation. Many a landscape which is understood as a wilderness is informed by culture, too (Küster 1995, 1998, 2010).

Landscape is always created in the mind—as a cultural achievement of humankind. This is what the European Landscape Convention refers to: landscape is an area as perceived by human beings, its character resulting from the impact or interplay of natural or human factors. If one speaks of landscape, the human being is given centre stage, namely as a perceiver and designer. What a human being sees, he or she could capture on canvas—like a landscape painter.

Various conditions of the landscape are known to us from descriptions of lands and forests, but also from prints and paintings of former times. These sources essentially convey the same objective images, differing only in their approach and appraisal. The destruction of forests is hence occasionally cast in a negative and gloomy light under economic aspects where a light-filled, Arcadian landscape is perceived from an aesthetic perspective (Burrichter et al. 1980). The subjective viewpoint of the individual onlooker provides the starting point in any case.

The works of landscape painters therefore do not depict fanciful, but real landscape types of their time with individual, artistic means of expression. These landscape types were already being portrayed in the seventeenth century, for example by the Italian painter S. Rosa or, so outstandingly, by Cl. Lorrain from the Lorraine, with grazing cattle as accessories. Many pasture landscape motifs are also found in seventeenth century Dutch landscape painting, thought to have been established by Jacob van Ruysdael. Inspired by van Ruysdael's landscape paintings, the English painter Thomas Gainsborough then depicted the pasture-informed "park landscapes" of his native country in the eighteenth century. They can either serve as an impressive backdrop for his sceneries, or be in the foreground as the presented subject, as is the case in his early landscapes, "Cornard Wood" (1748). This picture, which not only reflects the former condition of many forests, but also the various ways in which they were used, is of a historical expressiveness that verbal descriptions would be hard to equal. The pasture landscapes of the nineteenth century meanwhile arise before us with their remodelled and gnarled tree shapes from the pictures by the Romantics, M. von Schwind, L. Richter, Ch. Kröner, and others. That these Romantic painters should find enough inspiration in the diversity and expressiveness of common grazing lands at a time when Germany experienced the equivalent of the enclosure movement is virtually guaranteed (Figs. 4.1 and 4.2).

These few examples from three centuries in the history of western and central European art may serve to demonstrate how the expressive scenery of pasture landscapes has always cast a spell on the painters and printmakers of former times, and how widespread these landscape types were. The meaning of these

Fig. 4.1 Black-and-white reproduction of the colour painting "Cornard Wood" by Thomas Gainsborough (London 1748). The painting shows a typical thinning stage of wood pastures (from Burrichter et al. 1980)

Fig. 4.2 Similar thinning stage as in Fig. 4.1 in the "Borkener Paradies" wood pasture area with flowering blackthorn (*Prunus spinosa*) in the Ems River valley near Meppen in northern Germany (from Burrichter et al. 1980)

artworks hence extends beyond their purely artistic value as they simultaneously also document the history of the landscape and vegetation (Pott and Burrichter 1983; Ellenberg 1990, 1996; Ellenberg and Leuschner 2010).

Irrespective of how such pasture landscapes have inspired people of former times to artistic forms of expression, they also leave long-lasting impressions in modern man. This is because they are the epitome of an atmospheric landscape which, with its physiognomic diversity and small-scale interchange of pasture grass with colourful flowers, blooming forest edges, shrubby residual forest and bizarre tree shapes, has an incomparably stronger influence on pastoral life than the more or less pronounced monotony of our modern agrarian landscape. This diversity of the landscape, the apparently random, but in detail ever so regular, arrangement of vegetation units, the open land framed by trees and shrubs like scenery, jutting out and half covering the vista, lending it an appearance of vastness exactly because of this, all of it cloaks this landscape of contrasts in the magic of the primeval, and is yet anything but (Figs. 4.3 and 4.4). Its unique attractiveness results from the interplay of natural forces and anthropo-zoogenic influence (Pott and Hüppe 1991, 2008).

The idyll of these semi-natural pasture landscapes also inspired landscape gardeners to synthesize them. The product of their designs are the "English

Fig. 4.3 The rustic wood pasture area "Brögbern," north-west of Lingen in the Emsland region, provides a varied scene with a great diversity of structures, biotopes and species. It shows wooded cohesive parcels of beech-and-oak pasture forest, extensive cover of bushes and matgrass cattle tracks (from Pott and Hüppe 2008)

parks" and "English gardens", so highly popular to this day. In keeping with the model of the pasture landscape, an impression of randomness and apparent nature is to be created here, despite the clearly conceived planning and systematic arrangement of individual elements. A zoogenic pasture product hence becomes the subject of aesthetic purposes.

The Swedish botanist G. Romell and Germanist Jost Trier from Münster not only considered pasture landscapes as prototypical of the English parks, but also placed both in a direct historico-genetic context. While Romell (1967) regards the exclusively grazed pasture landscape as the starting product, Trier (1963, 1968) speaks of "terrains that were grazed and whose leaves were used", i.e. of combined grazing and pollarding landscapes from which the parks in England originally sprang. This means that such pasture forests and common pasturages are not only of interest for natural scientists as traditionally used relict landscapes, but also for art historians and garden designers as instructive and living study objects. Many of the landscape and forest parcels in the landed property of Italian Villas still show the traces of such intensive forest use today. Nowadays, they exhort us to interpret correctly in order to preserve such historic phenomena of forest use and forest treatment (Pott 2003).

Fig. 4.4 Woodland that has been thinned by grazing cattle, with grass pastures, clumps of bushes and residual forest as a "park landscape" in the "Borkener Paradies" region of the Ems River valley near Meppen in the year 1980 (from Pott and Hüppe 2008)

4.3 Nature or Culture?

The conscious perception of a cultivated landscape is fascinating. One should first of all analyse its components confronting us as either natural or cultural elements. But this is not the aim of capturing cultivated landscape: instead, all the details must be combined into an overall picture. This approach permits us to work out what constitutes the particular individual identity of a landscape.

Anyone wishing to understand the overall appearance of a landscape today will need to look into its history (Behre 1988, 2008; Pott 2014). Every cultivated landscape has been shaped over a very long time. It not only contains the traces of today's uses, but also those of various past uses. One could compare landscape with an archive that has been filled with documents of nature and culture for millennia. Knowing a specific landscape's history will help us arrive at better estimations of its future developments. Finding out about the consequences of earlier climate fluctuations, for example, will enable us to better predict the repercussions of current climate changes. The traditional uses of meadows and pastures, hedges, of pollarded willows and coppices generally supported the proliferation of rare animal and plant species (Pott 1996). If we wish to go on providing a habitat for such elements in our environment, the corresponding traditional uses will also need to be continued. But adherence to earlier utilization methods can also be important for other reasons: in the past, people used to source raw materials locally, for example by using wood for heating and construction (Hoskins 1988; Bork et al. 1998), by extensive use of the land for pasture and pollarding, by a corresponding agrarian economy of farming and heath management (Hüppe 1993), and by a varied forestry industry with regulated systems of use, including the regular clearing of coppices and meadows by fire to grow rye (Pott 1985, 1986).

Uncovering the history of a landscape will often reveal that not everything one regards as such is actually nature. Scenically attractive heathlands such as the Lüneburg Heath are frequently equated with nature, but have been brought about by intensive use: the forests that used to grow there were cleared and the open land grazed by domestic animals. In many places, the heathland farmers even used to remove the humus layer to obtain fertilizer for their meagre fields. This was hard work and in no way as "romantic" as poets and painters pictured it (Emanuelsson et al. 1985; Küster and Volz 2005; Krzywinski et al. 2009). In time, overexploitation of the land had dramatic consequences: once the sand had been bared of the humus, the wind would blow it into dunes that buried entire settlements (Figs. 4.5 and 4.6). And wherever hardly any humus remaned, only heather would be able to grow. It burgeoned into a splendour that everyone felt compelled to deem beautiful.

Fig. 4.5 *Juniperus communis* heath in the "Bockholter Berge" nature preserve near Münster in Westphalia in a complex with dwarfshrub heathland of *Calluna vulgaris* on acidic, sandy soils. These formations arise from acidophilous mixed oak forests after fires and grazing by sheep, goats or German heath

Fig. 4.6 Sand dunes and various succession stages of their first and re-colonization by *Corynephorus canescens*-meadows of Corynephoretea-vegetation

People wished to protect this "nature" they took pleasure in and loved to hike through. Such a "nature" could not be conserved by simply leaving it to itself, however, because humus would accumulate again over time, and the heather overgrown by forest pioneers and shrub species. In this manner, heathlands could turn into forests, and that by very natural means.

The landscapes that people created with their often notable diversity of species and habitats can hence also only be preserved by people. Their use must be continued, or as an alternative heathlands, for example, can be maintained by removing shrubs at regular intervals. This is no longer profitable in agricultural terms. Any maintenance modelled on the methods of an earlier heathland peasantry will cost a lot of money and human effort. So how much are today's cultural assets of wood pastures or heathlands worth to us, or any other landscape for that matter?

4.4 The Development of Today's Forests Since the Last Ice Age

Looking at the countless pollen diagrams available by now for the end of the Late Glacial and the post-glacial period in central Europe, the regional differences are considerable, and often marked by considerable local deviations from the essential features of the vegetation's and landscape's general development. These are attributable to human agency, whose influence characterized the post-glacial period, but not in the same manner everywhere. The anthropo-zoogenic impact on the progression of the vegetation and landscape was stronger than usually assumed so far (Pott 1985, 1986, 2000). But the basic outlines of the late- and post-glacial forest development are well-established by now and shall only be summarize briefly here for this reason:

In the late glacial Dryas period from around 11,500 BC, summer temperatures should certainly have reached or exceeded the 20 °C assimilation temperature conducive to plant growth, at least near the ground, long before the pine (*Pinus sylvestris*) gained a foothold. The predominant vegetation types were mats rich in grasses and sedges, coppices of low-growing willows, dwarf birches and sea buckthorn (*Hippophae rhamnoides*), as well as *Artemisia*-rich plant communities of the subarctic steppe type. The early appearance of *Ephedra* pollen (Burrichter and Pott 1987), amongst others, speaks for a continuous character of the climate in this period. One is compelled to date a possible migration of continental steppe elements to central Europe to this time. This is supported by the observation that insular *Stipa* and *Festuca* steppes are also interspersed in the southern margins of the Arctic tundra today and able to exist there.

The "birch-and-pine" period (Allerød) and later Preboreal around 9000–8000 BC brought extensive coverage of ice-free central Europe, with birch forests in the north-west and pine forests in the east and south-east. The vegetation was evidently comparable to subarctic forest steppes in character, but the existence of extensive open steppes is to be expected in drier areas like the Thuringian and Mainz basins, which probably already had less rainfall then. The blocks, gravel and shifting sand areas created by the periglacial climate will meanwhile have also provided many habitats where steppe plants enjoyed an advantage over forest growth. Neither should the open character of the pineland be underestimated: there is no reason to assume that the pine forestation of the lowlands equalled the destruction of subarctic steppe elements or impeded the transit of Pontic-Sarmatic plants from the Balkans and Danube region in the Allerød or Preboreal. Especially as the association of true steppe plants with light pine forest into a forest steppe is still as characteristic of the continental areas with cold winters as a vegetation type today as the forest-free steppe itself. A great many steppe plants from south-eastern Europe—resilient to the cold and equipped with devices for enduring great dryness and sudden temperature fluctuations—should have already migrated into large parts of Central Europe by the Allerød (Fig. 4.7).

A further consolidation of the forests is to be assumed in the "hazel period" (early warm period, Boreal around 7000–6000 BC), where a strong proliferation of Hazel (*Corylus avellana*) went hand in hand with an advance of pine trees into the previously birch-rich landscapes, while at the same time the replacement of the pines by mixed oak forests (including elms) was also starting in rich soils.

Especially in the western uplands with their ocean-influenced, humid climates, the pine and birch trees were supplanted by hazel bushes spreading north of the Alps from the west.

Fig. 4.7 The Scots pine (*Pinus sylvestris*) has a broad physiological and ecological amplitude and is nowadays usually found in extreme habitats, by dint of competition, as on the sand dunes here. It is a relict species in central Europe which, coming from the south and east, already covered the whole of Europe in the Preboreal period, before the climatically favoured broad-leaved tree species forced it into today's moorland and dry habitats, where the deciduous trees couldn't follow

Entire hazel forests must have existed for a time, especially in the Harz region, the Weser Uplands, the Eifel, the Rhenisch Slate Mountains, in Upper Hesse and the Black Forest (Fig. 4.8). Hazel bushes were also found in the lowlands, but probably not quite in such numbers. This situation, also applies to the east, where they were only able to flourish once the local climate had also become more Atlantic, after the English Channel had broken through. The Boreal proliferation of the hazel was meanwhile also unable to make as much headway in the south-eastern parts of central Europe because some of its potential growing areas were already occupied by the spruce (*Picea abies*). Coming from its refuges in the northern Dinarites (mostly Slovenia and Croatia) and south-eastern parts of the Alps, the spruce tree had spread around the high mountain ranges in the east and also become indigenous again in the regions to the west from there. The spruce trees thrived in the pine forests of the time, shading out the Scots pines (*Pinus sylvestris*). In this manner they spread across the eastern Alps from the south-east right through to Upper Bavaria, the Bohemian Forest, the Ore Mountains and Thuringian Forest, up to a number of upland areas in the Rhön Mountains, the Sauerland region and through to the Harz Mountains. The progressive distribution of hazel bushes from the west and advance of the spruce from the south-east laid the foundation for

Fig. 4.8 The hazel (*Corylus avellana*) needs warm temperatures and loves light. The bushes are suitable for coppicing and still cultivated in copses today. This is what central Europe's first deciduous forest formations must have looked like in the Boreal period

divergent developments of the vegetation and landscape in the west and east of Germany and central Europe that is still characteristic today.

The Atlantic period, which followed from around 6000 to 3200 BC, lasting approximately 3000 years, was attended by the appearance of a stable forest landscape largely characterized by elm (*Ulmus* spec.), oak (*Quercus*), lime (*Tilia*

Fig. 4.9 Xerothermic vegetation complex with dry grasslands on shallow rock and a natural drought tolerance limit of the forest at the "Klusenstein" cliff in the Sauerland region. The forest has been spreading everywhere since the Atlantic period and only rocky habitats such as this one have remained naturally forest-free for edaphic reasons

spec.), ash (*Fraxinus excelsior*), maple (*Acer* spec.) and alder trees (*Alnus glutinosa*). Although most of these species had already started appearing towards the end of the Boreal, they only reached their full distribution in the Atlantic period. The most significant climatic change that the mixed oak forest period brought in comparison with the early warm period (Boreal) was increased humidity. The spreading of dense, mixed deciduous forests markedly reduced the areas of xerothermic vegetation, which increasingly lost out in the competition for habitats, and rendered the transition of thermophilic floral elements through greater distances largely impossible. The restriction of photophilic and thermophilic species to special, insularly dispersed, forest-free sites sundered their connection to the main distribution areas (Fig. 4.9). This created the relict-like areal disjunctions still observable today.

With the summer temperatures 1.5–2 °C higher than today, on average, and more rainfall, there can be no doubt that the proliferation of broad-leaved species was linked with the general increase in precipitation, which is also confirmed by the appearance of a great number of low and high moors, many of them extensive. Given the highly diverse habitat conditions in central Europe's individual natural regions, various variants of the Atlantic mixed oak forest need to be assumed, more or less all of which have also been substantiated by way of pollen analysis:

- alder-rich variants in the marshy lowlands of the large river valleys and flat country
- elm-rich variants on nutrient-rich, fresh soils (most of all in the riparian zones)
- lime-rich variants in the loess regions and above ground moraine, as well as with higher proportions of ash (*Fraxinus*) and maple (*Acer*) in the higher reaches of the low mountain ranges
- birch-rich variants on the poor sandy soils of the north-west German geest (e.g. with pine, *Pinus sylvestris*)
- possibly spruce-rich variants in the Alpine foothills and eastern uplands.

The significant climatic change brought by the period of mixed oak forests (Atlantic period) in comparison with the Preboreal and Boreal, i.e. a higher humidity, also continued in the warm period (Subboreal) to follow. One profound consequence was a supplanting of the pine in the western parts of central Europe by biodiverse deciduous forests including oak, elm, lime, and later increasingly also beech trees (*Fagus sylvatica*). The spreading of sub-oceanic shade-tolerant species, first and foremost beeches, and progressive appearance of high moors point to a damp and cold climate with a growing proportion of snow and shortened growing season. Starting from the Subboreal, at the latest, the open landscapes with their communities of steppe plants can also be assumed to have been pushed out of the plain into small, insularly dispersed, exposed patch habitats in the arid regions, ensuring the final severance of connections to their areas of origin. The flora and vegetation were provided with their autonomous character since then.

The various hardwood and softwood species have hence migrated in our direction again from their refuge areas in the course of later climate improvements during the late ice age and thereafter. They came in stages and a very specific order, controlled by climate changes, over a period of approximately 9000 years from the first to the last migrating species. In the process, the beech tree (*Fagus sylvatica*) spread northwards from its ice-age refuges in the Mediterranean until it reached present limits, initially probably taking at least two routes to northern and central Europe from the south-west and south-east (Fig. 4.10).

The post-glacial proliferation of the beech tree in central Europe proceeded under simultaneous human influence almost everywhere, however. This means that there were constant overlaps between the natural course of the development and the moulding influence of people (Speier 2006). The selection of places for human settlement in prehistoric times was almost exclusively restricted to locations that could support beeches. It has long been known that the first notable interventions of human beings in the virtually continuous cover of deciduous forest after the Atlantic period took place at the beginning of the Neolithic Age, when the switch from the Mesolithic Age's land use to a more or less sedentary lifestyle and economy set in. Ever since then, the influence of farmers and cattle breeders on the remodelling and refashioning of the vegetation and landscape, needs to be recognized as decisive.

Fig. 4.10 Young beeches (*Fagus sylvatica*) are establishing themselves in a mixed oak forest in northern Germany after a lowering of the ground water table. This is how the spreading of the beech tree must have proceeded in the Atlantic period: after the climate change in the Subatlantic period, the shade-tolerant species is gradually overgrowing the light-dependent oaks and will henceforth dominate the tree layer (Pott 2010)

4.5 Man and Nature

Sedentary farming spread throughout central Europe as a new lifestyle in the sixth millenium BC. The so-called "neolithic revolution" had advanced there from the Balkans and Danube region in the south-east along with the people of the Linear Pottery culture. The economic and social switch to a sedentary rural lifestyle with agriculture and animal husbandry was simultaneously also attended by a transformation of the natural landscape into a cultivated landscape. The settlement areas of the older Linear Pottery culture in loess-rich plains show the first impact of farmers on the primeval landscape of the time. Based on the evidence from pollen studies for grain cultivation, the establishment of the first permanent settlements can be dated to around 5500–4500 BC. The post-glacial warm period, which led to better climate conditions in central Europe than we experience today, was evidently conducive to this process. The earliest farming cultures hence created the first open landscapes for their respective cultivation and use (Fig. 4.11). The settlements of succeeding Rössen groups (from around 4000 BC) or other Middle/Late Neolithic cultures became established in approximately the same time periods as the

Fig. 4.11 Reconstruction of a Neolithic Linear Pottery longhouse of the Rössen Culture in the Oerlinghausen open-air museum near Bielefeld. These thatched oak constructions could reach a length of 40 m and accommodated clans and their domestic animals. These buildings were not infinitely long-lived and devoured a tremendous amount of wood in the heretofore primeval landscape's first cleared islands

Linear Pottery culture, but extended from the primary settlement areas to the adjoining peripheral zones of limestone hills and highland slopes. They find their chronological and spatial pendant in the passage graves of Nordic Megalith architecture that have come down to us as impressive relicts of the so-called Funnelbeaker people and pay witness to the first settlements in north-western Europe's geest plains. Around 3000 BC, the largely dry and somewhat richer geest soils of the beech-and-oak-forest (Fago-Quercetum) constituted the nucleus of agricultural settlements and served the creation of farms and arable land. Many surviving Megalith graves of the Funnelbeaker cultures are striking testimonies to these first settlement periods. The northward advance of Neolithic farming cultures across the loess boundary did not take place until around 4000 BC, but very soon almost all the other areas of northern Europe had been "neolithicized". The Funnelbeaker culture was the carrier of this neolithicization in large parts of central and northern Europe. This and other Neolithic or later cultures can often be clearly observed in the pollen diagrams. They represent the beginning of human influence on natural ecosystems, and of the two-way interaction between humankind and nature! This evidently involved two different approaches, both of them described for the first time in Denmark: one with an emphasis on pollarding and with relatively little intervention in the forest, and then the so-called land seizure, where forest grazing resulted in a considerable thinning of the forest. *Fagus sylvatica*, for example, has never been able to claim its potential territory in the Pleistocene lowlands of north-western Germany for this reason. We can assume that the potential proportion of beech was higher on these Pleistocene Old and Young Drifts, and that the natural beech-and-oak forest (Fago-Quercetum) would have potentially claimed a greater area than today's oak-and-birch forest (Betulo-Quercetum) in the absence of human settlement activities. Even the prehistoric influence of human beings was therefore not exerted on a static vegetation but on a dynamic process; and one can no longer speak of a continuously natural vegetation since the later Atlantic period. Some authors consider this human impact on the natural landscape to be so drastic that a new term has been coined for the succeeding age of today, dominated as it is by humankind and human activity: the Anthropocene (Steffen et al. 2007).

Except in soils that were poor in nutrients or affected by ground or back water, the beech tree achieved absolute dominance almost everywhere at the time on various substrates (Fig. 4.12). Due to its edaphic and climatic advantages, it also developed its tremendous competitive power across small-scale locational differences, which still permits a rough tripartition into beech, mixed beech and beech-free forest today. The locational amplitude of the beech is so great that it forms a continuous area under current climatic conditions, from the plain to the montane level, where its main region lies, and from Sicily through to southern Sweden—except for habitats that are dry in summer. As an Atlantic-Submediterranean geoelement, *Fagus sylvatica* is even able to form the forest border in mountain ranges that are influenced by Atlantic climate, such as the Cévennes and Vosges, and in the Swiss Jura. The so-called maritime timber line lies where a median temperature of around 10 °C is still reached in July.

Today, *Fagus sylvatica* is found at nearly all altitudes of the low mountain ranges, from the flatlands to the highest slopes. The beech forests meanwhile are most common in the plain in the northern environs of the low mountain ranges, on limestone and loess. They are already typically developed at altitudes of 50 m above sea level, while the woodrush-beech forests (Luzulo-Fagetum) on silicate rocks do not reach altitudes lower than 160 m above sea level, and furthermore often still need to be regarded as transitional types between beech-and-oak forests and pure beech forests at these altitudes (Fig. 4.13). This phenomenon clearly demonstrates the superior competitive power of the beech at better locations. In the geest areas of the northern German plain, for example, the beech is only found in association with the sessile and common oak (*Quercus robur* and *Q. petraea*), most of all in the *Lonicera periclymenum*-beech forest (*Periclymeno-Fagetum*), but also in the oak-hornbeam forests of the *Stellario-Carpinetum* type (phytosociological terms as per Pott 1995). In north-western Germany, the loess loam soils and loess-containing substrates of the fertile plains are local domains for continuous lowland beech woods from the millet grass-beech (Milio-Fagetum) forest complex. This small-scale differentiation reaches back to the time of the beech invasion.

The great range expansion of the woody plants as described above started in the course of the post-glacial climate changes and development of the vegetation. The extermination of so-called megaherbivores, i.e. the largest grazing animals, since the land had been seized by human beings in the stone age, has meanwhile certainly

Fig. 4.12 A beech (*Fagus sylvatica*) forest with closed canopy and little understory, referred to as "hall forest" in Germany because trees of all ages make their way into the canopy, creating the impression of a Gothic cathedral in old-growth forests

been conducive to the development of forest thickets (Martin 1967; Vera 1997; Vera et al. 2006; Johnson 2009). People not only exterminated these large animals by hunting them, however, but also by claiming ever more land.

The megaherbivore theory is based on the assumption that large herbivores will have a crucial impact on the vegetation and landscape wherever they are found in

Fig. 4.13 Limestone-and-beech forest of the Galio odorati-Fagetum type (after Pott 1995), rich in spring geophytes like the hollowroot-birthwort (*Corydalis cava*)

great numbers. They are in particular able to break up previously continuous forests sufficiently for them to be supplanted by semi-open, park-like pasture landscapes. This is for example demonstrated by the African tree-savannahs, kept forest-free by elephants as a keystone species. The theory attributes the fact that central Europe, for example, was not a savannah-type park landscape in prehistoric times, but a continuous forest area, to people keeping the animal populations small by hunting, and/or to the extinction of the largest megaherbivores such as elephants and rhinos in central Europe by the end of the last ice age (Martin and Klein 1984).

In the Atlantic period, when people began to settle down, central Europe was dominated by extensive forests of oak, elm, lime, ash and maple trees, as we have seen, with beeches and hornbeams added later. The clearing activities and forest grazing opened the forests up over time, creating regionally differentiated, open and semi-open cultivated landscapes with small-scale structures of fields, pastures and meadows, along with a great number of forests, copses and hedgerows. Areas of open land are therefore anthropo-zoogenic in the forest climate of central Europe.

Our observations concerning the impact of grazing livestock on what used to be more or less continuous forests date from the time of Early Medieval forest use, which led to various remodellings of the landscape, and hence forest structures, depending on the type of grazing livestock and grazing intensity (Pott and Hüppe 1991, 2008).

The megaherbivore theory deems it likely and possible, however, that a mosaic of areas with various degrees of openness and various succession stages would also predominate in large parts of central Europe through the browsing of animals alone, without human influence. If this were to be the case, one would need to assume the continued existence of large herds of wild animals such as wild horses, bison and other herbivore species (Johnson 2009).

What speaks against the herbivore theory is the existence of species that depend on an undisturbed development of the forest over centuries. In addition, central Europe has hardly any endemic open land plant species and subspecies, in contrast to the Mediterranean or steppe landscapes in the east, which would indicate a relatively young age of the open vegetation, even if there are also contrary opinions to this (Walentowski and Zehm 2010). Pollen research has also failed to turn up any indications of open landscapes for the period in question (Birks 2005; Litt 2000). The pollen of open land species (such as grasses, for example) only start appearing more frequently with appreciable clarity from the beginning of the Neolithic Age and the introduction of agriculture and cattle breeding. If open landscapes had indeed been widespread before that, this would be demonstrable without fail.

Ever since the Neolithic period, the domesticating livestock species had furthermore started to compete for food with the aurochs, wisents and wild horses, which sought refuge in the riparian areas of the many-branched rivers, in swamps and moors. Switching to the forage plants of the forest was impossible for most wild animals because they are unable to digest and/or neutralize the substances and antigens contained in many herbs and woody plants. This was different for deer and elks (*Alces alces*), which feed on highly nutritious buds and shoots, and found plenty of grazing in the new growth of the forests. The elk, which loves forests with many shallow waters, has been slowly forced out into eastern Europe, however. Red deer (*Cervus elaphus*) and wild boar coped well with the loss of open pasture landscapes and became true forest animals. Some of the exterminated grazing animals are being reintroduced today and employed to maintain grassland landscapes and common pasturages for nature conservation. Especially the abandoned former wood pastures that are popularly referred to as "Urwald" (primeval forest—e.g. Bentheimer Urwald and Neuenburger Urwald, and the Hasbruch north of Oldenburg, Figs. 4.14 and 4.15) in Germany and have been left to themselves as protection areas and/or so-called "natural forest reserves" for quite a while, are showing a development today where shade-tolerant trees are growing up densely until they tower over the old, broad-crowned pasture trees and shade them out (Pott and Hüppe 1991, 2008; Behre 2010). The euphemistic term "natural forest reserve" is not quite accurate here because these are not natural forests, but the old forests should of course be allowed to continue developing naturally. Perhaps "forest reserves" would be more honest, with more or less dense, continuous timber forests being the natural successor.

Fig. 4.14 The "Hasbruch" near Oldenburg is a former wood pasture and pollarding forest whose multi-layered structure is still recognizable today: oaks were left free-standing to supply timber and acorns for the animals, while hornbeams were moulded into candelabra-type shapes by regular pollarding to provide leaf fodder. The evergreen *Ilex aquifolium* directly on the forest floor is the result of positive grazing selection (from Pott and Hüppe 2008)

4.6 Forest Landscapes or Park Landscapes?

The duration, intensity and impact of the anthropogenic influence in the Neolithic period were not consistent in the various natural regions of central Europe, however, but exhibited temporal and regional differences between the coastal regions, geest areas, mountains and uplands with their carbonaceous and loess soils, as well as the low and high mountain regions.

Geobotanical research into the intensity and continuity of settlements, into the evidence for the migration or cultivation of specific crops plants, and into the manner of these agricultural activities, was only started around 50 years ago. It is founded on the decisive methodological basis provided by the knowledge of, ability to register, and statistical confirmation of pollen from plants whose presence is largely or exclusively linked to forms of human settlement. Typical settlement indicator pollen for north-western Germany for example include those of grain species (cereals), plantain (*Plantago*), goosefoot (*Chenopodium*), mugwort (*Artemisia*), red sorrel (*Rumex acetosella*), stinging nettle (*Urtica*), cornflower (*Centaurea cyanus*), buckwheat (*Fagopyrum*), flax (*Linum usitatissimum*) and walnut

Fig. 4.15 Old oaks formerly subjected to copse management and coppicing grew through again once this was discontinued, creating the impression of an eerie "fairy tale forest" today, with the trees grown into bizarre shapes. "Ahlhorner Heide" heath in the "Wildeshausener Geest" nature park, north-western Germany (from Pott and Hüppe 2008)

(*Juglans*), as well as grasses, composites, etc. All these plants were part and parcel of the original native vegetation, but have been advanced by human beings. The evaluation of prehistoric cultivation methods additionally relies on indicators of a secondary nature that mostly point to a partial clearance of the landscape such as, for example, the decline of specific tree species and proliferation of grass and herb pollen. These indicators, among others, permit us to reconstruct the proportion of forested to open land.

The coastal marsh already featured high, verifiable frequencies of grass pollen in the Atlantic period, which originate from naturally wood-free tidal reeds or salt meadows and reflect the transgression and regression phases of the North Sea in their oscillating graphs. Permanently forest-free landscapes only start appearing recently along with human settlement.

Similar, naturally high grass pollen frequencies that occur before the appearance of settlement-indicating pollen are also found in the upland moors. These moors were originally sparsely wooded and covered with tall *Molinia* tussocks, as is still observable occasionally today.

The Atlantic period witnessed the development of deciduous forests with beech trees everywhere, which is reflected in a high frequency of tree pollen. Only the settlement activities of the Bronze Age and their expansion in the Iron Age served to reduce the forests and provide the first anthropo-zoogenic open landscapes. Land seizure is also evident in the Alpine foothills, where agriculture only plays a minor role today because of the specific climate situation, and livestock farming now predominates with its pasturages. But large parts of these foothills were put to agricultural uses right into the eighteenth century. This used to involve special regional forms of agriculture (e.g. ridge and furrow) where the areas in question needed to be wrested from the forest. But the landscape only opened up in very recent times here as well. Even the highest ranges of the Alps attest to oscillations of the timber line with a highly fluctuating share of tree pollen since the natural climatic change from the Boreal to the Atlantic period, with anthropo-zoogenically induced timber line depressions at the end of the Neolithic period and in the Bronze Age. The continually verifiable grass pollen frequencies largely stem from the alpine level, where we have been able to produce paleoecological evidence for a shifting of the timber line to over 2350 m above sea level in the warm period from fossilized Swiss pine wood dating from 6860 to 6140 years BP (Fig. 4.16). The pollen analysis data hence provide no indications for natural park landscapes in mostly forested, temperate central Europe throughout the entire geographic comparison (Pott et al. 1995).

Fig. 4.16 Larch and Swiss stone pine forest of the *Larici-Pinetum cembrae* type (after Pott 1995) that has been thinned out by alpine grazing and wood harvesting at the intra-alpine timber line in the Fimbatal valley (Val Fenga, Silvretta, Switzerland). Post-glacial, natural and climate-controlled fluctuations of the timber line by several hundred meters have been substantiated in this region for the warm phase of the Atlantic period around 5000 BC (from Pott et al. 1995)

This supports the assumption that large swathes of central Europe have been deforested between the Neolithic age and today. This still leaves partly unclear the proportion of natural open land in wet and humid habitats, e.g. upland and lowland moor landscapes, and in the riparian zones. As well, the exact share of natural dry grassland biotopes in the open land areas will be debated again and again. But what is certain is that the natural continental and Submediterranean dry grassland communities in the climatically controlled forest and grass steppes of eastern and south-eastern Europe are increasingly turning into natural open landscapes. This should always be taken into account for the discussion.

In the deciduous, nemoral broadleaf forest zones within the potential distribution area of the beech, primary xerothermic dry grasslands are limited to small-scale extreme habitats only. However, extensive secondary dry grasslands have often spread to open land areas through mowing and grazing in the course of the anthropo-zoogenic thinning of the forest. Left untouched, these areas will all gradually be reclaimed by the forest.

Even the prehistoric deforestation stages in central Europe of the Neolithic and Bronze Age are very beautifully demonstrable by combinations of modern pollen analysis and archaeobotanical data. These stages continue on into the Iron Age and into the era of the Roman Empire, with local reforestation of briefly neglected, formerly cultivated areas, not being observable before the Migration Period.

Many studies demonstrate widespread agricultural activity with open land at the end of the pre-Roman, Iron Age extending far beyond the Roman Empire's contemporary boundaries. Around 450 AD, however, at the high point of the migrations, there are increasingly unequivocal indications of reforestation processes that are linked to declining settlements and a correspondingly reduced agricultural activity. Only at the beginning of the Early and High Middle Ages around the turn of the millennium are the anthropo-zoogenic open land areas inexorably on the increase again, at the expense of the forest.

References

Behre K-E (1988) The role of man in European vegetation history. In: Huntley B, Webb T (eds) Vegetation history. Handbook of vegetation science. Kluwer-Verlag, Dordrecht, pp 633–672

Behre K-E (2008) Landschaftsgeschichte Norddeutschlands. Wachholtz, Neumünster, 308 p

Behre K-E (2010) Der Neuenburger Urwald – ein Denkmal der Kulturlandschaften. Brune-Mettcker Druck- und Verlagsgesellschaft, Wilhelmshaven, 136 S

Birks HJB (2005) Mind the gap: how open were European primeval forests? Trends Ecol Evol 20 (4):154–156

Bork H-R, Bork H, Dalchow C, Faust B, Piorr H-P, Schatz T (1998) Landschaftsentwicklung in Mitteleuropa. Klett-Perthes, Gotha, 328 p

Burrichter E, Pott R (1987) Zur spät- und nacheiszeitlichen Entwicklungsgeschichte von Auenablagerungen im Ahse-Tal bei Soest (Hellwegbörde). In: Köhler N, Wein N (eds) Natur- und Kulturräume, Festschrift Ludwig Hempel. Münstersche Geogr. Arbeiten, vol 27. Schöningh, Paderborn, pp 129–135

Burrichter E, Pott R, Raus T, Wittig R (1980) Die Hudelandschaft "Borkener Paradies" im Emstal bei Meppen. Abhandl Landesmuseum f Naturk 42(4), 69 p

Ellenberg H (1990) Bauernhaus und Landschaft in ökologischer Sicht. Verlag Eugen Ulmer, Stuttgart, 585 S

Ellenberg H (1996) Vegetation Mitteleuropas mit den Alpen, 5. Aufl. Verlag Eugen Ulmer, Stuttgart, 1095 S

Ellenberg H, Leuschner C (2010) Vegetation Mitteleuropas mit den Alpen, 6. Aufl. Ulmer Verlag, Stuttgart, 1333 S

Emanuelsson U, Bergendorff C, Carlsson B, Lewan N, Nordell O (1985) Det Skånska Kulturlandskapet. B.T.J. Datafilm, Lund, 248 p

Hoskins WG (1988) The making of the english landscape, 13th edn. Penguin-Books, London, 327 p

Humboldt Av (1845) Kosmos. Entwurf einer physischen Weltbeschreibung. Stuttgart-Tübingen (Reprint Eichhorn, Frankfurt 2004)

Hüppe J (1993) Die Entwicklung der Tieflands-Heidegesellschaften Mitteleuropas in geobotanisch-vegetationsgeschichtlicher Sicht. Ber d Reinh Tüxen-Ges 5:49–76

Johnson CN (2009) Ecological consequences of late Quaternary extinctions of megafauna. Proc Royal Soc Ser B 276:2509–2519

Krzywinski K, O'Connell M, Küster H (2009) Europäische Kulturlandschaften, 1. Aufl. Aschenbeck Media, Bremen, 216 S

Küster H (1995) Geschichte der Landschaft in Mitteleuropa von der Eiszeit bis zur Gegenwart. Verlag C. H. Beck, München, 424 S

Küster H (1998) Die Geschichte der Gärten und Parks. Insel-Verlag, Frankfurt, 320 p

Küster H (2010) Geschichte der Landschaft in Mitteleuropa. Von der Eiszeit bis zur Gegenwart. 4. Aufl. Verlag C. H. Beck, München, 448 S

Küster H, Volz W (2005) Natur wird Landschaft – Niedersachsen. Zu Klampen Verlag, Springe, 143 p

Litt T (2000) Waldland Mitteleuropa. Die Megaherbivorentheorie aus paläobotanischer Sicht. In: Großtiere als Landschaftsgestalter. LWF-Bericht 27, 2, Landesanstalt für Wald- und Forstwirtschaft, Bayer

Martin PS (1967) Prehistoric overkill. In: Martin PS, Wright HE (eds) Pleistocene extinctions: the search for a cause. UMI Books on Demand, Ann Arbor, MI, pp 75–120

Martin PS, Klein RG (1984) Quaternary extinctions. A prehistoric evolution. University of Arizona Press, Tuscon

Pott R (1985) Vegetationsgeschichtliche und pflanzensoziologische Untersuchungen zur Niederwaldwirtschaft in Westfalen. Abh Westf Mus Naturk 47(4):1–75

Pott R (1986) Palynological evidence of extensive woodland management in turn with agriculture in the area of the Hauberge in Siegerland, North Rhine-Westfalia, FRG. In: Behre KE (ed) Anthropogenic indicators in pollen diagrams. Verlag Balkema, Amsterdam, pp 125–134

Pott R (1993) Farbatlas Waldlandschaften. Ausgewählte Waldtypen und Waldgesellschaften unter dem Einfluss des Menschen. Ulmer Verlag, Stuttgart, 224 S

Pott R (1995) Die Pflanzengesellschaften Deutschlands. 2. Aufl., Ulmer Verlag, Stuttgart, 622 S

Pott R (1996) Biotoptypen. Schützenswerte Lebensräume Deutschlands und angrenzender Regionen. Ulmer Verlag, Stuttgart, 448 S

Pott R (1997) Von der Urlandschaft zur Kulturlandschaft – Entwicklung und Gestaltung mitteleuropäischer Kulturlandschaften durch den Menschen. Verhandl d Ges f Ökol 27:5–26

Pott R (1998) Effects of human interference on the landscape with special reference to the role of grazing livestock. In: Wallis de Vries MF, Bakker JP, van Vieren SE (eds) Grazing and conservation management, conservation biology series. Kluwer Academic, Boston, pp 107–134

Pott R (1999) Diversity of pasture-woodlands of North-Western-Germany. In: Kratochwil A (ed) Biodiversity in ecosystems: principles and case-studies of different complexity levels, Tasks of vegetation sciences, vol 34. Kluwer Academic, Dordrecht, pp 107–132

Pott R (2000) Die Entwicklung der europäischen Buchenwälder in der Nacheiszeit. Rundgespr d Kommission f Ökol 18:49–75

Pott R (2003) Biodiversität kulturhistorischer Wälder in Mitteleuropa. In: Colantonio-Venturelli R (ed) Paessagio culturale e biodiversitá. Villa Vigoni, Menaggio, pp 17–45

Pott R (2010) Klimawandel im System Erde. In: Pott R (ed) Berichte der Reinhold-Tüxen-Gesellschaft, vol 22. Reinhold-Tüxen-Gesellschaft, Hannover, pp 7–33

Pott R (2014) Allgemeine Geobotanik, Biogeosysteme und Biodiversität, 2nd ed. Springer, Berlin, 652 S.

Pott R, Burrichter E (1983) Der Bentheimer Wald – Geschichte, Physiognomie und Vegetation eines ehemaligen Hude- und Schneitelwaldes. Forstwiss Centralblatt 102(6):350–361

Pott R, Hüppe J (1991) Die Hudelandschaften Nordwestdeutschlands. Abh Westf Mus f Naturk 53 (1/2):1–314

Pott R, Hüppe J (2008) Naturschutzfachliche Bedeutung und Biodiversität kulturhistorischer Wälder und Hudelandschaften in Nordwestdeutschland. Abh Westf Mus f Naturk 70 (3,4):199–226

Pott R, Hüppe J, Remy D, Bauerochse A, Katenhusen O (1995) Paläoökologische Untersuchungen zu holozänen Waldgrenzschwankungen im oberen Fimbertal (Val Fenga, Silvretta, Ost-Schweiz). Phytocoenologia 25(3):363–398

Romell G (1967) Die Reutebetriebe und ihr Geheimnis. Studium Generale 20:362–369

Speier M (2006) Holozäne Dynamik der Europäischen Rotbuche (*Fagus sylvatica*) in der regionalen Waldentwicklung des Westfälischen Berglandes. Dechenania 159:5–21

Steffen W, Crutzen PJ, McNeill JR (2007) The Anthropocene: are humans now overwhelming the great forces of nature? Ambio 36:614–621

Trier J (1963) Venus-Etymologien um das Futterlaub. Bohlau Verlag, Köln, 207 p

Trier J (1968) Anger und Park. Veröffentl Inst und Lehrstuhl f Landschaftsbau und Gartenkunst, Berlin, vol 19, 17 pp

Vera FWM (1997) Metaforen voor de Wilderness. Minist. Van Landbouw, 's-Gravenhage, 426 p

Vera FWM, Bakker ES, Olff H (2006) Large herbivores: missing partners of Western European light demanding tree and shrub species? In: Danell K, Duncan R, Bergstroem R, Pastor J (eds) Large herbivore ecology, ecosystems, dynamics and conservation. Cambridge University Press, Cambridge, pp 203–231

Walentowski H, Zehm A (2010) Reliktische und endemische Gefäßpflanzen im Waldland Bayern – eine vegetationsgeschichtliche Analyse zur Schwerpunktsetzung im botanischen Artenschutz. Tuexenia 30:59–81

Sclerophyllous Versus Deciduous Forests in the Iberian Peninsula: A Standard Case of Mediterranean Climatic Vegetation Distribution

Rosario G. Gavilán, Beatriz Vilches, Alba Gutiérrez-Girón, José Manuel Blanquer, and Adrián Escudero

Abstract

Iberian Mediterranean forests are mainly dominated by *Quercus* species, which belong to different sections of the genus characterized by the condition of leaves, sclerophyllous vs. deciduous. There are also other Fagaceae-dominated forests such as beech forest, but they are considered relict and have a restricted distribution on mountains. Sclerophyllous oak forest covers the biggest part of the area, mainly in valleys or plateaus, while deciduous forests are distributed along mountain ranges where the higher amount of precipitation compensates for the typical dryness of Mediterranean climate. As examples we can mention *Q. canariensis* or *Q. pyrenaica* that have functional particularities, i.e. a marcescent leaves that only fall when the new ones appear, and have a quite restricted distribution so that they are considered endemic species. Although the widespread sclerophyllous forests of *Quercus rotundifolia* cover a large area, there are other types that are of ecological, chorological, and climatic interest. This is the case for cork oak forest (*Q. suber*) or olive woodlands (*Olea sylvestris*), that are also treated in this paper. We have also included some azonal vegetation types, mainly deciduous forests, present along riverbeds in extended areas of Mediterranean Iberia.

Keywords

Forest · Mediterranean climate · Marcescent · Quercus · Deciduous · Sclerophyllous

---

R.G. Gavilán (✉) · B. Vilches · A. Gutiérrez-Girón · J.M. Blanquer
Departamento de Biología Vegetal II, Facultad de Farmacia, Universidad Complutense, Madrid, Spain
e-mail: rgavilan@ucm.es

A. Escudero
Área de Biodiversidad y Conservación, Universidad Rey Juan Carlos, Móstoles, Madrid, Spain

5.1 Introduction

The Mediterranean part of the Iberian Peninsula forms the most extended area of the Mediterranean Basin together with north-western African territories. This climate is extended over the 80% of the Iberian Peninsula territory. The rest of Mediterranean areas of Europe and North Africa run as a more or less narrow fringe along the sea coast. It is the only territory in Europe with both continental and arid Mediterranean vegetation. It borders on temperate and summer rainy climate territories in the north, as it does in some other regions of the world that also have a Mediterranean climate. Iberian Mediterranean territories are situated to the south of Spanish high mountain ranges such as the Pyrenees and the Cantabrian Range; both mostly belong to the Eurosiberian region.

The central area of the Iberian Peninsula is occupied by a large and elevated plateau (670 m asl in average), divided by the Central System, a northeast-southwest mountain range that reaches 2592 m asl in Sierra de Gredos. This plateau is surrounded by other ranges like the Cantabrian Range in the north, Sistema Ibérico in the east, Betic ranges in the southeast, and Sierra Morena in the south. Further away there are other important ranges like Sierra Nevada in the southeast, where the highest point of the Peninsula is reached (3478 m asl). This plateau is the oldest area, where rocks (mainly metamorphic rocks) were created during the Carboniferous. In the Tertiary, big mountains such as the Betic ranges and the Pyrenees arose, the Meseta bent, and ranges inside it were formed. Outside it other ranges were formed, such as the Sistema Ibérico, due to the pressure of the Pyrenees on the plateau. After this period, sedimentary deposits appeared in the Ebro and Guadalquivir watersheds. That resulted in the present landscape becoming very hilly (Fernández-González 1986).

The climate of the Mediterranean area of the Iberian Peninsula is characterized by a severe dry season coinciding with the warmest and most favorable season for plant growth, at least in thermic terms. Rainfall is very variable and unpredictable or irregular in space and time. Due to the extension of the Iberian Peninsula, the continental climate is also an important ecological factor in vegetation distribution. In general, annual climatic amplitudes (extremes) reach their lowest values near the coast, while the highest appear in the center of the Peninsula.

If we take into account the global circulation of air, the Iberian Peninsula is situated in the so-called Ferrel cell, a secondary circulation feature in between high- and low-pressure areas, at mid-latitudes, and is influenced by western winds whose high-speed flowing aloft is called the Jet Stream. The two components of the Jet Stream, the Polar and the Subtropical Jets, have a different position in winter and summer. In winter, the warm-air masses shift to the equator and the Polar Jet flows to the south bringing cold and rain to the Iberian Peninsula. In summer, the Subtropical Jet shifts north; one of its warm-air masses, the Azores anticyclone, brings warm weather to the Iberian Peninsula. Its western- and southwestern parts are the territories most influenced by these phenomena, and their summers are very dry and winters very humid. The easternmost areas, closer to the Mediterranean Sea, are not so dry in summer due to the behavior of the Iberian Peninsula as a small continent. Inland, the air is progressively heated during summer, so low pressure areas are induced at the same time that water is evaporating over the surrounding seas. This humid sea-air is destabilized inland when it is heated, causing storms. On the other hand, winters are drier than in the western part due to the lesser influence of western winds. In summary, length and intensity of summer dryness increase from north to south and from northeast to southwest, similarly to temperature that increases from north to south (Gavilán 2005; Blanquer Lorite et al. 2014).

Five thermoclimatic belts were formerly defined for the Mediterranean Iberian Peninsula based on thermic indices and temperatures (Rivas-Martínez 1987), although more were also applied to other Mediterranean territories (Rivas-Martínez et al. 2002). The thermomediterranean belt runs near to the coast, and also reaches some inland areas through the Gualdalquivir valley. The climate is warm (no less than 18 °C of annual mean temperature) and without a freezing season. The Mesomediterranean belt reaches 800–900 m asl and is the largest geographically, with mean annual temperature over 13 °C. The Supramediterranean belt, higher in altitude, reaches 1700–2000 m asl. It can be considered a mountain belt although it covers the greatest part of the northern plateau. The mean annual temperature is between 8–13 °C. Oro- and Cryoromediterranean belts occupy small areas in mountains above 1700–2000, or more than 2000 m asl, respectively. The mean annual temperature is 4–8 °C in the oro- and less than 4 °C in the cryoromediterranean belt. The last only appears in the top of mountains of Sierra Nevada, if we consider as Submediterranean those alpine territories of the other Iberian Mediterranean mountains it means that Submediterranean is an intermediate between temperate and Mediterranean climates (Rivas-Martínez et al. 2002). With regard to precipitation (annual rainfall: P) certain ombrotypes (rainfall types) also have been defined. The semiarid type (200–350 mm) appears in the southeast of the Iberian Peninsula or in some rain-shadowed areas like in the Ebro watershed or in the Guadix and Baza depressions (Andalusia). It is linked to the thermo- and mesomediterranean belts. The dryness (350–600 mm) is more extended in the Peninsula and it is linked to the thermo-, meso- and supramediterranean belts. The subhumid (600–1000 mm), humid (1000–1600 mm) and hyperhumid (1600–2300 mm) ombrotypes occur usually in the supra-, oro- and cryoromediterranean belts, but also in some mesomediterranean areas like Sierra de Grazalema (Cádiz, Andalusia), the western part of the Sistema Central (Cáceres, Extremadura), Sistema Ibérico (Castilla), and Catalonian coast range (Rivas-Martínez et al. 1990).

The aim of our paper is to show the main characteristics of vegetation types that cover extended areas of the Iberian Peninsula, sclerophyllous and deciduous forest. Those features explained their particular phytogeographic distribution pattern, including floristic composition of woody and herb layers but also the canopy layer, the geographical or altitudinal distribution in this vast Mediterranean territory and, finally, the main climatic characteristics that differentiate them. The combination of all of these features in a multidimensional space help us to understand the complexity of those communities and the landscape they form.

5.2 Evergreen Sclerophyllous Forests (Table 5.1)

Oak species form the main vegetation types of the Iberian Mediterranean landscape. *Quercus rotundifolia*, *Q. ilex* and *Q. suber* are the most characteristics species of those forests. They cover extensive territories in a broad range of altitudes and variety of bioclimates. *Quercus coccifera* also forms typical shrublands and *Olea europaea* var. *sylvestris* forms woodlands; they occur in clay soils (montmorillonite) of southern Spain.

5.2.1 *Quercus ilex/rotundifolia* Forests

These forests are distributed in the Central Meseta, western Iberian Peninsula, Andalusia, Ebro Valley and Catalonia; and even in the Eurosiberian part where it is a relict formation. They grow well on a variety of substrates: calcareous or siliceous soils, and even on very restrictive ones such as gypsum, dolomitic and serpentine soils. They cover a wide range of altitudes, from sea level to over 2000 m asl in Sierra Nevada. Likewise, they are adapted to a wide range of precipitations, from semi-arid sites in Sierra Alhamilla (Almería, P = 350 mm/year, approx.) to humid sites in the Montseny (Girona, *Q. ilex* stands, P > 1000 mm/year).

The huge ecological amplitude of holm oaks is a reflection of their ability to withstand very variable thermic, substrate and moisture conditions (Granda et al. 2014). Under such different environmental scenarios, a complex series of holm oak forests have developed, based on forest structure, floristic and dynamic criteria. In general, three main types can be recognized.

These forests have been conserved as coppice stands and also as "dehesas." The coppice stands traditionally have been managed to obtain fire fuel, but now are related to recreational activities such as hunting (deer, wild boar). The total surface occupied by these stands is gradually growing as a consequence of the systematic abandonment of lands of marginal crops. The dehesa is a typical multi-purpose sustainable agroforestry system consisting of a mosaic of widely spaced scattered oaks (Blanco et al. 2005) whose abandonment encourages the installation of shrub communities (Gavilán and Fernández-González 1997). Dehesas have been traditionally used for cattle raising, being particularly extensive in the western part of the Iberian Peninsula, due to the inability of soils to sustain agriculture practices. They are not perceived as having a high diversity and are not usually awarded the appropriate conservation status. However, they are at greater risk of incorrect land-management decisions as when one attempts to restore this type of system with alien species which not only damage these poorly protected areas, but also other areas nearby that are of paramount interest for conservation (Plieninger and Wilbrand 2001; Gavilán et al. 2015).

Several types of grassland appear in these stands as a consequence of canopy and cattle management: those formed by tall grasses (*Agrostis castellana*, *A. pourretii*) in humid zones used mainly by bovine cattle; or those formed by perennial short

Table 5.1 Forest units of central Spain. For each unit the following information is presented: phytosociological name of forest association, the approximate altitudinal range (Alt.); Exposure (Exp.: north or/and south); Distribution range (DR: western or eastern part); Bioclimatic belt (Belt: meso [MM] and/or supramediterranean [SM]; see text); corresponding woody seral associations

Forest type/ Syntaxonomical unit	Alt.	Exp.	DR	Belt	Woody Seral communities	
					Scrublands	Shrublands
Sclerophyllous forest						
Junipero oxycedri-Quercetum rotundifoliae	500–1100	N/S	E	SM/MM	Cytiso scoparii-Retametum sphaerocarpae Cytiso scoparii-Genistetum floridae Lavandulo-Adenocarpetum aurei	Rosmarino-Cistetum ladaniferi Santolino-Cistetum laurifolii
Genisto hystricis-Quercetum rotundifoliae	600–900	N	W	SM/MM	Genisto hystricis-Cytisetum multiflori Cytiso multiflori-Retametum sphaerocarpae	Lavandulo pedunculatae-Genistetum hystricis Cisto ladaniferi-Genistetum hystricis
Pyro bourgeanae-Quercetum rotundifoliae	250–600	S	W	MM	Cytiso multiflori-Retametum sphaerocarpae	Genisto hirsutae-Cistetum ladaniferi
Deciduous forest						
Luzulo forsteri-Quercetum pyrenaicae	900–1300	N/S	E	SM	Cytiso scoparii-Genistetum floridae	Rosmarino-Cistetum ladaniferi Santolino-Cistetum laurifolii Halimio ocymoidis-Cistetum laurifolii
Genisto falcatae-Quercetum pyrenaicae	700–1000	N	W	SM	Genisto hystricis-Cytisetum multiflori Thymo-Cytisetum multiflori	Lavandulo pedunculatae-Genistetum hystricis
Arbuto unedonis-Quercetum pyrenaicae	450–800	S	W	MM	Cytisetum multifloro-eriocarpi	Halimio ocymoidis-Ericetum umbellatae Halimio

(continued)

Table 5.1 (continued)

Forest type/ Syntaxonomical unit	Alt.	Exp.	DR	Belt	Woody Seral communities	
					Scrublands	Shrublands
						ocymoidis-Cistetum psilosepali *Polygalo microphyllae-Cistetum populifolii*

grasses (*Poa bulbosa, Trifolium subterraneum*) appear in dehesa and are usually exploited by sheeps; annual grasslands can also grow in those open areas.

5.2.2 *Quercus rotundifolia* Forests on Acid Soils

These forests grow mainly in the mesomediterranean belt of the western half of the Iberian Peninsula, but sometimes also in supramediterranean zones where rainfall is low and sporadic, in the red-stone outcrops of the eastern ranges (Table 5.1). Under optimal conditions the trees can reach the height of 15 m. These forests have been historically transformed to dehesas, although some well-preserved stands remain in certain parts of western Spain and in central- and southern Portugal.

The canopy layer is almost exclusively composed of holm oaks. Only under more humid conditions do some smaller oaks appear, such as *Q. faginea* subsp. *broteroi* and *Q. pyrenaica* and, mainly, the evergreen sclerophyllous cork oak (*Q. suber*). On the driest edge, *Juniperus oxycedrus* or wild olives (*Olea sylvestris*) are found. The herbaceous layer is composed of few plants; standing out are *Hyacinthoides hispanica, Paeonia broteroi, Carex dystachya* and some spore-forming vascular plants (*Asplenium onopteris, Selaginella denticulata*). The sub-canopy layer forms an intricate thicket where many evergreen tall shrubs and climbers appear, such as *Daphne gnidium, Phillyrea angustifolia, Ph. media, Rhamnus alaternus, Arbutus unedo, Lonicera* spp., *Asparagus* spp., *Viburnum tinus* or *Pyrus bourgeana*.

The shrubby formations that are dynamically related to these forests are of the most typical vegetation elements of the Iberian Peninsula, the so called *jarales*. These formations are dominated by different rock-rose species such as *Cistus ladanifer, C. salvifolius* and *C. crispus*, some medium-size shrublets (chamaephytes) such as *Lavandula stoechas* group, *Thymus zygis* and *T. mastichina* and, finally, by some spiny *Genisteae* such as *Genista hystrix* or *G. falcata*. Large areas of the shallow and more eroded soils of the western half of the Iberian Peninsula are covered by these poor communities. The shrubs usually release allelopathic compounds to the soil in order to avoid competition from forest species. Where soils are deeper, conspicuous shrubby and retamoid (broom-like) formations dominated by *Retama sphaerocarpa*, *Adenocarpus decorticans*, *Cytisus multiflorus* or *C. scoparius* can grow. The mature forests on the acid soils of the thermo-Mediterranean zones of the Guadalquivir valley are rich with *Myrtus communis*.

5.2.3 *Quercus rotundifolia* Forests on Limestone

This forest type covers extensive territories on the meso- and supramediterranean belts on calcareous substrata, which also appear in southern Portugal. Most of their representations are extensive coppice stands. The herbaceous layer is richer than on acidic sites; including for example, *Paeonia coriacea* and *Thalictrum tuberosum*, and some orchids (*Cephalanthera* spp., *Epipactis* spp.). Seral shrublands are dominated by Labiatae such as *Salvia lavandulifolia*, *Lavandula latifolia*, *Thymus vulgaris* and *Rosmarinus officinalis*, and also by some Genisteae such as *Genista scorpius*, *G. cinerea* and *Retama sphaerocarpa*. These types of forest also appear in subhumid areas with continental climate in northern Spain. They are rich in nanophanerophytes (*Spiraea obovata* subsp. *hispanica*, *Buxus sempervirens*, *Berberis vulgaris*). On the thermomediterranean coastal fringe of eastern territories, there is a subtype of that forest which is enriched with thermophyllous and xeromorphic species such as *Chamaerops humilis*, *Rhamnus oleoides*, *Asparagus albus*, *Osyris quadripartita* and even some vines, such as *Rubia peregrina* subsp. *longifolia*. However, they are poorly conserved due to the intense agriculture practices in these warm areas. Some scattered and poorly conserved stands of this type of forest also grows in southern areas of Mallorca and even in Ibiza (Iriondo et al. 2009).

5.2.4 *Quercus ilex* Thermomediterranean Forests

Scattered *Quercus ilex* stands grow on the thermomediterranean coastal fringe of Catalonia, avoiding dry and semi-arid areas, and in northern Spain, on the coastal border of the Cantabrian Sea, in the Eurosiberian floristic province (Folch et al. 1984). *Quercus ilex* is a close relative of *Q. rotundifolia*, featuring larger leaves and taller trunks. The taxonomical status of both of these oaks is being resolved, although one can note floristic and structural differences between the two extremes of the species. More mesophytic environments, in combination with milder winter temperatures, lead to very complex and rich forests. The subcanopy layer is composed of many small broad-leaved evergreen trees such as *Viburnum tinus*, *Arbutus unedo*, *Pistacia lentiscus* and *Phillyrea latifolia*. Vines such as *Tamus communis*, *Smilax aspera*, *Lonicera implexa*, *Clematis* spp. and *Rubia peregrina*, which were scarce in *Q. rotundifolia* forests, become abundant.

This type of forest is also widely distributed in the Balearic Islands, in the Mediterranean Sea. The presence of numerous endemics, such as the geophyte

Cyclamen balearicum, the climber *Smilax balearica*, *Rhamnus ludovici-salvatoris* and *Paeonia cambessedesii*, readily distinguish these stands.

5.2.5 *Quercus suber* Forests

Cork oak is probably one of the best examples of passive pyrophytes or plants well adapted to fires in the Mediterranean Basin. Buds are protected by cork and sprout after wildfires. The bark of these trees is the basis of the well-developed cork industry. The forests are confined to acid soils under higher rainfall conditions (>750 mm), higher than the acidophilus *Q. rotundifolia* forests. The best formations are located in the ranges of the southwestern part of the Iberian Peninsula (Extremadura and western Andalusia and Portugal), and also in the northern part of Mediterranean Catalonia, in northern Spain. Some scattered stands appear on acid outcrops of the eastern ranges such as Sierra de Espadán in the vicinity of Valencia, Sierra Contraviesa in Granada, and even in northeastern Spain (Sil valley at Galicia).

The herb layer is richer than in *Q. rotundifolia* forests, with plants of deciduous forests such as *Pteridium aquilinum*, *Luzula forsteri* and *Brachypodium sylvaticum*, and endemic plants such as *Sanguisorba hybrida*, *Teucrium scorodonia* subsp. *baeticum* and *Ruscus hypophyllum*. The tree layer of the western Iberian forests may include other oaks like *Quercus pyrenaica* or *Q. faginea* subsp. *broteroi*. The secondary canopy layer of these forests is floristically unusual, being rich in lauroid plants (evergreen trees with elongated, hard, shiny leaves similar to those of laurel, *Laurus nobilis*), such us *Arbutus unedo*, *Phillyrea latifolia* and *Viburnum tinus*. In moister areas, these formations are enriched by other lauroid trees, such as *Prunus lusitanica* and *Laurus nobilis*.

On the other hand, the seral shrublands are characteristic of the Iberian Peninsula. They probably form one of the most diverse examples of the Mediterranean heaths: *Erica umbellata, E. lusitanica, E. australis,* and Cistaceae such as *Cistus populifolius* and *Halimium ocymoides*.

In recent decades, many of these forests have been replaced by *Eucalyptus* stands, for paper production. Eucalypts adapt well to these areas because of the climatic similarities between their native territories and the western part of the Iberian Peninsula. However, they produce soil acidification, and a strong impoverishment in species.

5.2.6 Other Sclerophyllous Formations

Olive is the most representative tree of the Mediterranean Basin. It has been extensively cultivated in the Iberian Peninsula, but the natural groves of its wild ancestor tree, *Olea europaea* var. *sylvestris*, can be found both on acidic and calcareous soils. Olives make up the mature forests of the thermomediterranean portion of the Guadalquivir Valley, where the compact clay soils are unfavorable for *Quercus*, but suitable for agriculture. Olives can also be found on sunny rock crests of some southern ranges (Guinochet et al. 1975; Rivas-Martínez et al. 1997). Furthermore, some relict stands appear on coastal cliffs of Asturias (northern Spain). These formations are rich in thermophilous species such as *Calycotome spinosa, Rubia peregrina* subsp. *longifolia*, and *Chamaerops humilis*. In Menorca (Balearic Islands), they are enriched with *Euphorbia dendroides, Phillyrea rodriguezii* and *Prasium majus*. These forests have almost disappeared, particularly those on rich soils, because of the intense human activity that has transformed the original landscape into an agricultural one.

Quercus coccifera formations are also widespread in the meso- and thermomediterranean parts of the Iberian Peninsula, on a wide range of substrates. They form extensive, impassable maquis or shrublands, usually related as seral communities of *Q. rotundifolia* forests since they share most plants: *Rhamnus lycioides, Pistacia lentiscus, Jasminum fruticans* and the semi-parasite *Osyris alba*. Certain scattered stands of *Quercus coccifera* on the humid and subhumid thermo-Mediterranean calcareous outcrops of Portugal (i.e. Serra da Arrábida) form real forests because these oaks develop there into trees. Below 350 mm of annual rainfall, *Q. coccifera* forms extended communities which are considered a typical semi-arid vegetation type (Peinado and Rivas-Martínez 1987).

5.3 Deciduous Forests

Deciduous forests have an optimal distribution in Eurosiberian territories, appearing rather scarcely in the Mediterranean area of the Iberian Peninsula (Sánchez-Mata 1989). Certain deciduous or semi-deciduous oak species such us *Quercus pyrenaica, Q. faginea* and *Q. canariensis*, form forests. Due to similarities in their floristic composition and structure to the Eurosiberian forests, they have been phytosociologically included in the class *Querco-Fagetea*. They develop in mountainous belts, usually supramediterranean, where the climatic conditions are suitable. However, in areas where the climate is humid enough or in soils with a higher water storage, they can reach lower areas in the meso- and even the thermomediterranean belts. Traditionally, they have been exploited in different ways: wood extraction, wood for heating, agriculture, grazing by cattle, meadow management for winter fodder, and even conversion to conifer plantation, and thus were partially or completely destroyed.

5.3.1 *Quercus pyrenaica* Forests

These forests are very common in the Mediterranean area of the Iberian Peninsula, but absent in the Pyrenees. They grow on siliceous substrata, in the sub-humid or humid supramediterranean belt (Vilches et al. 2013a, 2015). The leaves of this broad-leaved oak dry in autumn as do other deciduous trees but do not fall until spring when new leaves appear, in a phenomena called 'marcescence' and it is related to the Mediterranean climate regime (Fig. 5.1). In the canopy *Q. pyrenaica* is dominant but other trees may occur, such as *Populus tremula*, *Prunus avium*, *Sorbus aucuparia*, *Acer monspessulanum*, *A. granatensis*, *Castanea sativa* and *Ilex aquifolium*. The ground layer is formed by numerous forest herbs such as *Arenaria montana*, *Aristolochia paucinervis*, *Aquilegia hispanica*, *Luzula forsteri*, *Primula veris* subsp. *canescens* and *Veronica officinalis*. The scrub seral communities of these forests dominate the landscape in extended areas of the Iberian Peninsula due to the intense destruction of the forests. Shrub communities are composed of retamoid species (*Cytisus*, *Genista*, *Echinospartum*), depending on the area and on forest type. However, it is interesting to note the presence of *Cytisus scoparius* in almost all these communities (Table 5.1).

The greatest extension and diversity of *Quercus pyrenaica* forests are in the middle and northern areas of the Iberian Peninsula, where several phytosociological associations have been described (Vilches et al. 2013b). In northwestern areas (from Peña de Francia and Serra de Estrela to the north), under cold and humid climates, these forests are enriched by *Erythronium dens-canis*, *Potentilla montana* and *Omphalodes nitida*. Seral scrub communities are composed of *Genista florida* subsp. *polygaliphylla*, *Erica arborea* and *Cytisus striatus*, while shrublands include some heathers such as *Erica australis* subsp. *aragonensis* and *Calluna vulgaris*. In northeastern areas (Sierra de Ayllón, Sistema Ibérico, Sierra de Leyre), also under similar climatic conditions, Pyrenean oak forests may include *Pulmonaria longifolia*, *Festuca heterophylla*, *Stellaria holostea* and *Melampyrum pratense*.

Fig. 5.1 *Quercus pyrenaica* deciduous forest in central Spain in winter. The leaves of this broad-leaved oak do not fall until spring when new green leaves appear

The seral scrub communities are dominated by *Genista florida* subsp. *florida* and *Cytisus scoparius*; *Genista anglica, G. pilosa, Erica cinerea* and *E. vagans* form heath-lands; and *Cistus laurifolius* and *Halimium ocymoides* appear in other shrublands.

The sub-humid forests in these areas have a poorer ground layer and their seral communities only support other broom-like plants such us *Genista falcata, G. hystrix, Adenocarpus hispanicus, Cytisus striatum,* and some Cistaceae (*Cistus ladanifer* or *Halimium viscosum*). In western areas of Spain and Portugal these forests appear in both meso- and supramediterranean belts. Scrubs are dominated by *Cytisus multiflorus* and *C. eriocarpus*. The mesomediterranean forests share species with *Quercus rotundifolia* forests: *Ruscus aculeatus, Arbutus unedo, Phillyrea angustifolia, Carex distachya, Rubia peregrina* and *Asparagus acutifolius*. In the herb layer of the supramediterranean forests, *Festuca elegans* is rather common, as in the Sierra Nevada, but in these areas retamoid communities are dominated by *Adenocarpus decorticans* instead of *Cytisus multiflorus* or *C. eriocarpus* (Pajarón and Escudero 1993).

Quercus pyrenaica forests have been used for extracting wood or grazing, while agriculture was not developed, mainly due to the nutrient-poor soils. The *Q. pyrenaica* canopy has been replaced by *Pinus sylvestris* and *P. pinaster,* for economic purposes. *Castanea sativa* occasionally is present in the canopy; it has frequently been cultivated in the potential territory of these forests, probably even before the Romans conquered the Iberian Peninsula, although they extended its use. At present, chestnut is used for two purposes: timber production as coppice stands, and edible-nut production in high forest stands. Actually, the latter management causes intense disturbance due to the removal of plants from the ground layer to avoid competition, while coppicing maintains the natural herb-layer diversity of deciduous forests. Good examples of chestnut stands appear in the north of the Mediterranean Iberian Peninsula, in western areas of Sistema Central and Extremadura, and even in Andalusia.

5.3.2 *Quercus faginea s.l.* Forests

These forests usually appear on calcareous substrates in bioclimatic conditions similar to *Q. pyrenaica* forests, although they can also reach the mesomediterranean belt. *Quercus faginea* subsp. *faginea* dominates the canopy layer although other tree species can also be found (*Acer opalus, A. granatense, Sorbus aria* and *Crataegus monogyna*). The ground layer is covered with some beautiful plants such as *Cephalanthera rubra, Dictamnus hispanicus,* and *Paeonia humilis*. Seral communities are composed of *Rosa, Rubus,* and *Berberis* species, whereas perennial grasslands are dominated by *Brachypodium phoenicoides*. Although *Q. faginea* has traditionally been considered strictly Mediterranean, it also remains in the Eurosiberian part (Basque country and surrounding areas).

In Aragón (northeastern Spain) differences on floristic composition of canopy or seral communities separate different two forest types. The westernmost ones are more ombrophilous and include *Acer monspessulanum* in the canopy layer. *Spiraea hypericifolia* subsp. *obovata, Rosa agrestis, R. micrantha* and *Genista occidentalis* grow as seral scrubs, while the ground layer is very poor. In eastern areas, and extending to parts of Castilla and Catalonia, these forests are enriched with *Acer opalus* together with *Viola willkommii, Daphne laureola* and *Prunus mahaleb*. To the south, in Valencia (Penyagolosa-Montgó), they are also accompanied by *Fraxinus ornus*. *Buxus sempervirens,* present in these forests, forms the most important seral community when *Q. faginea* canopy disappears, dominating the landscape due to the evergreenness of its leaves. In Andalusia these forests develop in particular microclimatic conditions of humidity restricted to the supramediterranean belt. They are not very extended and their floristic composition is slightly different.

Quercus faginea forest lands traditionally have been used for agriculture and consequently been destroyed. Olive trees and vines are grown in mesomediterranean territories, while alfalfa *(Medicago sativa)*, wheat, potatoes and barley are more common in the supramediterranean belt. In mountainous areas, perennial pastures of *Brachypodium phoenicoides* are used for livestock, mainly sheep and horses. In any case, particular well-conserved stands are found in some montane areas of the eastern half of the Peninsula. Finally, in certain cases, these forests have been substituted by cultures of *Pinus nigra*, for wood production.

In the boundary between the Eurosiberian and the Mediterranean floristic regions, close to the Pyrenees, *Quercus humilis* (syn. *Quercus pubescens*) forests may grow. They have traditionally been considered sub-Mediterranean formations, because of their xerophytic position with respect to Eurosiberian beech forests and their contacts with Mediterranean *Quercus faginea* forests. Recently, they have been included in the Eurosiberian region due to the summer rainfall regime in the areas where they grow.

Quercus faginea subsp. *broteroi* forms almost exclusive forests in the meso- and thermomediterranean, humid and subhumid territories in central Portugal and locally in some narrow valleys of Montes de Toledo. They appear only on siliceous substrates. In these forests some endemic plants like *Ulex densus, U. jussiaei* or *Armeria welwitschii* can be found. The scrub seral community is composed by *Quercus coccifera*.

5.3.3 *Quercus canariensis* Forests

Quercus canariensis forests occupy some small areas close to the Iberian coast in the meso- and thermomediterranean belts under humid or hyper-humid rainfall regime and where fog and mist are very frequent. The most extended forests are located in Sierra del Aljibe (Cádiz) in the southern tip of the Iberian Peninsula but they can also be found in southern areas of Portugal and also in Catalonia. In the canopy layer, trees such as *Castanea sativa, Ilex aquifolium, Arbutus unedo* and *Quercus faginea* subsp. *broteroi* also appear. The ground layer supports plants such as *Rubia peregrina, Ruscus aculeatus, Teucrium scorodonia* and *Paeonia broteroi*.

It is interesting to note the presence of *Ruscus hypophyllum*, the endemic *Rhododendron ponticum* subsp. *baeticum*, and *Culcita macrocarpa*, and some epiphytic ferns such as *Davallia canariensis* and *Polypodium* spp. (Fig. 5.2). The secondary forest is composed of *Arbutus unedo* formations enriched with a dwarf oak (*Quercus lusitanica*). Finally, seral communities, or heaths, are mainly composed by the endemic Leguminosae, *Stauracanthus boivinii*.

5.3.4 *Fagus sylvatica* and *Betula celtiberica* Forests

Fagus and *Betula* forests, typical of the Eurosiberian region, are restricted to mountainous areas, usually in the supramediterranean humid and hyper-humid belts, mainly in central or northern areas and on siliceous soils.

Beech (*Fagus*) forests are found in some favorable topographically areas of Sistema Central and Sistema Ibérico. The canopy layer is formed by beech and other trees such as *Quercus petraea*, *Taxus baccata* and *Populus tremula;* the ground layer comprises *Sanicula europaea*, *Hepatica nobilis*, *Melica uniflora*, *Stellaria holostea*, *Vaccinium myrtillus*, and *Primula vulgaris*, among others. These forests are very interesting from a biogeographical point of view as a type of relict vegetation, but have been misused for conifer cultivation, sometimes very intensively (Izco 1984).

The distribution of birch (*Betula*) forests in the Mediterranean part of the Iberian Peninsula is restricted to particular sites, topographically protected (i.e. small

Fig. 5.2 Epiphytic *Polypodium* spp. on *Quercus canariensis* in Sierra del Aljibe (Cádiz)

valleys, near creeks). The most extended and best conserved are in central Spain, in Sierra de Guadarrama (Madrid) and in Somosierra (Guadalajara). In the southern ranges of the Peninsula, some small and in many cases highly threatened populations of *Betula fontqueri* exist as well (López González 2006).

5.4 Azonal Vegetation

Certain soils sustain communities that are clearly different from the above mentioned, in spite of growing under similar climates. These communities are basically controlled by some edaphic factors, such as water availability or the presence of particular soil elements.

5.4.1 Riparian Vegetation

Riparian vegetation in the Iberian Peninsula is mostly composed of broad-leaved trees of Eurosiberian origin. Accordingly, most of these line forests may be considered Eurosiberian islands in a Mediterranean landscape. It is usually possible to detect a strip of tree bordered by willow communities in contact with the river. Between this tree-line and the climax forests, several forest bands following a gradient of soil water content can be easily recognized along larger Iberian rivers. Ash (*Fraxinus*) and elm (*Ulmus*) groves are distantly located from the river. Poplar (*Populus*) forests and alder (*Alnus*) forests contact directly with willow communities. All of them present a very complex structure when they are well preserved, and contain many vines, spiny shrubs and abundant herbs (Sánchez-Mata and Fuente 1986).

Alnus glutinosa forests occur on the streams that run on siliceous material. They are mainly distributed in the western part of the Peninsula, including most of Portugal. As in other riparian forests, the herb layer is composed of numerous nemoral plants (*Scrophularia* spp., *Galium broterianum*), ferns (*Polystichum setiferum, Dryopteris filix-max* and *Osmunda regalis*) and vines such as *Humulus lupulus, Bryonia dioica,* and *Clematis campanuliflora*. Spiny and non-spiny shrubs (*Rubus* spp., *Rosa* spp.) are abundant in the gaps. Nowadays, most of these forests are reduced to narrow strips close to the river, except in some mountain gorges where we can still find well-preserved stands. In some of these stands one can also find other trees such as the endemic birch, *Betula celtiberica* subsp. *parvibracteata*, growing in some gorges of Central Spain: Montes de Toledo and Sierra Morena (Rivas-Martínez 1987).

Populus forests are widely distributed in the Iberian Peninsula. They occur on the lower parts of the larger rivers, in both siliceous and calcareous soils, and near many streams of the calcareous area. Lowland zones are dominated by *Populus alba,* usually accompanied by *Salix alba, S. triandra* and *Celtis australis,* whereas supramediterranean zones are dominated by *Populus nigra* and sporadically by *Populus tremula* and *Betula celtiberica*. Some well-conserved streams of the supramediterranean belt, in certain zones of the southern Iberian range, mainly in deeper calcareous canyons, present very complex riparian forests. Some of these trees are *Populus nigra, Ulmus glabra, Tilia platyphylos, Betula celtiberica, Fraxinus excelsior*, and even *Acer pseudoplatanus* and *Taxus baccata*.

Fraxinus angustifolia and *Ulmus minor* groves have been systematically destroyed because their rich soils have been historically transformed to agriculture (Ortega 1989). Furthermore, during the last 20 years, elm trees have been seriously devastated by Dutch elm disease, caused by *Ophiostoma ulmii*, whose spores are transferred by the elm bark beetle (*Scolytus multistriatus, S. scolytus, S. ensifer*) that lives in symbiosis with it.

Salix woodlands are usually associated with riparian forests. These 3–5 m tall communities are in contact with the water, and thus they are frequently disturbed by periodical floods. Some of these willows are *Salix atrocinerea* on acid soils, *S. salvifolia* on sandy beds and *S. eleagnos* on calcareous soils.

In thermo- and mesomediterranean areas with arid or semiarid climate, river forests can be dominated by different *Tamarix* species in sporadic water courses: *T. canariensis, T. africana, T. gallica* or *T. boveana* depending on whether soils are rich in sand, clay or gypsum. In similar environments, but with greater stoniness, *Nerium oleander* occupies and forms dense and conspicuous shrubby communities. Similarly, the endemic *Securinega tinctoria* grows on sandy soils of streams that are always dry during summer, in western areas of the central Iberian Peninsula.

References

Blanco E, Casado MA, Costa M, Escribano R, García M, Génova M, Gómez A, Gómez F, Moreno JC, Morla C, Regato P, Sainz H (2005) *Los bosques ibéricos: una interpretación geobotánica*. Editorial Planeta, Barcelona

Blanquer Lorite JM, Gutiérrez Girón A, Gavilán RG (2014) Importance of climatic data analysis for biodiversity studies in Mediterranean mountains of the Iberian Peninsula. *Lazaroa* 35:197–201

Fernández-González F (1986) *Los bosques españoles*. Unidades Temáticas Ambientales de la Dirección General del Medio Ambiente. M.O.P.U, Madrid

Folch R, Franquesa T, Camarasa JM (1984) *Historia Natural dels països catalans, 7: Vegetació*. Enciclopèdia Catalana S.S, Barcelona

Gavilán RG (2005) The use of climatic parameters and indices in vegetation distribution. A case study in the Spanish Sistema Central. Int J Biometeorol 50:111–120

Gavilán R, Fernández-González F (1997) Climatic discrimination of Mediterranean broad-leaved sclerophyllous and deciduous forests in central Spain. J Veg Sci 8:377–386

Gavilán RG, Sánchez-Mata D, Gaudencio M, Gutiérrez-Girón A, Vilches B (2015) Impact of the non-indigenous shrub species *Spartium junceum* (Fabaceae) on native vegetation in central Spain. J Plant Ecol 9:132–143. https://doi.org/10.1093/jpe/rtv039

Granda E, Escudero A, Valladares F (2014) More than just drought: complexity of recruitment patterns in Mediterranean forests. Oecologia 176:997–1007

Guinochet M, Guittonneau G, Ozenda P, Quézel P, Sauvage C (eds) (1975) La flore du bassin méditerranéen: essai de systématique synthétique. C.N.R.S, Paris

Iriondo JM, Albert MJ, Giménez Benavides L, Domínguez Lozano F, Escudero A (eds) (2009) Poblaciones en Peligro: Viabilidad Demográfica de la Flora Vascular Amenazada de España.

Dirección General de Medio Natural y Política Forestal (Ministerio de Medio Ambiente, y Medio Rural y Marino), Madrid

Izco J (1984) Madrid Verde. Comunidad Autónoma de Madrid, Madrid

López González G (2006) Guía de los árboles y arbustos de la Península, Ibérica. edn. Mundi-Prensa, Madrid

Ortega C (coord) (1989) El libro rojo de los bosques españoles. Adena-WWF España, Móstoles (Madrid)

Pajarón S, Escudero A (1993) Guía Botánica de las sierras de Cazorla, Segura y, Alcaraz. edn. Pirámide, Madrid

Peinado M, Rivas-Martínez S (eds) (1987) La vegetación de España. Univ. Alcalá de Henares, Alcalá de Henares

Plieninger T, Wilbrand C (2001) Land use, biodiversity conservation, and rural development in the dehesas of Cuatro Lugares, Spain. Agrofor Syst 51:23–34

Rivas-Martínez S (1987). Memoria del mapa de series de la vegetación de España, 1:400.000. Serie Técnica. ICONA. Ministerio de Agricultura, Pesca y Alimentación, Madrid

Rivas-Martínez S, Fernández-González F, Sánchez-Mata D, Pizarro JM (1990) Vegetación de la Sierra de Guadarrama. Itinera Geobot 4:3–132

Rivas-Martínez S, Asensi A, Díez-Garretas B, Molero J (1997) Biogeographical synthesis of Andalusia (southern Spain). J. Biogeogr 24:915–928

Rivas-Martínez S et al (2002) Vascular plant communities of Spain and Portugal. Itinera Geobot 15:5–992

Sánchez-Mata D (1989) Flora y vegetación del macizo oriental de la Sierra de Gredos (Avila). Diputación provincial de Avila. Institución Gran Duque de Alba

Sánchez-Mata D, de la Fuente V (1986) Las riberas de agua dulce. Unidades Temáticas Ambientales de la Dirección General del Medio Ambiente. M.O.P.U, Madrid

Vilches de la Serna B, De Cáceres M, Sánchez-Mata D, Gavilán RG (2013a) Indicator species of broad-leaved oak forests in the eastern Iberian peninsula. Ecol Indic 26:44–48

Vilches de la Serna B, Merle H, Ferriol M, Sánchez-Mata D, Gavilán RG (2013b) *Minuartio valentinae-Quercetum pyrenaicae*: a new Iberian broad-leaved oak forest in the eastern coastal mountains and seral plant communities. Lazaroa 34:209–217

Vilches de la Serna B, Sánchez-Mata D, Gavilán RG (2015) Marcescent *Quercus pyrenaica* forest in the Iberian Peninsula. In: Box EO (ed) Vegetation structure and function at multiple spatial, temporal and conceptual scales, pp. 257–283, Geobotany Studies. Springer

What Is the 'True' Mediterranean-Type Vegetation?

Mark A. Blumler

Abstract

It has been demonstrated that evergreen sclerophylls are not always well adapted to, and do not flourish in, strongly winter-wet summer-dry climates [Blumler (Warm-temperate deciduous forests. Springer, 2015); Blumler and Plummer (Warm-temperate deciduous forests. Springer, 2015)]. What vegetation then is best adapted to mediterranean climates? Therophytes may be natural competitive dominants where soil is fertile, summer drought is protracted and extreme, and precipitation is low enough that it does not percolate downward significantly. Fertility matters because of its influence on the seasonality of the water regime (i.e., the "mediterraneanness" of the water regime, in a sense). Empirical data, collected originally by Raunkiaer but reinforced by more recent studies, suggest a greater tendency of annuals to flourish in seasonally dry than in arid environments. Studies of vegetation dynamics indicate that annuals can be highly competitive and perhaps even "climax" where winters are wet (but not too wet), and summers are hot, long, and dry. Tall, laterally expanding clonal plants of the sort that Grime classifies as competitors, do not thrive under such conditions, being largely restricted to mesic habitats where the seasonal drought is moderated. This leaves Grime's competitive-ruderals, tall annual plants, as the most competitive species present in open habitats. Biome modeling and vegetation mapping would improve if the assumption of one biome per climate were relaxed and instead different life forms or functional types were mapped as overlays in a GIS.

M.A. Blumler (✉)
Departments of Geography and Biological Sciences, SUNY-Binghamton, Binghamton, NY, USA
e-mail: mablum@binghamton.edu

Keywords

Cloning and succession • Edaphic climate • Life-form spectra • Mediterranean environments • Therophytes

6.1 Introduction

The association of evergreen sclerophyllous shrubs with geographically disjunct regions of winter-wet, summer-dry Mediterranean climate is deeply entrenched in the biogeographical literature and for over 100 years has been considered a classic example of convergent evolution (Schimper 1898; Mooney and Dunn 1970). On the other hand, some studies have pointed out that evergreen sclerophyllous (chaparral/maquis) vegetation occurs also in non-Mediterranean regions, typically subtropical, and frequently with a summer-rain maximum (see Blumler 2005, 2015; Blumler and Plummer 2015). Paleobiologists have noted that the Mediterranean-type climate is geologically recent, while the major lineages of sclerophyllous shrubs are far older (Axelrod 1975; Blumler 1991, 2005, 2015). Finally, Box (1982) pointed out that the emphasis on evergreen sclerophylls has obscured the very high growth form diversity that is characteristic of Mediterranean regions.

I have shown that winter-deciduous trees, especially oaks, are more prominent than evergreen sclerophylls in the most summer-dry regions, though these trees usually occur in open formations with a herbaceous understory (Blumler 1991, 2005, 2015; Blumler and Plummer 2015). This is not to say that there is no convergence represented by sclerophylls, but that their climatic adaptation is to sub-Mediterranean, or more precisely, subtropical semi-arid regimes. Nutrients also matter, since evergreen sclerophylls are especially likely to dominate on infertile substrates (e.g. Beadle 1966; Specht 1969; Small 1973; Chapin 1980; Milewski 1983; Mooney 1983).

So, if evergreen shrubs are not characteristic of the most summer-dry regions, what then is the "true" Mediterranean-type vegetation? Box (1982) would argue that a diversity of vegetation types is characteristic, and I am in full agreement (e.g., Blumler 2005). At the same time, if one focuses attention on Mediterranean climate in its purest, i.e. most winter-wet, summer-dry form, one life form does come to the fore (though not to the complete exclusion of many others). As Christen Raunkiaer (1934) meticulously documented, the Mediterranean climate is characterized by annual plant species, and to a lesser extent geophytes (although he presumed that woody plants were the natural dominants). Unfortunately, subsequent scholars misstated his conclusions (admittedly, he was not entirely clear: *vide infra*), stressing aridity rather than seasonality of precipitation. Here, I argue that annuals are indeed the natural competitive dominants, as well as the most speciose of the life forms, in the regions of most pronounced winter-wet, summer-dry climate.

6.2 Seasonality of the Soil Water Regime

In the Mediterranean-climate portion of California, several plant communities dominate over extensive areas (Table 6.1), with little evidence that any are merely seral to some other type. Rather, they are differentiated in part on the basis of climate, and partly according to edaphics (e.g., Wells 1962), which in turn mediate climate through impacts on water availability. Blumler (1984) coined the term "edaphic climate" to refer to soils that can cause the water regime experienced by the plant to be either less mediterranean than the ambient climate, or equally or perhaps even more so. For instance, California's famous vernal pools—grassland depressions that are underlain by a hardpan—are waterlogged in winter but dry out completely during the summer. Their edaphic climate, then, is extreme mediterranean. And they are vegetated almost exclusively by annuals (Holland and Jain 1977). In contrast, chaparral typically occurs on coarse-textured, relatively infertile substrates (Hanes 1977; Cody and Mooney 1978). These free-draining soils are relatively dry during the rainy season, but dry out much more slowly than clay soils do during rainless periods. So the edaphic climate of chaparral soils is less mediterranean than climate records would suggest. Noy-Meir (1973: 37) pointed out that, in arid regions, sandy and rocky soils typically support taller, woodier vegetation than heavy-textured soils—the so-called "inverse texture effect"—because, during the dry season, some moisture is retained in sands long after it is lost through capillary action and evaporation from clays. The infertility of typical chaparral soils moderates the edaphic climate further, since it slows plant growth during the rainy-season, leaving more moisture in the soil as the summer drought sets in. In contrast, where N deposition from car exhaust fertilizes shrub-dominated soils in southern California, the growth of annual plants increases. This causes more of the rainy season moisture to be utilized and putting the shrubs under increased summer drought stress, while also increasing the flammable fuel load when the annuals set seed and die. The result is massive conversion from shrubs to annual grassland (Wood et al. 2006; Blumler 2011).

An Israeli example that illustrates the role of edaphic climate is Hillel and Tadmor's (1962) study of Negev Desert vegetation patterns. In the Negev, there are two main types of wadi: loess wadis, which are fertile, and gravel wadis, which are not but into which small amounts of water can infiltrate to a great depth and remain available year-round. Woody plants dominate the gravel wadis. The loess

Table 6.1 Some vegetation types that dominate extensive areas in Mediterranean-climate California

Mixed evergreen forest	Deciduous oak savanna
Chaparral	Pine woodland
Evergreen oak woodland	Coastal sage scrub
Evergreen oak savanna	Succulent/Sage Scrub
Oak-pine woodland	Bunchgrass prairie
Deciduous oak woodland	Annual grassland

Munz (1959), Baker (1972), Major and Barbour (1977), Mattoni and Longcore (1997)

wadis store sufficient water that on average 250–500 mm is available to plants each year, providing "...growing conditions similar to those in some Mediterranean-type habitats of northern Israel" (Hillel and Tadmor 1962: 40). But the Negev wadis experience much greater year-to-year fluctuations than places in Israel that literally receive 250–500 mm or more annual rainfall, because desert precipitation is more sporadic than that of wetter regions (Lydolph 1985). During years of good precipitation, annual grasses and legumes form "lush stands" in the loess wadis (Hillel and Tadmor 1962: 39). But when the rains fail, or come late, the dominants are the rhizomatous perennial Bermuda grass (*Cynodon dactylon*) and shrubs. Bermuda grass also occurs in the Mediterranean part of Israel but is of minor importance there, as is true of rhizomatous species in general—a point to which I shall return below.

In the eastern Mediterranean and the summer-dry Fertile Crescent, fertile soils are also often rocky. Hard limestone gives rise to fertile soils, but because of irregular solution weathering, soil depth varies over short distances and there are many rock outcrops. This creates complexity at a microsite scale; while the deep soil areas have a pure mediterranean edaphic climate, the rocks moderate summer drought by mulching against evaporation and crevices often enable deep infiltration. Litav (1967) and Blumler (1992b, 1993, 1998) showed that perennials tend to be strongly associated with rock outcrops on such substrates, while annuals tend to dominate open ground, especially in the absence of disturbance.

6.3 Raunkiaer's Research

Raunkiaer (1907) noted that annuals are the best protected from the unfavorable season and suggested, therefore, that they are important in steppes and deserts. Subsequently, however, he emphasized seasonality of precipitation. Raunkiaer (1908: 143) identified "a Therophyte climate in the regions of the subtropical zone with winter rain"; he stated that the therophyte climate, in Europe, "is that of the Mediterranean countries" (Raunkiaer 1914: 343) and also noted that annuals are more prevalent in southern Spain than in the French Mediterranean because the latter is close to the border with Europe's hemicryptophyte climate. Raunkiaer (1934: 567) also found that, while in Italy the therophyte percentage is elevated, it is not so high in the north as in the central and southern parts. Although he had no good data from southern lowlands, he suggested that the therophyte percentage there would be higher than in central Italy.

Southern France receives considerable precipitation in summer, so its climate is sub-Mediterranean. Generally speaking, seasonality of precipitation, in this case concentration of the rainfall in the cool season and drought and heat in summer, increases as one travels south and/or east across the Mediterranean Basin and into the Near East (Blumler 1993; Blumler and Plummer 2015). However, this statement simplifies what is actually a very complex reality (Trewartha 1961; Blumler 2005).

Table 6.2 Life forms of the Sierra Nevada, Spain

Elevation	Chamaephytes	Hemicryptophytes	Therophytes
0–633 m			51%
633–1500 m			29%
1500–2530 m	20%	51%	19%
>2530 m	25%	67%	5%
>3000 m	50%	50%	0%

Raunkiaer (1934: 560)

Raunkiaer distinguished the therophyte climate of the Mediterranean region from the hemicryptophyte climate of central Europe, and much of his research was oriented towards locating the boundary between the two. He observed, and documented repeatedly, an analogous transition that occurs up mountain slopes within the Mediterranean: typically, annuals decrease with altitude, while hemicryptophytes replace them rather suddenly as the most prevalent life form and then gradually decrease, while chamaephytes increase towards the alpine belt (e.g., Table 6.2). This is not surprising, since the high-altitude summer is comparatively short and cool, reducing the drought stress. Moreover, "winter" annuals, as dominate low elevations, would not be able to grow in the subfreezing winter temperatures of high elevations and so would need to complete their life cycle rapidly during the short springtime window between snowmelt and summer drought. Raunkiaer pointed out that in comparing different Mediterranean places, one should compare lowlands only, if possible, so that the confounding effect of elevation is removed.

Raunkiaer (1934) compiled spectra for a large number of Mediterranean islands (Table 6.3). These data show that the therophyte percentage increases as one goes south in the Mediterranean; what happens from west to east is less clear (Fig. 6.1). (Contour lines on this and subsequent maps were determined with Inverse Distance Weighting). There are exceptions to the north-south trend, such as the Balearics: Raunkiaer suggested that they have more high-elevation land than the much smaller islands that comprised the bulk of his data set. This is plausible, since Majorca reaches 1400 m; by comparison, on Madeira, with maximum elevation of 1800 m, Raunkiaer found that the lowlands have 51% therophytes, while the highlands have only 28%. Greek islands also typically are more rugged and elevated than the Italian ones—for instance, Samos rises to over 1400 m, which may explain its comparatively low therophyte percentage. Life-form spectra from regional floras (including the ones Raunkiaer analyzed, plus a few published after he died) show a tendency for an increase in therophytes towards both the south and east, until one reaches the desert (Fig. 6.2).

Raunkiaer was vague about what he thought happened to the therophyte percentage once one got into true desert to the south of the Mediterranean Sea. He stated that "The Mediterranean therophytic climate ... extends from the southern limit of the Sahara to the foot of the Alps, and from the Atlantic to the mountain masses and plateaux of central Asia" (Raunkiaer 1914: 352). Thus, while he

Table 6.3 Percent therophytes on Mediterranean islands

Location	%
Madeira lowlands	51
Madeira highlands	28
Madeira total	39
Porquerolles	48
Balearics	36
Ligurian Islands	34
Tuscan Islands	44
MaddalenaIslands	54
Pontine Islands	56
NeapolitanIslands	48
Pantellaria	61
Malta	51
Lampedusa/Linosa	63
Tremitic/Pelagosa	53
Zakynthos	46
Thasos	50
N. Sporades	52
Aegina	60
Karpathos	50
Samos	33

Raunkiaer (1934)

included the aseasonal, hyper-arid central Sahara and the summer-rain southern Sahara, he excluded the summer-rain deserts and semi-deserts of inner Asia east of the Pamirs. He had a single life-form spectrum from Timbuktu, on the southern side of the Sahara, with a moderately high percentage of therophytes, albeit not so high as much of the southern Mediterranean. He also had a flora of Aden, which is hyper-arid (mean annual precipitation 40 mm) but has no pronounced seasonality to the precipitation. Aden has a therophyte percentage only slightly above the 13% in Raunkiaer's (1914) "normal spectrum". Raunkiaer assumed that the therophyte percentage remained higher than "normal" right across the Sahara, but whether he believed it to be even greater than in the southern Mediterranean is uncertain. Given the values for Aden and Timbuktu, it seems unlikely that he did.

6.4 Subsequent Life-Form Analysis

Subsequent research and discussion have sometimes been sloppy, with a pronounced tendency to misstate Raunkiaer's conclusions, emphasizing aridity rather than precipitation seasonality. For instance, Cain (1950) stated that Raunkiaer believed there is a therophytic climate of tropical and subtropical deserts. But Raunkiaer mentioned only the subtropics with winter rain, comprising deserts, steppes, and Mediterranean regions. Cain also mislocated one of Raunkiaer's sites: "Yekaterinoslaw" (Ekaterinaslav) is in the Ukraine, but Cain thought it was

6 What Is the 'True' Mediterranean-Type Vegetation?

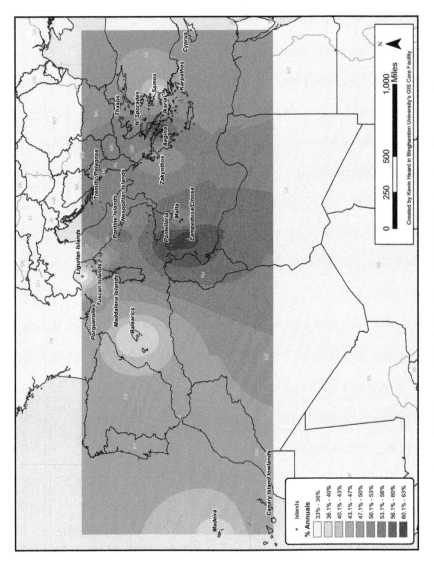

Fig. 6.1 Percent therophytes in the floras of Mediterranean islands (Raunkiaer 1934). IDW used to determine contour lines

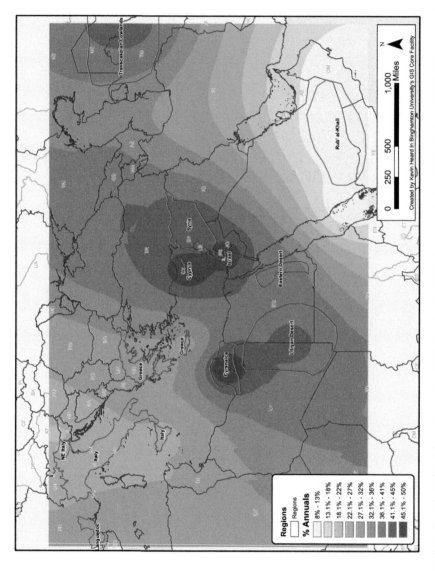

Fig. 6.2 Percent therophytes in regional floras (Raunkiaer 1934; Mandaville 1986; Danin and Orshan 1990; El-Ghani 1998). IDW used to determine contour lines

in the Near East. Raunkiaer was careful and thorough, and preferred to use floras, whereas subsequent investigators frequently have relied upon single samples (e.g., Cain 1950; Daget et al. 1977; Batalha and Martins 2002). Raunkiaer also understood that scale matters, in part because of the issue of elevation and its effect upon climate. Others seem to have ignored the issue. Cain classified his data set according to the Köppen system (though not entirely accurately) and paid little attention to seasonality, an approach that influenced several subsequent researchers. He noted that Table Mountain, in the mediterranean region of South Africa, has a lower therophyte percentage than the normal spectrum. However, Table Mountain is known for its infertility, which tends to reduce the therophyte component (Milewski 1983).

Sarmiento and Monasterio (1983) gathered life-form spectra for tropical savannas in South America and Africa. These tend to have summer-wet, winter-dry climates and also tend to have therophyte percentages intermediate between the normal spectrum and Mediterranean values. Unfortunately, Sarmiento and Monasterio were unable to report any life-form spectra from east Africa, which often features a bimodal climate (two rainy seasons and two short dry periods), or from the Sahel, with its extreme rainy season and protracted dry season.

In a global survey, Blumler (1984) concluded that annuals are especially characteristic of winter-wet, summer-dry climates but also prominent in other seasonal, or sometimes even bimodal regimes, such as the Sonoran Desert of Arizona; true deserts were less likely to have high numbers of therophytes, unless (as is often the case) the rainfall pattern was seasonal.

To the east of the Sahara, the zone of hyper-arid desert continues as the Rub' al-Khali of Arabia. As is true of the Sahara, winter rain increases towards the north, and summer rain towards the south. Mandaville (1986: 147) reported the flora of the Rub' al-Khali—37 species in an area the size of Texas!—and noted that "Annual plants are virtually absent, a situation contrasting sharply with that in Arabia to the north where more than half the desert species are therophytes".

Danin and Orshan (1990) reported therophyte percentages for Israel, finding no significant difference between the Mediterranean part of Israel and its desert, and the region transitional between the two: all are 49–54% therophytes. They also divided Israel into 27 regions and correlated life forms with regional mean annual precipitation. They found that annuals are least important in the driest regions, increase to a plateau of over 50% over a range from 200 to 500 mm annual rainfall, and then decrease again at higher precipitation amounts. However, there is a large gap between the wettest region, Mt. Hermon (43% annuals), with almost 1000 mm mean annual precipitation, and the next wettest region (Upper Galilee, mean precipitation = 665 mm, 52% annuals). The decrease at high precipitation is almost if not entirely due to Mt. Hermon, which also is the only region that extends up to high altitudes (up to 2300 m, vs. no more than 1200 m elsewhere). So the decrease that Danin and Orshan associated with high precipitation is more likely due to elevation.

Batalha and Martins (2002) reported very low therophyte percentages from the cerrado of Brazil; this perhaps should not be surprising given the extremely low fertility of cerrado soils. They also collected other life-form spectra from the literature, including Raunkiaer, Cain, and Sarmiento and Monasterio, as well as several more recently published reports. Cain's errors were propagated, unfortunately, and like him they did not distinguish single samples from floras; they also did not account for elevation, for instance, and included the recently published flora of Mt. Kyllini, Greece (Dimopoulus and Georgiadis 1992), which rises to almost 2400 m and therefore, not surprisingly, has only 24% annuals. Like Sarmiento and Monasterio, they had no data from the Sahel, where much of the vegetation is classified as annual grassland (Breman et al. 1980; Blumler 1984; Hiernaux and Turner 1996; Turner 1998). They also reported no data from the less seasonal, sometimes bimodal, savannas of east Africa. There, most savannas are dominated by perennial grasses (McNaughton 1985), which suggests that they are not particularly therophyte-rich. Northern Australia also has native annual-dominated grasslands and savannas associated with summer-wet, winter-dry climate (Blumler 1984) but no life-form spectra to my knowledge.

One of the few regions from which many life-form spectra have emerged in recent years is the Indian subcontinent (e.g. Reddy et al. 2011). Unsurprisingly, given the seasonal, monsoon precipitation regime, these studies tend to show moderately high levels of therophytes. However, they also tend to be based on single samples or transects rather than whole floras, so they are not analyzed here. Finally, Van der Merwe and van Rooyen (2010) reported life-form spectra for three neighboring South African associations that experience 200–400 mm, 100–400 mm, and <100–200 mm annual rainfall, respectively. Therophytes were, respectively, 26.7%, 29.5%, and 10.9% of their floras, illustrating that where fertility is not terrible, South Africa does have therophytes. Although the percentages are not remarkably high, it is also true that the precipitation regime is comparatively less seasonal (Blumler 2005). Note that the driest of the three associations has a percentage of therophytes that is less than in the normal spectrum.

6.5 California Therophyte Percentage

Raven and Axelrod (1978) emphasized the many annuals in the California flora—the frontispiece to their volume includes four color photos of spectacular wildflower displays of predominantly annual forbs (e.g. Figs. 6.3 and 6.4)—and noted that annuals constitute 28% of the flora. This figure compares well with Italy (26%) and Greece (31%) (Raunkiaer 1934), while it is much less than the value for Israel (Danin and Orshan 1990). But California is more mountainous than Israel, and perhaps even more so than Italy and Greece. Cain (1950) included spectra from two Mojave Desert sites, Death Valley (42% therophyte) and Salton Sink (47%); both are very dry but are also highly seasonal (winter-wet, summer-dry) deserts.

Herbert Baker (1972) carried out a mammoth seed-weight study, which entailed gathering seeds from thousands of California species, carefully assigned to plant communities as classified by Munz (1959). Table 6.4 shows the percentages of

Fig. 6.3 One of four photos of wildflower displays that adorn Raven and Axelrod's (1978) frontispiece. This display of mostly annual forbs from Bear Valley, Colusa County, is evocative of the descriptions of parts of lowland California by John Muir, Frederick Clements, and others

Fig. 6.4 Another of Raven and Axelrod's photos, illustrating their emphasis on annual plants

Table 6.4 Percent therophytes in Baker's (1972) seed collections

	N	% Therophytes
Desert		
Sagebrush scrub	152	42
Shadscale scrub	40	30
Creosote bush scrub	255	45
Alkali sink	58	38
N. Juniper woodland	63	44
Pinyon-Juniper woodland	171	39
Joshua tree woodland	162	45
Mediterranean		
Mixed evergreen forest	339	23
Coastal strand/dune	131	32
Coastal sage	328	41
Chaparral	594	33
N. Oak woodland	211	55
S. Oak woodland	183	55
Foothill woodland	535	50
Coastal prairie	189	56
Valley grassland	333	68

therophytes that Baker encountered in each of the mediterranean-type and desert communities. Montane and North Coast communities are excluded from Table 6.4 because neither has a true mediterranean climate (not surprising, then, that their therophyte percentages were much lower). Wetland communities, which would not have mediterranean edaphic climates, were also excluded. Although Baker's results are not as reliable as if one were to examine entire floras, they do fit the pattern proposed here. California's winter-rain deserts have high percentages of therophytes, but not as high as most of its mediterranean-climate communities. Among the latter, mixed evergreen forest is comparatively low, but it is found only on mesic sites such as north slopes, riparian edges, and coastal slopes receiving summer fog. Chaparral is also a bit low, but as explained above, has a moderated mediterranean edaphic climate. Coastal strand, of course, is sandy, and experiences moderated summer temperatures. Of the remainder, Valley grassland is found where summer temperatures are hottest and often, though not always, on the most fertile soils. The Oak woodland types, which typically support a herbaceous understory, also occur for the most part where summers are hot and soil relatively fertile (Blumler 2015).

Schiffman (2007) analyzed floras from 12 grassland nature preserves, including 1 from Coastal prairie, and reported the percentage of native annual forbs in each. For the 11 Valley grassland preserves, the median was about 51% native annual forbs; and this figure does not include the native annual grasses. In addition, of course, these sites all have significant numbers of introduced annuals.

Overall then, these results, and especially the comprehensive and reliable studies of Mandaville (1986) and Danin and Orshan (1990), confirm Blumler's (1984) conclusions. These results also support his suspicion that hyper-arid but aseasonal deserts do not have a particularly high percentage of therophytes. Danin and Orshan suggested that the decrease in annuals with increasing aridity in Israel may correspond to the point at which vegetation becomes restricted to the wadis (cf. Monod 1931). Low-elevation stations in the summer-dry parts of the Mediterranean consistently report more than 50% therophytes (Fig. 6.5). It appears that summer-rain, winter-dry regions also have high therophyte percentages, but not as high as in the Mediterranean region and California. However, without reliable life-form spectra from the Sahel, northern Australia, and central Chile, this conclusion remains tentative. For that matter, more spectra from selected parts of the Mediterranean itself, and better control on elevation, are needed before the patterns described here can be regarded as fully verified.

6.6 Summer-Dry Vegetation Dynamics

6.6.1 California

Clements (1934) observed "relict" stands of the bunchgrass purple needlegrass (*Nasella pulchra*) in the Central Valley and convinced most California ecologists that Valley grassland was a bunchgrass prairie before the arrival of the Spanish. Biswell (1956) demurred, pointing out that purple needlegrass responds well to fire and that Clements' populations were located along railroad lines, where annual burning was practiced. As Schiffman (2007) pointed out, Clements himself recognized the abundance and diversity of annuals in the vegetation (Clements and Shelford 1939), though as she put it,

> Like other observers, Clements noted the abundance of native annuals and then glossed over their identities as if they were unimportant. Despite his clear acknowledgement of their tremendous percent cover, these plants' transient nature indicated to him that they had little real ecological value. (Schiffman 2007: 53).

In recent decades, California ecologists and biogeographers have questioned increasingly the notion that perennials dominated the grassland (Wester 1981; Blumler 1992a, b, 1995; Hamilton 1997; Mattoni and Longcore 1997; Schiffman 2007; Minnich 2008). Some early studies had reported that perennials increase when grazing and fire are prevented, but subsequent studies have tended to show that any such increase is temporary (e.g., Bartolome and Gemmill 1981). Bartolome and Gemmill suggested that cattle grazing weakens the taller annuals, so when the livestock are removed purple needlegrass has plentiful establishment sites for its seedlings; but after a few years, the annuals thicken up, precluding any further establishment of the perennial. Certainly, early descriptions of the grassland suggest that annual wildflowers were abundant if not dominant (Blumler 1992b, 1995;

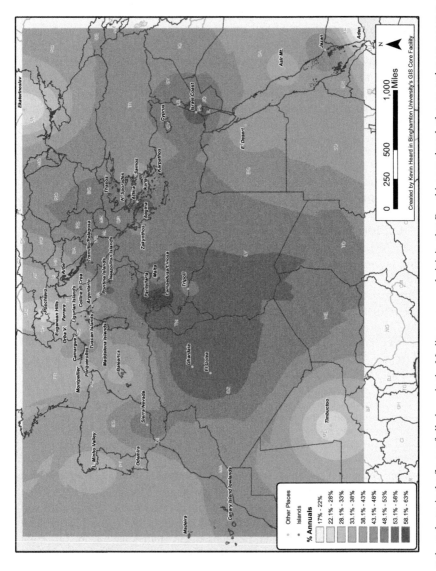

Fig. 6.5 Percent therophytes in the floras of all places in the Mediterranean and vicinity that Raunkiaer analyzed, plus several more recent studies (Raunkiaer 1934; Danin and Orshan 1990; El-Dermadesh et al. 1994; El-Ghani 1998; Al-Yemeni and Sher 2010). Single samples excluded. IDW used to determine contour lines

Mattoni and Longcore 1997; Schiffman 2007; Minnich 2008). Probably the most authoritative nineteenth century observer, and one who talked with many of the prominent California range men of the time, was William Brewer. He believed that the striking vulnerability of the California grasslands to Mediterranean plant invasion reflected the fact that it was naturally vegetated by annuals and that annual plants can fluctuate rapidly in abundance (Brewer 1883).

6.6.2 The Mediterranean and Near East

Working especially in the vicinity of their home base of Montpellier, Braun-Blanquet (1932; Braun-Blanquet et al. 1951) and others of his school of phytosociology described the successional sequences for the area. In southern France, "old-field" succession proceeds much as it does in England or the eastern US (e.g. Escarre et al. 1983; Kazakou et al. 2006): annuals give way especially to species that Grime (1977, 1979) would classify as "competitors", i.e. tall perennial herbs and woody plants capable of rapid lateral expansion. Such species would be "guerrilla" clones in Lovett Doust's (1981) classification, with long internodes between the successive shoots, in contrast to "phalanx" species that send up new shoots along short internodes, including bunchgrasses, bulbs with bulbils, and so on. [There also are intermediate forms, such as *Festuca rubra*, a bunchgrass with rhizomes, the latter being its main means of lateral expansion (Skalova et al. 1997)]. Even annual grasses tiller, and thus expand laterally to some extent if given room to do so. Many of Lovett Doust's guerrilla species are prostrate and thus would not be competitors in Grime's system.

Meiners et al. (2015) observed that competitor species in the eastern US dominate in "mid-succession", i.e. as the herbaceous phase comes to an end and the transition to forest takes place. Common competitors are the goldenrods (*Solidago*), brambles (*Rubus*), aspen (*Populus tremuloides*), black locust (*Robinia pseudoacacia*), and so on. In southern France the species most often described as coming to dominate old fields is the tall rhizomatous grass *Brachypodium phoenicoides*. Phytosociologists have extrapolated the French results to the rest of the Mediterranean Basin. In southern France, though, there is a significant amount of summer rain that is available to perennials (but not to winter annuals). So perennials have a much greater competitive advantage there than where summers are hot and bone dry.

In fact, guerrilla species become marginalized at best in true winter-wet, summer-dry Mediterranean environments (Blumler 2000). *B. phoenicoides* does not occur in Israel at all; perennial *B. sylvaticum* is present, but as its name suggests, it is usually found in shade. In contrast, the annual *B. distachyon* is not only present but common and widespread in open environments. In the US, *B. sylvaticum* is invasive in Oregon, with its sub-Mediterranean climate, while *B. distachyon* is invasive in California.

6.7 In Areas Without Guerrilla Clones, the Annual Plant Is King

In the western US, it is widely believed that natural grazing pressure must have been light, as suggested by the lack of sod-forming grasses. Comparisons are made frequently with the Great Plains, grazed by bison and much vegetated by rhizomatous, stoloniferous and other sod-forming grasses (Mack and Thompson 1982; Painter 1995). Rhizomatous and sod-forming grasses do exist in California and the remainder of the West, but they are restricted to mesic sites such as mountain meadows, north slopes, forest shade, wetlands, subirrigated bottomlands, and coastal slopes that experience summer fog. In addition to these mesic habitats, rhizomatous grasses also occur on sands. Though not really mesic, sandy substrates do have a much moderated edaphic climate. More to the point, it is not only prostrate, grazing-tolerant, sod-forming species that are so restricted, but also species that would fall into Grime's competitor class, such as the tall California bottomland dominant, creeping wild rye (*Leymus triticoides*). Furthermore, this distribution pattern does not hold only for the grasses, but for all other taxa as well: just as bunchgrasses and annual grasses replace the eastern sod-forming perennial grasses in mediterranean California, so too do "bunchforbs" (perennial) and annual forbs replace rhizomatous forbs, and "bunchshrubs" replace laterally spreading shrubs. In the eastern US, goldenrods and asters (formerly *Aster* spp., now split into three genera) with extensive rhizomes are common and diverse in old fields; in California asters are diverse but always in mesic places such as forest shade and wetlands, while goldenrod is neither diverse nor able to flourish in open grasslands except where there is subirrigation (e.g. Blumler 1992a). That is not to say that Asteraceae are unimportant! Rather, there is a remarkable diversity of asteraceous annuals, bunch forbs, and bunch shrubs in Valley grassland and other truly mediterranean environments. In mediterranean California, species of all phylogenetic backgrounds capable of rapid lateral spread are restricted to mesic habitats (or sand), where there is additional moisture during at least part of the summer.

The same pattern holds in Israel and other summer-dry parts of the Mediterranean and Near East. One does not encounter rhizomatous composites, legumes, or shrubs, let alone grasses, except on mesic sites and around rock outcrops (short-rhizomatous species, corresponding to the phalanx type, do occur). A recent Tunisian study of vegetation change in exclosures (Tarhouni et al. 2015) reported that the cover of hemicryptophytes and chamaephytes increased initially but subsequently decreased; after 20 years therophytes dominated. Protection from grazing had caused a decrease in Grime's S, CSR, and CS types, but an increase in RS. No mention is made of C types because there were none.

The environmental conditions that favor guerrilla clones are a topic that deserves separate, full-length treatment; here I will merely sketch some possibilities by way of explanation. Klimes et al. (1997) reported that clonal species are disproportionately common in cool, wet, and nutrient-poor environments. Mack and Thompson (1982) suggested that the bunchgrass form offers some resistance to heat. While this may have merit, I would suggest that guerrilla-clonal species—which are hemicryptophytes in Raunkiaer's system—must produce rhizomes or other propagating structures close to the surface, and the surface is likely to be extremely dry during the Mediterranean summer. In contrast, bunchgrasses, bunchforbs, bulbs, and other phalanx-type clonal species may have more ability to put roots down deeply, where they can tap lingering summer soil moisture (bulbs, by going dormant, may avoid the issue). In fact, Raunkiaer reached a similar conclusion:

> In the lowlands of the Mediterranean region, where the greatest danger is the excessive desiccation during the dry summer, the greatly heated soil surface is hardly any protection to the perennating shoots [of hemicryptophyes] with their rejuvenating buds. (Raunkiaer 1934: 551)

Regardless, the absence of guerrilla-clonal species changes the vegetation dynamics. Instead of replacing tall, competitive annuals by invading from the side, perennials must somehow compete with them as seedlings. How are they going to do that when their growth rate is slower and in almost all cases their seeds are smaller? Wild barley typically produces seeds of 35 mg or so; no perennial C3 (cool-season) grass comes close to that, except the xerophytic and thus slow-growing *Lygeum spartum* (Blumler 1992a). Grime (1979) classified cultivated barley (*Hordeum vulgare*) as a competitive-ruderal, and it is clear that he would place wild barley (*H. spontaneum*) and other tall Mediterranean annual grasses in that category also. In an environment where there are no competitors, competitive-ruderal species comprise the most competitive type of plant out there. I have argued that they can outcompete perennials for soil water in seasonally dry environments, through rapid growth during the rainy season that leaves very little for perennials to subsist on during the drought (Blumler 1992a, 1993, 1998, 2000).

Blumler (1992a, 1993, 1998) predicted that, on fertile soil in summer-dry regions such as Israel, tall annuals would come to dominate the vegetation of exclosures. Results from an exclosure at Neve Ya'ar in Galilee (578 mm mean annual precipitation) supported that prediction: wild oats (*Avena sterilis*) and wild barley increased to 96% of the cover on open soil in sun, while perennials decreased. Thus, "....traditional models of successional sequences in summer-dry regions are topsy-turvy with respect to reality" (Blumler 1993: 289). Other Israeli studies are consistent with these results, albeit not always so interpreted (Litav 1965; Litav et al. 1963; Noy-Meir et al. 1989). Noy-Meir et al. (1989) compared the vegetation at 15 pairs of sites experiencing differing levels of grazing across fence rows; the tall annual wild-cereal grasses, wild oats, wild barley and wild wheat (*Triticum dicoccoides*), showed the greatest positive response to reductions in grazing pressure. One site pair with perennial dominance on the ungrazed side was extremely rocky (35% rock cover—and considering that a plant establishing immediately adjacent to a rock will extend outwards from it for some distance, this translates to more than 50% "rock plus rock edge habitat"). If one removes that pair from the data set, there is no significant effect of grazing on the relative cover of annual plants (Table 6.5). No matter what the grazing pressure, annuals dominate.

Table 6.5 Life form, relative cover, and grazing intensity at 14 paired sites along fence rows in the Upper Jordan Valley, Israel

Life form	Relative cover		
	No grazing	Light-moderate grazing	Heavy grazing
N	8	13	7
Wild cereals	43.5	19.8	2.9
Other annuals	30.2	56.9	71.4
Total annuals	73.8	76.8	74.3
Perennials	26.6	23.5	25.4

Blumler (1993), from data of Noy-Meir et al. (1989), excluding the pair with the highest rock cover of 35%

6.8 Discussion

One of our most insightful ecologists, Mark Westoby (1980), made a similar argument to that presented here, concluding that annuals could become highly competitive under seasonal drought. Nonetheless, he accepted the received wisdom that perennials were the original dominants in the Near East and suggested that annuals became dominant only after agropastoralists had removed the perennials. But if annuals can out-compete perennials today, why would they not have been able to do so in the past? Even Michael Zohary (1982) eventually came around to accepting that annuals dominate some parts of Israel naturally, although in his earlier, much more influential publications he had insisted on a traditional phytosociological view (Zohary 1962, 1973).

I do not mean to give short shrift to geophytes. Raunkiaer was correct in saying that they represent a larger proportion of the flora in Mediterranean regions than is normal. This is not surprising since, among the perennial growth forms, the geophyte is best adapted to survive seasonal drought—better adapted than the stem succulent, for instance. Raven and Axelrod (1978) pointed out that monocots rarely evolve the annual habit. In the native California flora, annual monocots are found only in the grass, rush, and sedge families, and almost all of those are grasses. On the other hand, monocots do have a pronounced tendency to evolve the geophytic habit. The wonderful spring flower displays of Mediterranean regions feature annual forbs and monocot bulbs primarily, though some dicotyledonous geophytes also play a role.

It may seem trivially obvious that seasonality of precipitation would favor annual plants; but sometimes the trivially obvious is overlooked. In this case, traditional succession theory, with its classification of annuals as ruderals, pioneers and other bit players but no more than that, has surely had a major influence on perception. Also, if not trivially obvious it must seem at least plausible that, along a gradient of increasing precipitation seasonality, the ecological amplitude of annuals would expand. And certainly there are many studies from Mediterranean regions that report strong competition by annuals vs. perennials, even if the authors of those studies do not necessarily go as far as I do.

6.9 Biome Modeling: The Way Forward

Throughout his career, Elgene Box has evinced a refreshing attention to the diversity of growth forms in Nature. He developed a long list of growth forms to enter into his models of global and Mediterranean vegetation, respectively (Box 1981, 1982). In the latter paper, he actually apologized for using "only" 90 growth forms in his model, due to the constraints of computer time! He pointed out that forms that are rare today may prove to be pre-adapted to the coming climate changes (Box 1982). Thus, I am confident that he would recognize that annuals deserve to be studied just as much as any other life form, and that their adaptations are not necessarily simple. One implication of this paper is that Elgene Box's work continues! There are additional growth forms, such as tall, laterally expanding clonal herbs, to be added to his list.

My conception of edaphic climate, and its salience in Mediterranean regions, aligns with Box's (1982: 180) conclusion that "...the real climatic criterion is not summer precipitation but rather summer water availability..." His model, and most others, depend necessarily on climate statistics because of the need to keep things simple, and because soil data are neither so readily available nor so easily quantified. Greater attention to climatic measures of precipitation seasonality might improve the models.

The traditional "one region, one biome" model applied to vegetation mapping is no longer necessary. In this computer era, GIS mapping enables overlay of vegetation elements/life forms to replicate better the complexity in Nature. For annuals and Mediterranean climates, a graduated representation, as on the maps produced here, is better than the all-or-nothing representation of traditional cartography. Currently, the tendency is to give the nod to the trait(s) associated with the tallest plant type, i.e. to assume that the tallest plants that can establish will dominate. But clearly, that assumption does not hold well in mediterranean and other seasonally dry climates. Even if it were so, the other "players" in the ecosystem should not be ignored. Multiple overlays with each layer representing a growth form or functional type are likely to produce a more realistic picture of vegetation pattern.

Acknowledgements It is an honor to have the privilege of making a contribution to this celebration of E. O. Box's stellar career. He understood that vegetation modeling, or the explanation of global vegetation patterns, was a significant research avenue in its own right, and not only because of concern about future climate change. As such, within the USA he was a lone voice for a number of years. It also was a privilege to use this opportunity to harken back to one of plant geography's important pioneers, Christen Raunkiaer. Kevin Heard of SUNY-Binghamton's GIS Core created the maps.

References

Al-Yemeni M, Sher H (2010) Biological spectrum with some other ecological attributes of the flora and vegetation of the Asir Mountain of South west, Saudi Arabia. Afr J Biotechnol 9:5550–5559

Axelrod DI (1975) Evolution and biogeography of Madrean-Tethyan sclerophyll vegetation. Ann Mo Bot Gard 62:280–334

Baker HG (1972) Seed weight in relation to environmental conditions in California. Ecology 53:997–1010

Bartolome JW, Gemmill B (1981) The ecological status of *Stipa puchra* (Poaceae) in California. Madroño 28:172–184

Batalha MA, Martins FR (2002) Life-form spectra of Brazilian cerrado sites. Flora 197:452–460

Beadle NCW (1966) Soil phosphate and its role in molding segments of the Australian flora and vegetation with special reference to xeromorphy and sclerophylly. Ecology 47:992–1007

Biswell HH (1956) Ecology of California grasslands. J Range Manag 9:19–24

Blumler MA (1984) Climate and the annual habit. Unpublished M.A. Thesis, University of California, Berkeley

Blumler MA (1991) Winter-deciduous versus evergreen habit in mediterranean regions: a model. In: Standiford RB (tech. coord) Proceedings of the symposium on oak woodlands and hardwood rangeland management, Davis, CA, October 31–November 2, 1990, pp 194–197. USDA, Forest Service, Gen. Tech. Rep. PSW-126, Berkeley

Blumler MA (1992a) Seed weight and environment in Mediterranean-type grasslands in California and Israel. Unpublished Ph.D. Dissertation, University of California, Berkeley

Blumler MA (1992b) Some myths about California grasslands and grazers. Fremontia 20(3):22–27

Blumler MA (1993) Successional pattern and landscape sensitivity in the Mediterranean and Near East. In: Thomas DSG, Allison RJ (eds) Landscape sensitivity. Wiley, Chichester, pp 287–305

Blumler MA (1995) Invasion and transformation of California's valley grassland, a Mediterranean analogue ecosystem. In: Butlin R, Roberts N (eds) Human impact and adaptation: ecological relations in historical times. Blackwell, Oxford, pp 308–332

Blumler MA (1998) Biogeography of land use impacts in the near East. In: Zimmerer KS, Young KR (eds) Nature's geography: new lessons for conservation in developing countries. University of Wisconsin Press, Madison, pp 215–236

Blumler MA (2000) Vegetation dynamics in seasonally dry environments. In: Khoetsian A (ed) Proceedings of the IGU, biogeography study group scientific conference, biogeographical and ecological aspects of desertification processes in arid and semiarid environments, Yerevan, 23–29 May, 2000

Blumler MA (2005) Three conflated definitions of Mediterranean climates. Middle States Geogr 38:52–60

Blumler MA (2011) Invasive species, in geographical perspective. In: Millington AC, Blumler MA, Schickoff U (eds) Handbook of biogeography. Sage, London, pp 510–527

Blumler MA (2015) Deciduous woodlands in Mediterranean California. In: Box EO, Fujiwara K (eds) Warm-temperate deciduous forests. Springer, New York, pp 257–266

Blumler MA, Plummer JC (2015) Deciduous woodlands in the near eastern Fertile Crescent, and a comparison with California. In: Box EO, Fujiwara K (eds) Warm-temperate deciduous forests. Springer, New York, pp 267–276

Box EO (1981) Macroclimate and plant forms: an introduction to predictive modeling in phytogeography. Dr W. Junk, The Hague

Box EO (1982) Life-form composition of Mediterranean terrestrial vegetation in relation to climatic factors. Ecologia Mediterranea 8:173–181

Braun-Blanquet J (1932) Plant sociology. McGraw-Hill, New York

Braun-Blanquet J, Roussine N, Negre R (1951) Les Groupements Végétaux de la France Méditerranéenne. CNRS, Montpellier

Breman H, Cisse AM, Djiteye MA, Elberge WT (1980) Pasture dynamics and forage availability in the Sahel. Isr J Bot 28:227–251

Brewer WH (1883) Pasture and forage plants. In: U.S. Census Office, Report on the productions of agriculture as returned at the tenth census, vol 3. U. S. Government Printing Office, Washington, DC, pp 959–964

Cain SA (1950) Life forms and phytoclimate. Bot Rev 16:1–32

Chapin FS (1980) The mineral nutrition of wild plants. Annu Rev Ecol Syst 11:233–260

Clements FE (1934) The relict method in dynamic ecology. J Ecol 22:39–68

Clements FE, Shelford VE (1939) Bio-ecology. Wiley, New York

Cody ML, Mooney HA (1978) Convergence versus nonconvergence in Mediterranean-climate ecosystems. Annu Rev Ecol Syst 9:265–321

Daget P, Poissonet J, Poissonet P (1977) Le statut thérophytique des pélouses méditerranéennes du Languedoc. Colloq Phytosociol 6:81–89

Danin A, Orshan G (1990) The distribution of Raunkiaer life forms in Israel in relation to the environment. J Veg Sci 1:41–48

Dimopoulus P, Georgiadis T (1992) Floristic and phytogeographical analysis of Mount Killini (NE Peloponnisos, Greece). Phyton 32:283–305

El-Dermadesh MA, Hergazy AK, Zilay AM (1994) Distribution of the plant communities in Tihamah coastal plains of Jazan region, Saudi Arabia. Vegetatio 112:141–151

El-Ghani MMA (1998) Environmental correlates of species distribution in arid desert ecosystems of eastern Egypt. J Arid Environ 38:297–313

Escarre J, Houssard C, Debussche M, Lepart J (1983) Evolution de la végétation et du sol après abandon cultural en region méditerranéenne: étude de succession dans les garrigues du Montpellierais (France). Acta Oecol 4:221–239

Grime JP (1977) Evidence for the existence of three primary strategies in plants and its relevance to ecological and evolutionary theory. Am Nat 111:1169–1194

Grime JP (1979) Plant strategies and vegetation processes. Wiley, Chichester

Hamilton JP (1997) Changing perceptions of pre-European grasslands in California. Madroño 44:311–333

Hanes TL (1977) Chaparral. In: Barbour MG, Major J (eds) Terrestrial vegetation of California. Wiley, New York, pp 417–469

Hiernaux P, Turner MD (1996) The effect of clipping on growth and nutrient uptake of Sahelian annual rangelands. J Appl Ecol 33:387–399

Hillel D, Tadmor N (1962) Water regime and vegetation in the Negev Highlands of Israel. Ecology 43:33–41

Holland R, Jain S (1977) Vernal pools. In: Barbour MG, Major J (eds) Terrestrial vegetation of California. Wiley, New York, pp 515–533

Kazakou E, Vile D, Shipley B, Gallet C, Garnier E (2006) Co-variations in litter decomposition, leaf traits, and plant growth in species from a Mediterranean old-field succession. Funct Ecol 20:21–30

Klimes L, Klimesova J, Hendriks R, van Groenendael J (1997) Clonal plant architecture: a comparative analysis of form and function. In: de Kroon H, van Groenendael J (eds) The ecology and evolution of clonal plants. Backhuys Press, Leiden, pp 1–29

Litav M (1965) Effects of soil type and competition on the occurrence of *Avena sterilis* L. in the Judean Hills (Israel). Isr J Bot 14:74–89

Litav M (1967) Micro-environmental factors and species interrelationships in three batha associations in the foothill region of the Judean Hills. Isr J Bot 16:79–99

Litav M, Kupernik G, Orshan G (1963) The role of competition as a factor determining the distribution of dwarf shrub communities in the Mediterranean territory of Israel. J Ecol 51:467–480

Lovett Doust L (1981) Population dynamics and local specialization in a clonal plant *Ranunculus repens*. J Ecol 69:743–755

Lydolph PE (1985) The climate of the earth. Rowman & Allanheld, Lanham, MD

Mack RN, Thompson JN (1982) Evolution in steppe with few large, hooved mammals. Am Nat 119:757–773

Major J, Barbour MG (eds) (1977) Terrestrial vegetation of California. John Wiley, New York

Mandaville JP (1986) Plant life in the Rub' al-Khali (the empty quarter), south-central Arabia. Proc R Soc Edinb B Biol Sci 89:147–157

Mattoni R, Longcore TR (1997) The Los Angeles coastal prairie, a vanished community. Crossosoma 23(2):71–102

McNaughton SJ (1985) Ecology of a grazing ecosystem: the Serengeti. Ecol Monogr 55:259–294

Meiners SJ, Pickett STA, Cadenasso ML (2015) An integrative approach to successional dynamics: tempo and mode of vegetation change. Cambridge University Press, New York

Milewski AV (1983) A comparison of ecosystems in Mediterranean Australia and southern Africa: nutrient-poor sites at the Barrens and the Caledon coast. Annu Rev Ecol Syst 14:57–76

Minnich RA (2008) California's fading wildflowers: lost legacy and biological invasions. University of California Press, Berkeley

Monod T (1931) Remarques biologiques sur le Sahara. Rev Gén Sci Pures Appl 42:609–616

Mooney HA (1983) Carbon-gaining capacity and allocation patterns of mediterranean-climate plants. In: Kruger FJ, Mitchell DT, Jarvis JUM (eds) Mediterranean-type ecosystems: the role of nutrients. Springer, Berlin, pp 103–119

Mooney HA, Dunn EL (1970) Convergent evolution in Mediterranean-climate evergreen sclerophyll shrubs. Evolution 24:292–303

Munz PA (1959) A California flora. University of California Press, Berkeley

Noy-Meir I (1973) Desert ecosystems: environment and producers. Annu Rev Ecol Syst 4:25–51

Noy-Meir I, Gutman M, Kaplan Y (1989) Responses of Mediterranean grassland plants to grazing and protection. J Ecol 77:290–310

Painter EL (1995) Threats to the California flora: ungulate grazers and browsers. Madroño 42:180–188

Raunkiaer C (1907). The life forms of plants and their bearing on geography. In: Raunkiaer C (ed) (1934) The life forms of plants and statistical plant geography. Clarendon Press, Oxford, pp 2–104

Raunkiaer C (1908) The statistics of life forms as a basis for biological plant geography. In: Raunkiaer C (ed) (1934) The life forms of plants and statistical plant geography. Clarendon Press, Oxford, pp 111–147

Raunkiaer C (1914) On the vegetation of the French Mediterranean alluvia. In: Raunkiaer C (ed) (1934) The life forms of plants and statistical plant geography. Clarendon Press, Oxford, pp 343–367

Raunkiaer C (1934) Botanical studies in the Mediterranean region. In: Raunkiaer C (ed) The life forms of plants and statistical plant geography. Clarendon Press, Oxford, pp 547–620

Raven PH, Axelrod DI (1978) Origins and relationships of the California flora. Univ Calif Publ Bot 72:1–134

Sarmiento G, Monasterio G (1983) Life forms and phenology. In: Bourliere F (ed) Ecosystems of the world: tropical savannas. Elsevier, Amsterdam, pp 79–103

Schiffman PM (2007) Species composition at the time of first European settlement. In: Stromberg M, Corbin J, D'Antonio C (eds) California grasslands: ecology and management. University of California Press, Berkeley, pp 52–56

Schimper AFW (1898) Pflanzengeografie auf Physiologischer Grundlage. Gustav-Fischer-Verlag, Jena

Skalova H, Pechackova S, Suzuki J, Herben T, Hara T, Hadincova V, Krahulec F (1997) Within population genetic variation in traits affecting clonal growth: *Festuca rubra* in a mountain grassland. J Evol Biol 10:383–406

Small E (1973) Xeromorphy in plants as a possible basis for migration between arid and nutritionally-deficient environments. Bot Notiser 126:534–539

Specht R (1969) A comparison of the sclerophyllous vegetation characteristic of Mediterranean type climates in France, California, and southern Australia. I. Structure, morphology, and succession. Aust J Bot 17:277–292

Sudhakar Reddy C, Hari Krishna P, Meena SL, Ruchira B, Sharma KC (2011) Composition of life forms and biological spectrum along climatic gradient in Rajasthan, India. Int J Environ Sci 1:1632–1639

Tarhouni M, Ben Hmida W, Neffati M (2015) Long-term changes in plant life forms as a consequence of grazing exclusion under arid climatic conditions. Land Degrad Dev 28:1199–1211

Trewartha G (1961) The earth's problem climates. University of Wisconsin Press, Madison

Turner MD (1998) Long-term effects of daily grazing orbits on nutrient availability in Sahelian West Africa: 2. Effects of a phosphorus gradient on spatial patterns of annual grassland production. J Biogeogr 25:683–694

Van der Merwe H, van Rooyen MW (2010) Life-form spectra in the Hantam-Tanqua-Roggeveld, South Africa. S Afr J Bot 77:371–380

Wells PV (1962) Vegetation in relation to geological substratum and fire in the San Luis Obispo Quadrangle, California. Ecol Monogr 32:79–103

Wester LL (1981) Composition of native grasslands in the San Joaquin Valley, California. Madroño 28:231–241

Westoby M (1980) Elements of a theory of vegetation dynamics in arid rangelands. Isr J Bot 28:169–194

Wood YA, Meixner T, Shouse PJ, Allen EB (2006) Altered ecohydrologic response drives native shrub loss under conditions of elevated nitrogen deposition. J Environ Qual 35:76–92

Zohary M (1962) Plant life of palestine. Ronald Press, New York

Zohary M (1973) Geobotanical foundations of the middle east, vol 2. Gustav-Fischer-Verlag, Stuttgart

Zohary M (1982) Vegetation of Israel and adjacent areas. Dr L. Verlag, Wiesbaden

Characterization of Vegetation on the Plains of European Russia

Irina N. Safronova and Tatiana K. Yurkovskaya

Abstract

There are well-defined latitudinal zones on the plains of European Russia. The Tundra Zone is divided into four subzones. Communities of herbaceous perennial vascular plants dominate in the High Arctic tundra subzone. Prostrate dwarf shrubs *Salix* spp. and *Dryas punctata* take a great part in the vegetation of the Arctic tundra subzone. The Northern hypoarctic tundra is presented by hemiprostrate dwarf shrub-lichen-moss communities. The subzone of Southern hypoarctic tundra is formed by shrub communities with *Betula nana* and *Salix* spp. Mires are very important in the vegetation structure of the European Arctic. The Taiga Zone of Eurasia is mainly concentrated in Russia. Spruce forests are the zonal taiga type of European Russia. All forests in the Northern taiga subzone are characterized by sparse stand (= Bestand german), a considerable participation of birch and hypoarctic species. In the Middle Taiga Subzone, the forest canopy exhibits high density; the herb layer is well developed, with a thick cover of moss. In forests of the Southern Taiga Subzone, the diversity of the herb-dwarf shrub layer greatly increases because of the participation of nemoral species. The Subtaiga Subzone of European Russia is characterised by a combination of *Piceeta composita*-dominated forests, where nemoral tree species are represented. Mires play a large role in development and existence of taiga ecosystems. The Broadleaved Forests Zone of European Russia is the eastern edge of the European broadleaved forests. They are represented by *Tilio-Querceta* with *Fraxinus excelsior* and *Tilia cordata* with *Quercus robur*. The vegetation of the Forest-steppe zone is more or less mesophytic, including forests and shrub thickets, steppe meadows and meadow steppes. Mires are few. Steppes of European Russia are a part of a vast Eurasian steppe zone in

---

I.N. Safronova (✉) • T.K. Yurkovskaya
Komarov Botanical Institute, Saint-Petersburg, Russia
e-mail: irasafronova@yandex.ru; yourkovskaya@hotmail.ru

which herbaceous communities of xerophytic microthermic perennial plants dominate. A specific feature of the Northern Steppe Subzone is the participation of many xeromesophytes and mesoxerophytic forbs. In the Middle Subzone, forbs are fewer and comprise more xerophilous species. The Southern Subzone is characterized not only by bunch grasses in their communities, but xerophilous dwarf semishrubs. Extreme heterogeneity is a feature of the vegetation. The Desert Zone occupies a small part of the Caspian Lowland. Caspian deserts are the western edge of the extensive Caspian-Turanian Desert area. In European Russia, only the Northern Subzone is occurs on the Caspian Lowland and Turan Plains. The communities of xerophilous and hyperxerophilous micro- and mesothermic plants of different life forms, mostly dwarf semishrubs, semishrubs and shrubs represent this desert type of vegetation.

Keywords
Latitudinal zones • Subzonal types • Characterization of vegetation • Mires types • European Russia

The plains of European Russia extend from the Islands of the Arctic Ocean on the north to the Caspian Sea on the south, and from the Baltic Sea on the west to the Ural Mountains on the east. Their main feature is well-defined latitudinal zones: tundra, taiga, broadleaved forest, forest-steppe, steppe and desert (Fig. 7.1). Regionality can be detected in all zones, associated with an increase of continentality from the west to the east.

7.1 Tundra Zone

The tundra zone is a circumpolar one (Walker 2003). It is divided into four subzones, which can be classified in two groups: Arctic ("High Arctic") and Hypoarctic ("Low Arctic") Tundra. All four subzones are expressed in European Russia, from 82° N on the archipelago of Franz Josef Land to 66° N on the continent. The predominant part of this space is located on the islands.

Characteristic features of the tundra vegetation type are: the lack of a tree layer; the significant role of small woody plants (shrubs, stlaniks, dwarf shrubs) that grow slowly, are long-lived, and often evergreen; include many herbaceous perennial plants (rhizomatous, hummocky, cushion-like plants). Mosses and lichens exhibit large cover. An Important peculiarity of the tundra zone is perforated vegetation (with holes) (Yurtsev 1991).

7 Characterization of Vegetation on the Plains of European Russia

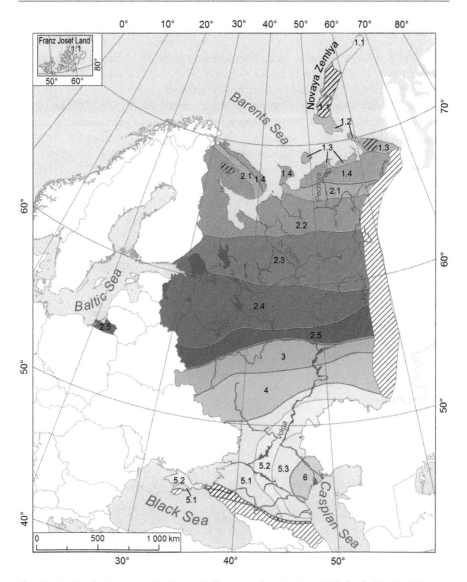

Fig. 7.1 Latitudinal zones and subzones in European Russia: 1.1—High Arctic Tundra Subzone, 1.2—Arctic Tundra Subzone, 1.3—Northern Hypoarctic Tundra Subzone, 1.4—Southern Hypoarctic Tundra Subzone, 2.1—Forest-Tundra Subzone, 2.2—Northern Taiga Subzone, 2.3—Middle Taiga Subzone, 2.4—Southern Taiga Subzone, 2.5—Subtaiga Subzone, 3—Broadleaved Zone, 4—Forest-Steppe Zone, 5.1—Northern Steppe Subzone, 5.2—Middle Steppe Subzone, 5.3—Southern Steppe Subzone, 6—Desert Zone. Shading are using for Mountains (Safronova and Yurkovskaya 2015)

7.1.1 High Arctic Tundra Subzone

High Arctic Tundra Subzone comprises communities of herbaceous perennial vascular plants (Fig. 7.2). It occupies the archipelago of Franz-Joseph Land and the northern island of the Novaya Zemlya Archipelago.

There is a wide variety of high arctic tundra communities, in spite of the fact that a considerable area of the islands has an ice covering. Communities have a polygonal and "spotty" structure. Petrophytic herb-lichen-moss tundra communities are predominant. Generally, cover reaches 50–60% on slopes of plateaus and on sea terraces. Cover of vascular plants is usually 10–30%. Common species are *Papaver polare, Saxifraga cespitosa, S. cernua, S. oppositifolia, Alopecurus alpinus, Phippsia algida, Deschampsia alpina, Cerastium arcticum,* and *C. regelii* ssp. *cespitosa.* We also find tundra communities with an abundance of *Salix polaris.*

Distribution of mosses and lichens is spotty. Overall they have about the same cover (30%), but small depressions are dominated by mosses (*Calliergon giganteum, Ditrichum flexicaule, Orthothecium chryseum, Sanionia uncinata, Campylium* sp., *Timmia* sp., *Cirriphyllum cirrosum, Philonotis* sp., *Bryum tortifolium, Aulacomnium turgidum, Rhacomitrium lanuginosus, Distichium capillaceum, Tomentypnum nitens, Rhacomitrium canescens,* and *Hylocomium splendens* var. *obtusifolia* and etc.). Names of vascular species follow S. K. Cherepanov (1995). Names of moss species follow M. S. Ignatov and O. M. Afonina (Ignatov and Afoniona 1992; Ignatov et al. 2006).

Fig. 7.2 High Arctic Tundra (Franz Jozef Land). Irina Safronova photo's

7.1.2 Arctic Tundra Subzone

The southern island of the Novaya Zemlya Archipelago and Vaygach Island are located in the Arctic Tundra Subzone. The larger part of the Islands of the Novaya Zemlya Archipelago is occupied by low mountains, the vegetation of which is similar to the vegetation of the plains. Prostrate dwarf shrubs such as *Salix* spp. and *Dryas punctata*, dominate a great part in the vegetation of Arctic Tundra Subzone, in polygonal and "spotty" communities of *Salix polaris* and *S. reptans*. The presence of *S. nummularia* characterizes the vegetation of this subzone. Such herbs as *Saxifraga hieracifolia, S. hirculus, Alopecurus alpinus, Deschampsia borealis, D. brevifolia, Dupontia* spp., are common.

7.1.3 Northern Hypoarctic Tundra Subzone

This subzone occupies Kolguev Island and a narrow strip of mainland along the coast of the Pechora Sea, from 52° E in the bay called "Kolokolova lip," up to 61° E on the east side of the "Hypudir lip". Northern hypoarctic tundra is represented by the hemiprostrate dwarf shrub-lichen-moss communities (*Arctous alpina, Vaccinium uliginosum* ssp. microphyllum, *V. vitis-idaea* ssp. *minus*) and the grass-lichen-moss communities (*Arctagrostis latifolia, Dupontia psilosantha, D. fisheri, Deschampsia borealis, D. brevifolia*). There are also shrub communities of *Betula nana* and *Salix lanata*.

7.1.4 Southern Hypoarctic Tundra Subzone

The subzone of southern hypoarctic shrub tundra includes the Northern coast of the Kola Peninsula, Peninsula Kanin, Malozemelskaya Tundra and Bolshezemelskaya Tundra and Yugorsky Peninsula, with a low range named Pai-Khoi, that has similar vegetation.

In the tundra of the Kola Peninsula, plants include *Betula nana, Calluna vulgaris, Empetrum hermaphroditum, Carex bigelowii*. In these communities boreal species such as bilberry, *Vaccinium myrtillus*, frequently occur. An edaphic variant of the southern hypoarctic tundra, the *Empetrum hermaphroditum* communities are especially widespread. The occurrence of *Empetrum* communities is associated with thin soil cover and rocks of the Barents Sea coast of the Kola. *Betula nana* and *Salix* spp. shrub tundra dominants are mainly concentrated adjacent to the "throat" of the White Sea. These communities dominate to the east of the White Sea, as well (Bogdanovskaya-Guieneuf 1938; Gribova 1980; Koroleva 2006a, b; Chinenko 2008, 2013a, b).

7.1.5 Mires of the European Arctic[1]

Mires (the term, which is common for all types of mires massifs or include all types of mires massifs) are very important for the vegetation structure of the European Arctic. They often dominate in Arctic landscapes, except at the highest latitudes (the Subzone of High Arctic Tundra). Distribution of mire types correlates well with latitudinal zonality and permafrost.

All mires of the tundra zone have frozen soils, which thaw out only in summer. The depth of the melt increases to the south and reaches a maximum in the palsa (= Hugelmoore in German).[2] Hollows of palsa can thaw out to the full depth.

Polygonal mires are the most notable type. Their range has been studied in the European Arctic, comparatively recently. They were known only from Asian Russia and Alaska (Katz 1971; Botch and Masing 1983), but Gribova and Yurkovskaya found polygonal mires between 61° E and 65° E on Malozemelskaya tundra and Bolshezemelskaya tundra, Yugor Peninsula, Vaigach Island, in the basin of the Cheyacha-, Sibirchatayacha-, Korotaicha-, and Black Rivers, and the lower part of the Kara River (Gribova and Yurkovskaya 1984). The western location of polygonal mires is located in the Nenets Ridge. Those mires are especially numerous in the valley of the Neruta River. The western boundary of polygonal mires still needs to be defined. The southern points of the polygonal mires are found near the lake of Warsh and, in the east, in the upper reaches of Kolva River (about 68°05″ N and 58° E). Their distribution is associated with a region of continuous permafrost, so that they are confined to the subzone of Northern Hypoarctic Tundra. The flora of polygonal mires is poor. There are about 25 species of vascular plants, more than 15 species of lichens and the same number of mosses. Arctic and hypoarctic species dominate. Arctic species *Carex rariflora, C. rotundata, C. stans, Hierochloe pauciflora* grow in hollows and streambeds, whereas *Salix arctica, S. reptans, Sphagnum lenense, Cetraria cucculata* grow in the slope and hill-polygons. Hypoarctic species predominate; *Eriophorum medium* and *E. russeolum* grow in hollows; *Empetrum hermaphroditum, Ledum decumbens, Salix glauca, S. pulchra, Vaccinium vitis-idaea* ssp. *minus, V. uiginosum* ssp. *microphyllum, Rubus chamaemorus, Calamagrostis lapponica,* and sometimes *Betula nana,* occur on hills and flat polygons. A lot of lichens and mosses belong to Hypoarctic species: *Cetraria nivalis, Calliergon stramineum, Aulacomnium turgidum, Dicranum angustum, D. elongatum,* and *Sphagnum lindbergii*. Boreal species of lichens and mosses are abundant. Boreal species of vascular plants: *Andromeda*

---

[1]Russian terminology in Mirescience (Moorkunde deutsch) correspond to the German terminology. This article used the terminology accepted in international publications in English (see Joosten et al. 2017). Mire (Moor) is a natural formation, occupying part of a land surface and characterised by a peat layer, logged with water and covered by specific vegetation (Sirin et al. 2017).

[2]Palsas are mounds of peat with a permafrost core, surrounded by seasonally melting fens—may coalesce, forming contorted ridges and swales, occupying several hundred hectares. They can exist as single mounds with fen flarks or small pools, as groups of mounds along a depression, or as massifs with a complicated structure.

polifolia, Oxycoccus microcarpa, Festuca ovina, and *Carex chordorrhiza* penetrate to the Subzone of Northern Hypoarctic Tundra through the mires.

Palsa are only occasionally observed in the southern part of the Subzone of Southern Hypoarctic Tundra. Fen and grass-moss mires are common throughout the Arctic. Among them the grass-sedge type of mires, with *Carex stans* and *Dupontia fisheri,* are most characteristic of the coasts of the islands and on the continent (Zubkov 1932; Aleksandrova 1956). Inland mires are dominated by *Eriophorum medium, Carex stans, C. rariflora;* in places *Comarum palustre, Epilobium palustre, Polemonium palustre, Calamagrostis neglecta* occur (Rebristaya 1977).

7.2 Taiga Zone

The Taiga Zone of Eurasia is mainly concentrated in Russia. Its north-western bounder lies in Scandinavia and the South-Western end extends to the Baltic States until the Kaliningrad region, North-East of Poland and North of White Russia. To the east Taiga Zone crosses Russia like a broad stripe until the Pacific Ocean.

From our point of view, taiga is not only dark coniferous forests. It is the area of all boreal ecosystems. Taiga ecosystems are a complex network of irregularly alternating forests and mires, lakes and river valleys. Their development and existence occur in close interaction (Yurkovskaya 2007a).

Mechanisms of regulation and self-regulation that are unique to taiga vegetation, create an optimal spatial structure for vegetation cover (Pignatti et al. 2002). Accordingly, the ratio of forests, mires and other ecosystems in a particular region of the taiga is not accidental and mostly determines stability of the vegetation in this zone.

Spruce forests are the zonal taiga type in European Russia. They are distributed unevenly. In West Karelia, spruce forests account for 28% of the total forested area. To the east, in the Arkhangelsk region, spruce forest accounts for 63% of the total forested area, and in the Republic of Komi, 54% (Chertovskoi 1978).

Five latitudinal Subzones, from Forest Tundra in the north to Subtaiga in the south, are present in the Taiga Zone of European Russia (Aleksandrova and Yurkovskaya 1989).

The structure of the forests changes from north to south, including the density and height of the stand, species composition, and other parameters. Some of these changes are shown in the Table 7.1 (Yurkovskaya and Pajanskaya-Gvozdeva 1993; Yurkovskaya 2007b; Yurkovskaya and Elina 2009).

7.2.1 Forest-Tundra Subzone

We consider the forest-tundra as the most northern subzone of the taiga zone. The same opinion was shared by geobotanists V. B. Sochava (1980), M. L. Ramenskaya (1974) and others. However, there are other opinions (Zinserling 1932; Walter and Box 1976; Aleksandrova 1980).

The structure of the forest changes from north to south; this involves density and height of the stand, species composition, and other parameters. *Betula czerepanovii*

Table 7.1 Structure and composition of spruce and pine forests (Yurkovskaya 2007b)

Parameters	Northern taiga	Middle taiga	Southern taiga
Spruce forests			
Height of spruce, m	17	20	25
Density of canopy	0.4 (0.2–0.6)	0.5 (0.4–0.7)	0.6 (0.4–0.8)
Species number of vascular plants	23	28	60
Constancy of some species (%)			
Picea abies	0	50	75
Picea abies × *P. obovata*	100	50	25
Calamagrostis arundinacea	14	30	70
Convallaria majalis	0	17	70
Empetrum nigrum s. L.	100	5	0
Pine forests			
Height of pine, m	16 (14–17)	20 (18–23)	22 (20–25)
Species number of vascular plants	20	20	50
Constancy of some species (%)			
Calamagrostis arundinacea	0	30	70
Convallaria majalis	0	30	90
Empetrum nigrum s. l.	85	10	0

Orlova [*Betula pubescens* ssp. *czerepanovii* (Orlova) Hämet-Ahti] forms the northern border of forests in the northwest of Russia.

The dominance of this birch is the main feature of the forest-tundra of the Kola Peninsula. We, as phytogeographers of northwestern Europe, include the birch forest-tundra in the taiga zone (Hämet-Ahti 1963; Hämet-Ahti and Ahti 1969).

The Forest-Tundra Subzone of the Kola Peninsula is a band 20–100 km wide from the northwest to the southeast. In the northwest it almost reaches the Barents Sea; in the central and eastern parts of the Kola Peninsula, birch forests penetrate deep into the Tundra Zone along the river valleys.

In birch forests and woodlands, *Betula czerepanovii* with *Empetrum hermaphroditum, Cladonia* spp., and less common taxa are widespread. Often green mosses, mainly *Pleurozium schreberi,* occur directly under the birch canopy, and lichens cover the open space.

Birch forests with *Vaccinium myrtillus* and *Empetrum hermaphroditum* are most common in the western part of the Forest Tundra Subzone, particularly in the area of Kola Bay. To the east they are much rarer. This is a park-forest, with a more or less well-developed understory of *Juniperus sibirica*, a dense shrub layer, *Empetrum hermaphroditum, Vaccinium myrtillus,* and a poorly developed moss cover.

Birch forests with *Vaccinium myrtillus* and *Chamaepericlimenum suecica* occur in an oceanic climate. They are found in the west and in the coastal parts of the White Sea. The spruce, *Picea obovata*, forms sparse forests at the border of forests on the plains to the east of the "throat" of the White Sea. *Betula czerepanovii* is present very often. Sparse forests of *Betula czerepanovii* can be found occasionally up to the western macroslope of the Polar Ural region (Ramenskaya 1974; Yurkovskaya et al. 2012). Larch, *Larix sibirica*, occurs near the Pechora River and forms the border of a forest to the east of the Ural Mountains. The western location of the small larch forest occurs in the basin of the Sula River, the left tributary of the Pechora River, almost 67° N. From the southern border of the forest-tundra to the north, the range of woodlands fully shifts from the watersheds to the valleys of streams and rivers. The watersheds there are occupied by South Hypoarctic Shrub Tundra, mainly *Betula nana* communities (Katenin 1970; Norin 1972).

7.2.2 Northern Taiga Subzone

All forests in the Northern Taiga Subzone are sparse, have a considerable population of birch (*Betula pubescens, B. cherepanovii*) and of hypoarctic species, such as *Empetrum hermaphroditum, Ledum palustre, Vaccinium uliginosum, Betula nana*, which are found only in mires, to the south.

In the west of the Northern Taiga Subzone, in Fennoscandia, which is part of the Kola-Karelian region, spruce forests are less common than pine forests. This domination by pine is one of the features of the vegetation cover of Fennoscandia, separating it sharply from the vegetation cover of the East European Plains, where spruce forests dominate (Yurkovskaya and Elina 2009). In the Kola-Karelian region, pine forests occur to the north, forming a band along the border with Norway. The distribution of these forests is determined not so much of zonal-geographic components but as a result of ecological, mainly edaphic conditions such as the predominance of sandy, rocky and skeletal soils.

Pine forests are interspersed with sparse birch forests only in Fennoscandia. Four groups are distinguished among Northern Taiga pine forests: (1) Lichen, (2) Moss-Lichen, (3) Moss and (4) Sphagnum. The most constant species are *Arctostaphylos uva-ursi, in the* Lichen type; *Empetrum hermaphroditum, Vaccinium vitis-idea,* in the Lichen, Moss-Lichen, and Moss types; and *Ledum palustre* in the Sphagnum type (Yakovlev and Voronova 1959; Vilikainen 1974; Gorshkov and Bakkal 1996; Neshataev and Neshataeva 2002; Kucherov and Zverev 2012).

To the east of the 36° E line of longitude, the spruce forests of *Picea × fennica* and *P. obovata* predominate. Most common are spruce forests with *Empetrum hermaphroditum* and *Vaccinium myrtillus. Larix sibirica* also occurs in these communities (Kucherov and Zverev 2010; Yurkovskaya 2014). Pine forests of *Pinus sylvestris* occur on sandy soils (Lashchenkova 1954; Martynenko 1999). Further to the east, closer to the Urals, *Pinus sibirica* takes the place of the spruce forests.

7.2.3 Middle Taiga Subzone

Middle Taiga, according to Russian geobotanists, is characterized by typical taiga vegetation: the canopy is characterized by high density, with a well developed herb layer, and a complete cover of moss. But the species diversity of middle taiga forests increases little in comparison with northern taiga. This is because of the absence of hypoarctic species which are typical for northern taiga. and the rarity of

Fig. 7.3 Spruce forest with Vaccinium myrtillus. Galina Elina photo's

boreo-nemoral and nemoral species. *Piceetum myrtillosum* forests dominate Middle Taiga (Fig. 7.3). *Piceetum vacciniosum* occurs on sandy loam soils. In addition to bilberry *(Vaccinium myrtillus)* and cowberry *(Vaccinium vitis-idaea)*, we always see *Trientalis europea, Majanthemum bifolium, Orthilia secunda* as components of boreal forests. In the moss layer, *Hylocomium splendens* dominates, accompanied by the permanent species *Dicranum polysetum* and *Pleurozium schreberi* (Zinserling 1932; Lavrenko 1980a). There are spruce forests with *Equisetum sylvaticum* and *Sphagnum* spp. (*Piceetum sylvatici-equisetoso-sphagnosum*); and sedge-sphagnum understoreys (*Piceetum globulari-caricoso-sphagnosum*) with *Sphagnum girgensohnii*. The cowberry pine forests (*Pinetum vacciniosum*) are the predominating type of pine forests (Yakovlev and Voronova 1959; Kucherov and Zverev 2012).

There are some changes in the tree composition and in the herb layers from the west to the east within the subzone. *Larix sibirica* appears at about 37° E, near Lake Vodlozero (Zinserling 1933; Kravchenko 2007). To the east, Larix sibirica plays a more prominent role in the vegetation of this subzone, forming larch and pine-larch forests on lime-containing rocks (Kucherov and Zverev 2011). The role of wet tall-grass spruce forests with *Aconitum septentrionale*, *Caccalia hastate* and others increases from west to east. More often as we travel north, we encounter the northern vine, *Atragene sibirica,* in taiga forests.

Single findings of *Abies sibirica* are noted at 43° E longitude in the basin of the Vaga River, a left tributary of the Northern Dvina River, and at 52° E longitude *Abies sibirica* occurs in the fir-spruce forest. Between 48° E and 52° E, on the Timan Ridge, *Pinus sibirica* occurs in swampy pine forests.

7.2.4 Southern Taiga Subzone

The Subzone of the Southern Taiga, extends between 55° N and 61° N latitudes, along the western border of Russia; it becomes narrower by half near the Ural Mountains. In forests of this subzone the diversity of the herb-dwarf shrub layer increases greatly due to the presence of nemoral species, and the diminished role of the moss layer. Even in the north of the southern taiga subzone, average numbers of vascular plant species double when compared with forests of the Middle and the Northern Taiga Subzones (Gnatyuk and Kryshen 2001; Yurkovskaya and Elina 2009).

The area of spruce forests of *Picea abies* is much wider in the Southern Taiga Subzone than in the north. They dominate almost to 35° E longitude, and are replaced to the east by the hybrid form of *Picea × fennica*. *Piceetum oxalidosum* and *Piceetum nemoriherbosum* form the dominant type of spruce forest. Near the Ural Mountains, the forests are fir-spruce and spruce-fir, sometimes with an understory of *Tilia cordata*.

In the Subzone of Southern Taiga in European Russia, lichen pine forests (*Pineta cladineta*) are often enriched by xeromesophytes such as *Koeleria grandis, Silene chlorantha, Gypsophila fastigiata, Dianthus arenarius, Viola rupestris* s.l., and *Jasione montana*; and by mesoxerophytes such as *Koeleria glauca, Pulsatilla patens,* and *Veronica spicata* (Kucherov and Zverev 2012). Except for these species in the Southern Taiga Subzone, pine forests there do not differ from the pine forests of the Middle Taiga (Fig. 7.4).

Fig. 7.4 Pine forest. Tatiana Yurkovskaya photo's

7.2.5 Subtaiga Subzone

Subtaiga is often considered an independent zone of broadleaved-coniferous forests. We concur with those researchers who include Subtaiga in the Taiga Zone, treating it as the southern Subzone (Semenova-Tyan-Shanskaya and Sochava 1956; Isachenko and Lavrenko 1980). Is characterized by a variety of *Piceeta composita* forests, in which nemoral tree species are represented only in the understory or in the second layer, never in the first. The Southern Taiga forest types can co-occur here especially *Piceetum nemoriherbosum*. Generally the boundary between the Subzone of Southern Taiga and Subtaiga is not clear, as was noticed by Zinserling (1932).

The southern part of the subzone is formed by oak-spruce and pine-broadleaved forests, with fragments of broadleaved forests and boreal spruce forests, the distribution of which depends upon environments conditions. The southern border of the broadleaved-coniferous forests in general coincides with the southern boundary of the distribution of spruce species, *Picea abies, P. obovata* and hybrid forms. This border corresponds to the border region of the Moscow glaciation on the East European plain; that glaciation had a significant influence on the differentiation of modern vegetation.

Broadleaved-spruce forests on the East European plain are confined to well-drained watersheds and gentle slopes of hills. They develop mainly on sod-podzolic, rarely weakly podzolic, and sod-carbonate soils. Broadleaved-pine subtaiga forests occupy specific habitats. Broadleaved-pine forests are confined to the extensive ancient alluvial plains and terraces of big rivers with sandy and sandy loam soils, and watersheds with a relief of fluvioglacial origin, where moraine loams are overlain by sediments of a lighter mechanical composition.

7.2.6 Taiga Mires

Mires play a large role in the development and existence of taiga ecosystems. Mires form a network, which causes a natural fragmentation and heterogeneity of the vegetation cover of the taiga, increasing its sustainability in relation to natural and anthropogenic perturbations.

From the north to the south—"palsa", ribbed fen ("aapa"), sphagnum bogs and fen replace each other successively (Figs. 7.5, 7.6, and 7.7) (Yurkovskaya and Kuznetsov 2010).

Palsas are associated with sporadic permafrost distribution. Their geographical range is well known and may be called arcto-boreal. In the taiga, palsas are concentrated in the Forest-Tundra Subzone. The most northern site is located on the island of Kolguev (Bogdanovskaya-Guieneuf 1938). The southern border of palsa coincides with the border of the forest-tundra and northern taiga. Palsa reaches the north of the Northern Taiga Subzone only near the Ural Mountains (Yurkovskaya 1992).

7 Characterization of Vegetation on the Plains of European Russia

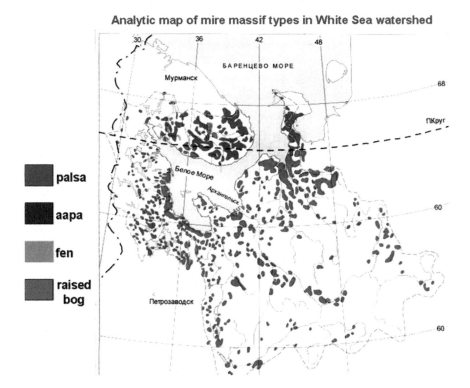

Fig. 7.5 Analitic map of mire massif types in White Sea watershed

Aapa mires are the next type to the south. Analysis and interpretation of space images confirms their pan-boreal distribution. In European Russia they are found in the Forest-Tundra, Northern- and Middle Taiga Subzones.

The classic range of Aapa mires in Europe is in northern Fennoscandia. In European Russia, they are most numerous in the Kola-Karelian region. On the Kola Peninsula, they are found in the Forest-Tundra Subzone and the Northern Taiga Subzone and occurs as the Lapland Aapa mires type. In Karelia, they reach 62° N latitude.

In the East, beyond the Vyg R. and the lake Onega up to the Urals, aapa mires occur sporadically, but are always present. They are the most common mire type only in the basin of the middle Pechora River. Aapa of the Onega-Pechora type extends to the Urals. In Karelia, the aapa are dominated by *Sphagnum papillosum*, with *S. warnstorfii* and *S. fuscum*. In the Onega-Pechora region, they are dominated by *Sphagnum magellanicum*. There are also differences in the composition of moss cover in the hollows and differences in the flora of vascular plants (Yurkovskaya 1992; and many others).

Raised bogs meet sporadically in the Kola-Karelian region. They don't spread over large areas except in the southeastern coast of the White Sea and on the plains to the east from the Onega Lake. Raised bogs prevail on the East-European plain. *Calluna vulgaris* is a geographically differentiating species for raised bogs. It is dominant in the Southern-White Sea type and absent in the Pechora-Onega one. Both types are

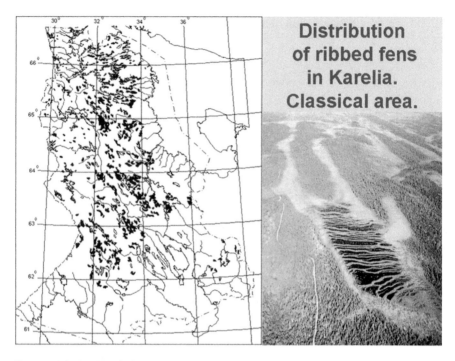

Fig. 7.6 Distribution of ribben fens in Karelia

Fig. 7.7 Palsa in Komi Republik. Nadezda Goncharova photo's

characterized by the absence of *Sphagnum rubellum;* that role is taken on by *Sphagnum nemoreum*. Both types are also characterized by the absence of *Sphagnum cuspidatum*. White Sea bogs type occurs in the Northern Taiga. Onega-Pechora bogs associated with the mainland of the northeast and occur within the Northern- and Middle Taiga.

Large areas of the raised bogs of the Western Russian type occur in the southwest of the taiga zone. They are characterized by the dominance of continental and suboceanic species, on the ridges are *Calluna vulgaris, Chamaedaphne calyculata, Sphagnum fuscum,* and *S. magellanicum*; in hollows we find *Scheuchzeria palustris* and *Rhynchospora alba, Sphagnum cuspidatum, S. balticum, S. majus* and *S. rubellum*. Southern taiga constitutes optimal conditions for them, but they also occur in the Middle Taiga and in Subtaiga. The more continental part of the Southern Taiga Subzone and Subtaiga Subzone is covered by bogs, with *Sphagnum magellanicum* the dominant.

Almost everywhere in the forest, from the Northern taiga to Subtaiga, are small bogs with a predominance of *Ledum palustre* and *Eriophorum vaginatum* in herb-dwarf shrub layer. Forming the matrix in the north is *Sphagnum fuscum,* and *Sphagnum magellanicum* in the south.

7.3 Broadleaved Forests Zone

This zone stretches in a narrow strip from the western border of Russia to the western macroslope of the Ural Mountains. On the Eastern slopes of the Ural Mountain, the broadleaved forests are absent. They are absent in Siberia up to the Far East, where broadleaved forests of the Manchurian type, with *Quercus mongolica,* appear. This gap in the Broadleaved Forests area is associated primarily with extreme continental climate (Box et al. 2001).

In European Russia broadleaved forests are represented by *Tilio-Querceta* where *Fraxinus excelsior* and *Tilia cordata* are associated with *Quercus robur*. Beech, *Fagus sylvaticus,* the most common species in the forests of Central Europe, is only found in the Kaliningrad region. Hornbeam (*Carpinus betulus*) has the occurrence. Maple (*Acer platanoides*) is widely distributed in the broadleaved forest, in addition to linden, oak and ash. Maple, grows in disturbed forests and most often forms secondary forest communities, as does linden.

Pine-broadleaved and pine forests with oak are common on light soils. Many xeromesophytes and mesoxerophytes contribute to herbaceous cover, for example, *Centaurea sumensis, Salvia verticillata, Aster amellus,* and *Inula hirta*. Increasing continentality from west to east results in a gradual reduction of European nemoral species (Lavrenko 1980b).

The Volga River is an important border. As one goes from west to east many tree, shrub and herbaceous species gradually disappear; for example, *Fraxinus excelsa,* the ash tree.

A small grove of oaks occur far to the north, in the Southern Taiga Subzone, up to 60° N; and linden forests occur further to the North to about 62° N, in the Middle Taiga Subzone. The most cold-resistant tree is elm, *Ulmus laevis*. Individual elm groves are found in the Middle Taiga Subzone of Karelia and in the Arkhangelsk region (Kravchenko 2007). Southern broadleaved forests penetrate far into the steppe region in ravines, the so-called ravine forests.

7.4 Forest-Steppe Zone

The Forest-Steppe Zone is the first zone on the Russian Plain, as one moves from north to south, where evaporation exceeds the annual precipitation. A feature of this climate zone is the instability of moisture and the alternation of wet and dry years. There are 500–600 mm/year of precipitation in the north and 300–400 mm in the south. The mean monthly July temperature is ≥ 20 °C to ≥ 22.5 °C. Humid western winds predominate during winter and summer. The sum of temperatures above 10 °C is 2200–2800 °C.

Vegetation is more or less mesophytic, including forests and shrub thickets on gray forest soils, and steppe meadows and meadow steppes on chernozem. Mires are few. Forest-Steppe Zone stretches from the border of Russia with Ukraine, to the Altai Mountains. The northern border of this zone coincides with the July isotherm of 20 °C, and approximately with the northern limit of continuous permafrost grey forest soils. The border extends sharply to the northeast from 51° N on the Seim River, to 54° N in the upper reaches of the Don River, and then the border moves gradually to the east-northeast almost to 55° N, at the western foothills of the Ural Mountains.

The southern boundary of the Forest-Steppe Zone coincides with the southern limit of ordinary chernozem. Beginning at the valley of the Oskol River, just north of 50° N, it passes at this latitude to the Black Kalitva River, then to the mouth of the Buzuluk River. Further, it rises in the northeast to Atkarsk and Khvalynsk, and crosses the Volga River at about 53° N. In Zavolzhye, it reaches east to the Samara River, then turns southeast and rounds the southern Ural Mountains (Malysheva and Malakhovsky 2005; Parshutina 2012).

Quercus robur (oak) forests, with *Tilia cordata*, are common for the Forest-Steppe Zone in European Russia (Fig. 7.8). *Carpinus betulus* (hornbeam) is missing here. *Acer campestre* (maple) disappears in the Don River basin and *Fraxinus excelsior* (ash), in the Volga region.

Meadow steppes consist mainly of mesoxerophytes and a large admixture of mesophytes. They are characterized by closed cover and a rich flora, 40–50 species in every community. Mesophytes, principally typical meadow plants, dominate in the vegetation. Rhizomatous mesophytes predominate among grasses, whereas bunch grasses are rare. Mosses often present significant cover.

Not only are forbs abundant in the meadow-steppe communities, but also the sedge *Carex humilis,* rhizomatous and long-rhizomatous grasses such as *Bromopsis riparia, B. inermis, Poa angustifolia, Phleum phleoides,* and *Helictotrichon schellianum.* Feather-grasses are represented by their most mesophytic species (xeromesophytes), such as *Stipa pulcherrima, S. tirsa, S. dasyphylla,* and *S. pennata.* The mesoxerophytes, *Stipa zalesskii* and *Helictotrichon desertorum,* play a noticeable role to the east of the Volga River, in Zavolzhye.

Fig. 7.8 Oak forest with herbs. Yuri Semenishchenkov photo's

At present natural vegetation on the watersheds is almost completely destroyed by ploughing, and fallow vegetation predominates. Its vegetation cover is very heterogeneous, and represents different stages of succession.

7.5 Steppe Zone

Steppes in European Russia are a part of a vast Eurasian steppe zone. The northern steppe border goes from 50° N at the national border with Ukraine, in a northeasterly direction, crosses the Volga River at 52° N, and reaches 53°30′ N. In the south, the Steppe Zone is limited by the Black Sea, the Caucasus and the desert zone on the Caspian Lowland.

Within the steppe zone yearly precipitation is 350–400 mm in the north. It is reduced to 250 mm in the south. The January temperatures increase from −14 and −16 to −10 and −11 °C, in the same direction, July temperatures from +19 and +21 to +24 and +25 °C; sum of temperatures above 10 °C goes from 2200–2500 to 3000–3400 °C. Soils are dominated by chestnut, including dark chestnut, chestnut and light chestnut. In the northern part of the zone, chernozem is typical. There is also solonetz, solonchak and sands.

Steppe type of vegetation develops in the temperate climatic belt of Eurasia and is characterized by herbaceous communities of xerophytic microthermic perennial plants (Lavrenko 1940, 1954, 1956; Lavrenko et al. 1991; Karamysheva 1993; Safronova 2002). Communities of feather grasses predominate (Fig. 7.9). In European Russia, these are communities of *Stipa tirsa, S. ucrainica, S. zalesskii, S. lessingiana, S. sareptana, S. pennata, S. capillata, S. korshinskyi*.

Fig. 7.9 Feather-grass steppe. Irina Safronova photo's

Zonal ecological (including edaphic) variants comprise communities of bunch grasses such as *Agropyron* spp., *Festuca*, spp., *Koeleria* spp., etc., (including the genus *Stipa* but not on plakors); communities of tufted sedges (*Carex* spp.) or forbs of *Allium, Galatella, Tanacetum*, etc. Dwarf semishrubs of *Artemisia, Camphorosma, Kochia, Thymus*, etc., and shrubs of *Amygdalus, Caragana, Cerasus, Spiraea*, etc., occur in the steppe type of vegetation in addition to herbaceous plants. Dwarf semishrub is the life form that usually dominates in the desert zone. In the steppe zone, dwarf semishrub communities are confined to rocky-gravelly and saline substrates. Some of them, such as *Artemisia nitrosa, A. marschalliana, A. lessingiana, Thymus marschallianus*, etc., are steppe species; others are found in steppe and desert zone, e.g., *Artemisia pauciflora, Atriplex cana, Camphorosma monspeliaca, Artemisia santonica,* and *A. lerchiana.*

The Eurasian Steppe Zone is divided into three latitudinal subzones, which are well developed in European Russia (Bohn et al. 2000a, b; Safronova 2010). In each subzone one or more species of the genus *Stipa* reach a phytocoenotic optimum. Subzonal steppe types differ in community structure. In every subzone communities dominated by *Stipa pennata, Festuca beckeri, Koeleria glauca* occur, and they are confined to sandy soil; communities dominated by *Stipa capillata* occur on sandy loam soils; communities of *Festuca valesiaca*—occur on saline and gravelly soils.

7.5.1 Northern Subzone

In the Northern Subzone, feather grass steppe communities are the zonal type on the dark chestnut and ordinary chernozem soils (Lavrenko 1978, 1980a). A specific feature of Northern Steppe is the participation of many species of xeromesophytes and mesoxerophytic forbs. The feather grasses are represented by one or another species, as the dominant. *Stipa tirsa* and *S. ucrainica* communities predominate in the western part of the Eastern Europe. *Stipa zalesskii* and *S. tirsa* communities can be found nearly throughout Northern Steppe Subzone. The phytocoenotical role of *S. tirsa* is more important to the west of the Don River, and *S. zalesskii* to the east of the Volga River. *Stipa pulcherrima* occurs throughout the Subzone in the feather grass communities.

The border between two large regional types, Black Sea steppe type and Volga-Kazakhstan steppe, exists along the Volga River and the Western macroslope of the Ergeny Hills.

7.5.2 Middle Subzone

In the middle subzone forbs are less numerous and abundant than in the northern steppes and include more xerophilous species. This is the so-called dry steppe on chestnut soils. In this subzone *Stipa ucrainica* and *S. lessingiana* steppes reach their phytocoenotic optimum. Communities of *Stipa ucrainica* are typical mainly to the west of the Volga River. Communities of *Stipa lessingiana* are widespread. They occur in various environments, whereas communities of *Stipa capillata* are prevalent on sandy-loamy soils; those of *Stipa pennata* are prevalent on sandy soils, coenoses of *Stipa sareptana*—on saline soils; coenoses of *Stipa korshinskyi* are prevalent on carbonate soils, coenoses of *Stipa zalesskii* are prevalent in places with the additional moisture. The permanent floristic components are compact-bunchgrasses such as *Stipa lessingiana, Koeleria cristata, Agropyron desertorum*—on loamy soils, *Agropyron fragile, Koeleria glauca, which occur* on sandy soils and on sands, and the sedge, *Carex supina*. On saline soils the obligatory component is a long-rhizomatous grass, *Leymus ramosus*. Dwarf semishrubs are represented by a few species, which are locally abundant, *Artemisia austriaca, A. lerchiana, A. marschalliana, A. pauciflora,* and *Kochia prostrata*.

7.5.3 Southern Subzone

Steppe communities of the southern subzone, the desert steppe, on the light chestnut soils are characterized in their composition not only by bunch grasses, but xerophilous dwarf semishrubs, mostly in the genus *Artemisia* subgenus *Seriphidium*. These steppes are the most xerophytic and are considerably poorer in species than those listed above.

In Russia they occur on the Ergeny Hills and in the Caspian lowland. Communities of *Stipa sareptana* are the most characteristic of this subzone, and occur in various environments. Communities of *S. lessingiana* are associated with light soils such as sand and sandy loam; those of *S. capillata* are associated with sandy-loam.

The uniqueness of the vegetation cover of desert steppes in European Russia is its extreme heterogeneity. This is due to the wide distribution of saline soils in the Caspian Lowland. Vegetation complexes include grass-dominated communities and dwarf semishrubs communities (Larin et al. 1954; Safronova 1975).

7.6 Desert Zone

The Desert Zone occupies a small part of the Caspian Lowland. Caspian deserts form the western edge of the extensive Caspian-Turanian desert area that stretches almost 3000 km from the Ergeny Hills in the west to the mountains of the Dzungarian Alatau and Tarbagatai in the east. To the east of the Caspian Sea, the desert area is divided into three latitudinal subzones. European Russia has only the Northern Subzone.

The communities of xerophilous and hyperxerophilous micro- and mesotermic plants of different life forms, mostly dwarf semishrubs, semishrubs and shrubs characterize the desert type of vegetation on the Caspian Lowland and Turan plains. The dominant biomorphs in the Caspian-Turanian desert zone are dwarf semishrub species of *Artemisia* in the subgenus *Seriphidium*. Communities of *Artemisia* spp.

occur in a variety of habitat types in the plains and low hills: on clayey, loamy, sandy, gravelly and stony soils; takyrs occur on sandy, saline soils (Fig. 7.10). Zonal-ecological variants are represented by semishrubs and shrub communities (Korovin 1961; Blagoveshchensky 1968; Safronova 1996). Communities of shrubs such as *Atraphaxis, Calligonum, Caragana, Ephedra, Haloxylon, Salsola*, etc.; and dwarf semishrubs such as *Astragalus, Convolvulus, Krascheninnikovia, Salsola,* etc., are confined to habitats with better moisture such as sand, rocky slopes of hills, and saline depressions with a shallow groundwater level.

Grasses are characteristic of communities on loamy, sandy and rocky-gravelly substrates throughout the Turanian Desert, in every latitudinal Subzone—Northern, Middle and Southern. Some of these grasses such as *Stipa sareptana, S. lessingiana, Agropyron fragile,* and *Poa bulbosa*, are also distributed in both desert and steppe zones; other species such as *Stipa richteriana* only occur in deserts (Rachkovskaya et al. 1990; Ladygina et al. 1995; Safronova 1998; Rachkovskaya et al. 2003).

Fig. 7.10 Dwarf semishrub desert. Irina Safronova photo's

7.6.1 Northern Subzone

The desert border of the Northern Subzone in the European Caspian region occurs northward to 47° N from the Kuma River along the Eastern macroslope of the Ergeny Hills, then turns northeast, crosses the Volga River at about 47°30′ N, and extends further to the east, just north of 48° N latitude (Ivanov 1958, 1961; Bohn et al. 2000a, b; Safronova 2002).

The desert climate of the northern Caspian Lowland is characterized by low precipitation, 150–250 mm/year and high evaporation, 800–900 mm/year. The mean January temperature is −7 to −10 °C. The mean temperature of July is +24 to +26 °C. The sum of temperatures above 10 °C is 3400–3600 °C.

The relief is generally low. Most of the Caspian Lowland lies below the level of the World ocean (−27 m on the coast of the Caspian Sea); brown desert soils are characterized for the Northern Subzone. Sands and solonchaks occupy large areas.

The vegetation is characterized by a number of features: (1) the significant part of the Caspian Lowland is occupied by sand and soil of light texture; (2) two ecological variants (psammophyte and hemipsammophyte) are prevail; (3) pelitophyte deserts, formed on the plains with loamy soils, occupy small areas; (4) the feather-grasses (*Stipa sareptana* and *S. lessingiana*) occur in wormwood desert communities on brown loamy-sand and sandy soils, they are absent on loamy brown soils; (5) species richness of the communities on the Caspian Lowland is very poor, on loamy and sandy loam soils up to 5–10 species, on sands, up to 20; (6) the wormwood deserts are dominant in the Caspian Lowland: *Artemisia lerchiana* communities occupy a variety of habitats; *Artemisia arenaria* communities are typical on sands; *Artemisia pauciflora* communities are associated with solonetz, *Artemisia santonica* communities grow on meadow-saline or solonchak soils; (7) spatial structure of the vegetation cover is very heterogeneous.

Halophytic vegetation is typical of the coast of the Caspian Sea and along the salt depressions in the sandy areas. It is represented by communities of perennial and annual saltwort, halophilic grasses and wormwoods on solonchaks and solonetz. Perennial saltworts, *Halocnemum strobilaceum*, *Halimione verrucifera*, *Kalidium caspicum*, *Salsola dendroides*, *Suaeda physophora*, *Limonium suffruticosum* are the most common taxa. A prominent role is played by halophilic grasses, such as *Aeluropus pungens*, and *Puccinellia distans*. Communities of annual saltworts characterized by *Climacoptera crassa*, *Salsola brachiata*, *Bassia sedoides*, *Frankenia hirsuta*, *Suaeda acuminata*, *S. altissima*, and *Salicornia perennans*, are widespread. Communities of annual saltworts occupy a huge space; they are associated not only with saline soils, but also fallow saline soils.

Due to strong anthropogenic influences, steppe species such as *Artemisia austriaca*, *Festuca beckeri*, *Festuca valesiaca*, *Koeleria gracilis* penetrate into the desert. At the same time, under the influence of pasture, hayfields and fires, wormwoods often disappear from feather-grass-wormwood desert communities and grasses become abundant, creating a "steppe" aspect. Sometimes in the composition of the communities, except for the abundant grasses, a large role is played by weed species, mainly annuals such as *Ceratocephalus testiculata*, *Descurainia sophia, Lagoseris sancta, Lepidium perfoliatum, Sisymbrium altissimum*, are great. Thus, under the influence of human activity, there is a strong transformation of the vegetation of the Desert Zone of European Russia.

7.7 Conclusion

There is a well-defined series of latitudinal zones and subzones on the plains of European Russia. It is our position that the Forest-Tundra is the northern part of the Taiga Zone. Another opinion is that the Subtaiga is an independent zone of broadleaved coniferous forests. We join with those researchers who include Subtaiga in the Taiga Zone, as the Southern Subzone. We consider the Forest-Steppe as an independent zone.

The predominant part of the Tundra Zone is located on the Arctic islands. Despite this, the vegetation of the zone exhibits great variety.

Taiga ecosystems are a complex network of irregularly alternating forests and mires, lakes and river valleys. They develop and interact in close association. It should be stressed again that mires play the greatest role in the development and existence of the Taiga Zone.

The density and height of the forests, their species composition, and other parameters change from the north to south. The vegetation structure also changes from west to east. In the west, in all Subzones, the pine forests predominate over spruce. In the centre of European Russia, in all Subzones spruce forests dominate. In the area adjacent to the Ural Mountains, the structure of taiga becomes more complicated; together with monodominant spruce and pine forests the major vegetation types are bidominant fir-spruce, larch-pine, and pine (*Pinus sibirica*)-spruce forests.

The role of *Picea abies,* and its hybrid form with *P. obovata,* differs in the complex of subzonal spruce forests. The area of *Picea abies* spruce forests in the Southern Taiga Subzone is much wider than in the north. They dominate almost to 35° E longitude, being replaced to the east by the hybrid form of *Picea × fennica*.

At present, natural vegetation in the watersheds in the Forest-Steppe Zone has been almost completely destroyed by ploughing so that fallow vegetation predominates. Vegetation cover is very heterogeneous, and represents different stages of restoration. In the Steppe Zone, the natural steppe communities are only preserved on steep slopes, on saline soils, and on outcrops of rocks. The Steppe Zone is occupied by fields, large areas—by fallows of different ages. Nature reserves are very important to the restoration and conservation of steppe vegetation.

Acknowledgements The work received financial support from the Russian Foundation for Fundamental Researches, the grant numbers are 14-04-00362 and15-05-06773.

References

Aleksandrova VD (1956) Vegetation of the South island of Novaya Zemlya between 70° 56' and 72° 12' n.l. In: Vegetation of the far north of the USSR and its reclamation. Moscow-Leningrad, pp 187–306

Aleksandrova VD (1980) The Arctic and Antartic: their division into geobotanical areas. Cambridge University Press, Cambridge, 247 p

Aleksandrova VD, Yurkovskaya TK (1989) Geobotanical zoning of the non-Chernozem zone of the European part of the USSR. Leningrad, 64 p

Blagoveshchensky EN (1968) About the desert type of vegetation. Probl Des Dev 5:14–24

Bogdanovskaya-Guieneuf ID (1938) Natural conditions and reindeer pastures of the island Kolguev. Proc Inst Polar Agric Ser Reindeer Herding 2:5–161

Bohn U, Gollub G, Hettwer C (eds) (2000a) Karte der natürlichen vegetation Europas/Map of the natural vegetation of Europe. Maßstab/Scale 1: 2,500,000. Karten/Maps/zusammengestellt und bearbeitet von/compiled and revised by Udo Bohn, Gisela Gollub, Christoph Hettwer. Bundesamt für Naturschutz/Federal Agency for Nature Conservation. 9 blatts/sheets. Bonn–Bad-Godesberg

Bohn U, Gollub G, Hettwer C (eds) (2000b). Karte der natürlichen vegetation Europas/Map of the natural vegetation of Europe. Maßstab/Scale 1: 2,500,000. Legende/Legend/zusammengestellt und bearbeitet von/compiled and revised by Udo Bohn, Gisela Gollub, Christoph Hettwer. Bundesamt für Naturschutz/Federal Agency for Nature Conservation, I–XVI. Bonn–Bad-Godesberg, 153 p

Botch MS, Masing VV (1983) Mires ecosystems in the USSR. Mires Swamp Bog Fen Moor. Reg Stud 4b:95–152

Box EO, You H-M, Li D-L (2001) Climatic ultra-continentality and the abrupt boreal-nemoral forest boundary in northern Manchuria. Separate print, pp 183–200

Cherepanov SK (1995) Vascular plants of Russia and adjacent States (the former USSR). St.-Petersburg, 991 p

Chertovskoi VG (1978) Spruce forests of the European part of the USSR. Lesanaya Promyshlennost', Moscow, 176 p

Chinenko SV (2008) Comparison of local flora of the Eastern part of the Northern coast of the Kola Peninsula with local flora of adjacent regions. Bot J 93(1):60–81

Chinenko SV (2013a) Comparison of local flora of the Eastern part of the Murmansk coast of the Kola Peninsula and adjacent regions: composition of taxa and life forms. Bot J 98(1):10–24

Chinenko SV (2013b) Comparative analysis of coenotic flora of the Eastern part of the Murmansk coast of the Kola Peninsula. Bot J 98(2):134–166

Gnatyuk EL, Kryshen AM (2001). Study of spatial differentiation of the flora of Central Karelia using statistical methods. Biogeography of Karelia. In: Proceedings of Karelian Scientific Center of Russian Academy of Science, Petrozavodsk, vol 2, pp 43–58

Gorshkov VV, Bakkal IJ (1996) Species richness and structure variations of Scots pine forest communities during the period of 5 to 210 years after fire. Silva Fennica 30(2–3):329–349

Gribova SA (1980) Tundra. In: Vegetation of the European part of the USSR. Nauka, Leningrad, pp 29–69

Gribova SA, Yurkovskaya TK (1984) To geography of polygonal mires in the European part of the USSR. Geogr Nat Resour 2:41–46

Hämet-Ahti L (1963) Zonation of the mountain birch forests in Northernmost Fennoscandia. Ann Bot Fen 34(4):1–127

Hämet-Ahti L, Ahti T (1969) The homologies of fennoscandien mountain and coastal birch forests in Eurasia and North America. Vegetatio XIX(V):1–6

Ignatov MS, Afonina OM (1992) Check-list of mosses of the former USSR. Arctoa 1:1–85

Ignatov MS, Afonina OM, Ignatova EA (2006) Check-list of mosses of East Europe and North Asia. Arctoa 15:1–130

Isachenko TI, Lavrenko EM (1980) Botanical-geographical zoning. In: Vegetation of the European part of the USSR. Nauka, Leningrad, pp 10–22
Ivanov VV (1958) Steppe of Western Kazakhstan in connection with the dynamics of their cover. Nauka, Moscow-Leningrad, 288 p
Ivanov VV (1961) On the border of steppes and deserts of the South-East of the European part of the USSR. In: Proceedings of the Institute of Biology, vol 27, pp 105–110
Joosten H, Tanneberger F, Moen A (eds) (2017) Mires and peatlands of Europe status, distribution and conservation. Schweizerbart Science Publishers, Stuttgart
Karamysheva ZV (1993) Botanical geography of steppes of Eurasia. In: Steppes of Eurasia: problems of conservation and restoration. Comp. articles in memory of E. M. Lavrenko. St.-Petersburg-Moscow, pp 6–29
Katenin AE (1970) Zonal position and general patterns of vegetation. In: Plant ecology and biology of Eastern European forest-tundra: Part 1, pp 27–36
Katz YaN (1971) Swamps of the Earth. Moscow, 295 p
Koroleva NE (2006a) Treeless plant communities of the Eastern Murman coast (Kola Peninsula, Russia). Veg Russ 9:20–42
Koroleva NE (2006b) Zonal tundra on the Kola Peninsula: reality or a mistake? Proc Mosc State Tech Univ 9(5):747–756
Korovin EP (1961) Vegetation of middle Asia and Southern Kazakhstan, vol 1. Tashkent, 452 p
Kravchenko AV (2007) A compendium of Karelian Flora (vascular plants), Petrozavodsk, 403 p
Kucherov IB, Zverev AA (2010) Siberian larch forests in the north-east of European Russia. I. Subarctic and subalpine open woodlands. J Biol 3(11):81–108
Kucherov IB, Zverev AA (2011) Larch forests of Northern European Russia. II. Middle and northern taiga forest. Proc Tomsk State Univ Biol 1(13):28–50
Kucherov IB, Zverev AA (2012) Lichen pine forests of middle and northern taiga of European Russia. Bull Tomsk State Univ Biol 3(19):46–80
Ladygina GM, Rachkovskaya EI, Safronova IN (eds) (1995) The vegetation of Kazakhstan and Middle Asia (Desert area): explanatory text and the legend to the map. St.-Petersburg, 130 p
Larin IV, Shiffers EV, Levina FJ et al (1954) Basic regularities of the distribution of vegetation and geobotanical zoning of the Northern Caspian area within the Volga–Ural interfluve. In: Issues of improving of forage in the steppe, semi-desert and desert zones of the USSR. Moscow-Leningrad, pp 9–30
Lashchenkova AN (1954) Pine forest. In: Productive forces of the Komi ASSR, vol 3, part 1. Syktyvkar, pp 126–156
Lavrenko EM (1940) Steppes of the USSR. In: Vegetation of the USSR, vol 2. Moscow-Leningrad, pp 1–206
Lavrenko EM (1954) Steppes of the Eurasian steppe region, their geography, dynamics and history. In: Problems of botany, vol 1. Moscow-Leningrad, pp 155–191
Lavrenko EM (1956) Steppes and agricultural lands on the site of steppes. In: Vegetation of the USSR: the explanatory text to the "Geobotanical map of the USSR, M 1: 4,000,000", vol 2. Moscow-Leningrad, pp 595–730
Lavrenko EM (1978) Vegetation of steppes and deserts of the Mongolian People's Republic. Probl Desert Dev 1:3–19
Lavrenko EM (1980a) Steppe. In: Vegetation of the European part of the USSR. Leningrad, pp 203–272
Lavrenko EM (ed) (1980b) Vegetation of the European part of the USSR. Leningrad, 426 p
Lavrenko EM, Karamysheva ZV, Nikulina RI (1991) Steppes of Eurasia. Leningrad, 145 p
Malysheva GS, Malakhovsky PD (2005) Zonal and subzonal boundaries steppes of the Volga Upland. In: Biological resources and biodiversity of ecosystems of the Volga region: past, present, future. Materials of the international meeting devoted to the 10th anniversary of the Saratov branch of Institute of problems of ecology and evolution named by A. N. Severtsov, 24–28 April 2005, Saratov, pp 87–89

Martynenko VA (1999) Light coniferous forests. In: Forest of the Republic of Komi. Moscow, pp 105–131

Neshataev YV, Neshataeva YV (2002) Syntaxonomic diversity of pine forests of the Lapland nature reserve. Bot J 87(1):99–121

Norin BN (ed) (1972) Soils and vegetation of the Eastern European forest-tundra, Part 2. Leningrad, 336 p

Parshutina LP (2012) On the southern border of forest-steppe within the Voronezh region. News Samara Sci Cent Russ Acad Sci 14:1(6):1634–1637

Pignatti S, Box E, Fujiwara K (2002) A new paradigm for the XXI-th century. Annali di Botanica II:31–58

Rachkovskaya EI, Safronova IN, Khramtsov VN (1990) On the zonality of vegetation of deserts of Kazakhstan and Middle Asia. Bot J 75(5):17–26

Rachkovskaya EI, Volkova EA, Khramtsov VN (eds) (2003) Botanical geography of Kazakhstan and Middle Asia (desert region). St.-Petersburg, 424 p

Ramenskaya ML (1974) To the typology of forest-tundra and mountain birch forests. In: Botanical studies in the subarctic. Apatity, pp 18–34

Rebristaya OV (1977) Flora of East of Bolshezemelskaya tundra. Leningrad, 334 p

Safronova IN (1975) On the zonal division of vegetation cover of Volga-Ural Interfluve. Bot J 60 (6):823–831

Safronova I (1996) Species of Artemisia subgenus Seriphidium in the West Turan and their ecology. In: Hind DJN (Editor-in-Chief) Proceedings of the international compositae conference, Kew, 1994, vol 2. Rojal Botanic Gardens, Kew, pp 105–110

Safronova IN (1998) Some actual issues of the botanical geography of the Northern Caspian region. In: Problems of botanical geography. On the 80th anniversary of the Department of Biogeography of St. Petersburg State University, St.-Petersburg, pp 100–106

Safronova IN (2002) About the Caspian subprovince of the Sahara-Gobi desert region. Bot J 87 (3):57–62

Safronova IN (2010) About subzonal structure of vegetation of the steppe zone of the European part of Russia. Bot J 95(8):1126–1133

Safronova IN, Yurkovskaya TK (2015) Zonal regularities of vegetation cover on plains of European Russia and their cartographic representation. Bot J 100(11):1121–1141

Semenova-Tyan-Shanskaya AM, Sochava VB (1956) Coniferous–broad-leaves forests. In: Vegetation of the USSR. Moscow-Leningrad, pp 346–365

Sirin A, Minayeva T, Yurkovskaya T, Kuznetsov O, Smagin V, Fedotov Y (2017) Russian federation (European part). In: Mires and peatlands of Europe status, distribution and conservation. Schweizerbart Science Publishers, Stuttgart, pp 590–617

Sochava VB (1980) Geographic aspects of the Siberian taiga. Novosibirsk, 256 p

Vilikainen MI (1974) Types of pine forests of Karelia. In: Pine forests of Karelia and increasing of their productivity. Petrozavodsk, pp 22–31

Walker DA (ed) (2003) Circumpolar Arctic Vegetation map. S. 1:7,500,000. CAVM. Team, Anchorage

Walter H, Box EO (1976) Global classification of natural terrestrial ecosystems. Vegetatio 32 (2):72–81

Yakovlev FS, Voronova VS (1959) Types of forests of Karelia and their natural zoning. Petrozavodsk, 190 p

Yurkovskaya TK (1992) Geography and cartography of the vegetation of the mires of European Russia and adjacent territories. St. Petersburg, 256 p

Yurkovskaya TK (2007a) Spatial and temporal relations between mires and forests in the boreal ecosystems. In: Peatlands of Western Siberia and carbon cycle: past and present. Proceedings of the second international field symposium, Khanty-Mansiysk, 24 August–2 September 2007. Tomsk, pp 47–48

Yurkovskaya TK (2007b) The structure of the vegetation of the tundra and taiga on the map of the reconstructed vegetation of Europe. In: Geobotanical mapping. Saint-Petersburg, pp 13–22

Yurkovskaya TK (2014) Light-coniferous forests of Russia in the analytical maps. In: Botany: history, theory, practice (on the 300th anniversary of the Komarov Botanical Institute of RAS): proceedings of the international scientific conference/Ed. by D. V. Geltman. SPb.: Publishing house Etu "LETI", pp 235–240

Yurkovskaya TK, Elina GA (2009) Reconstracted vegetation of Karelia on geobotanical and paleomaps. Petrozavodsk, 136 p

Yurkovskaya TK, Kuznetsov OL (2010) Mires ecosystems of the basin of the White Sea. In: White Sea. The natural environment of the White sea basin area, vol 1 (Chapter 14). Moscow, pp 278–300

Yurkovskaya TK, Pajanskaya-Gvozdeva II (1993) Latitudinal differentiation of vegetation along the Russian-Finnish border. Bot J 78(12):72–98

Yurkovskaya TK, Polozova TG, Snitko NP (2012) Natural birch forests on the analytical map of Russia. Bot J 97(10):1259–1275 (in Russian)

Yurtsev BA (1991) Problems of determination of tundra vegetation type. Bot J 76 (1):30–41 (in Russian)

Zinserling YD (1932) Geography vegetation of the North-West European part of the USSR. In: Proceedings of the Institute of Geomorphology, Leningrad, vol 4, p 376

Zinserling YD (1933) On the North-Western boundary of the Siberian larch (*Larix sibirica* Ledb). In: Proceedings of the Botanical Institute of the AS of USSR. Ser. 3. Geobotany, vol 1. Leningrad, pp 87–97

Zubkov AI (1932) Tundra of Gusinaya Zemlya. Proc Bot Mus USSR Acad Sci 25:57–99

Degeneration and Regression Processes in *Carpinus betulus* Forests of Trentino (North Italy)

Franco Pedrotti

Abstract

The goal of this note is to illustrate the processes of degeneration and regression underway in the forests of *Galio laevigati-Carpinetum betuli* in the hills of Gocciadoro, Central Alps, North Italy. In 1969, the woods of *Galio-Carpinetum betuli* still had good structural and floristic characteristics. Today, over 45 years later, the situation has changed and the association has undergone phenomena of degeneration and regression. In particular, today we observe the following: disappearance of almost half of the species, impoverishment of the species composition, reduction of the degree of cover of some species, increase in the degree of cover of some species (nitrophylous species), appearance of Neophytes (*Robinia pseudacacia* and *Ailanthus altissima*). From the phytocoenotic point of view, the consequence has been the disappearance of the *Galio laevigati-Carpinetum betuli* association, which today remains only in a small area of little more than 100 m^2.

Keywords

Common hornbeam forests · *Galio laevigati-Carpinetum betuli* · Degeneration processes · Regression processes · North Italy

8.1 Introduction

The European or common hornbeam (*Carpinus betulus*) in Trentino (Central Alps, Northern Italy) forms woods of a certain importance only in Val d'Adige, Valsugana, Valle del Chiese and Val Rendena with the *Galio laevigati-Carpinetum*

F. Pedrotti (✉)
Department of Botany and Ecology, University of Camerino, Camerino, Italy
e-mail: franco.pedrotti@unicam.it

Fig. 8.1 Forest of *Galio laevigati—Carpinetum betuli* of Gocciadoro (Trento); the herbaceous layer is greatly reduced due to the anthropogenic impact (Photo F. Pedrotti, April 2017)

betuli association. In Val d'Adige, this association is present in just one location, the hills of Gocciadoro near Trento, formed of sandstones and quartziferous phyllites (Fig. 8.1). In the remaining part of the Val d'Adige, formed of calcareous rocks, the forest vegetation is composed of *Fraxino orni-Ostryetum carpinifoliae*. Thus the hills of Gocciadoro are an ecological enclave in the context of the entire Val d'Adige.

In Trentino, human activity has caused degeneration and regression in the forests of *Galio laevigati-Carpinetum betuli* that have profoundly modified their original floristic and structural makeup and will lead in a short time to the disappearance of this association (Pedrotti 2010).

The goal of this note is to illustrate the processes of degeneration and regression underway in the forests of *Galio-Carpinetum betuli* in the hills of Gocciadoro.

8.2 The Hills of Gocciadoro

The hills of Gocciadoro are on the eastern orographic slope of the Val d'Adige, near the city of Trento, and occupy a very limited area 2 km long and 1 km wide; they run north-south, and reach an altitude of 320 m. They are furrowed by a fairly deep valley, the Valletta del Rio Salé (Fig. 8.2). The western slope of the hills is formed in part by quartziferous phyllites (Lower Archeozoic-Paleozoic) and in part by ignimbrites and andesite lavas (Lower Permian); the tops and the slopes with south-

8 Degeneration and Regression Processes in *Carpinus betulus* Forests... 171

Fig. 8.2 The hills of Gocciadoro (Trento); C—*Galio laevigati-Carpinetum betuli*, developed on western slopes of hills and in the Salé Valley; Q—*Luzulo niveae-Quercetum petraeae* developed on western slopes

western exposure are formed of ignimbrites and andesite lavas; in the valley of the Rio Salé there are also outcroppings of sandstones of Val Gardena (Middle Permian) (Largaiolli et al. 1964; Largaiolli 1971).

8.3 Vegetation, Climate, Soil

The hills of Gocciadoro are for the most part covered by woods of the *Galio laevigati-Carpinetum betuli* and *Luzulo niveae-Quercetum petraeae* forest associations. In the past, the *Galio laevigati-Carpinetum betuli* grew throughout the Gocciadoro area; today it remains only in the small site in the Valletta del Rio Salé, while the remaining part has been substituted by *Galio laevigati-Carpinetum betuli lamietosum orvalae*. In the clearings of the woods, there is a synanthropic shrubby association, *Lamio orvalae-Sambucetum nigrae*. The border vegetation is formed by *Chelidonio majoris-Alliarietum officinalis*.

Luzulo niveae-Quercetum petraeae is present on the tops of the hills and on the rocky and xeric slopes with western and south-western exposure, on the ignimbrites and on the andesite lavas. The woodland has clearings with shrubby vegetation (*Prunetum mahaleb*) and herbaceous vegetation of xeric meadows (*Tunico-Koelerietum gracilis*) (Pedrotti 1963), which have been reduced greatly because of the progressive anthropization of the environment.

On the escarpments where the woods have been cut and in some clearings of *Galio laevigati-Carpinetum betuli* and *Luzulo niveae-Quercetum petraeae*, there is an increasing spread of *Robinia pseudacacia*, forming a shrubwood provisionally called *Robinietum* s.l. (Fig. 8.3).

Thus one can describe the following phytosociological units present on the hills of Gocciadoro:

Fig. 8.3 Woods of the hills of Gocciadoro (Trento); C—*Galio laevigati-Carpinetum betuli*; Q—*Luzulo niveae-Quercetum petraeae*; R—Degeneration of *Luzulo niveae-Quercetum petraeae* caused by invasion of *Robinia pseudacacia*; D—regression of *Luzulo niveae-Quercetum petraeae* caused by tree cutting (Photo P. Pupillo, October 2017)

Galio laevigati-Carpinetum betuli (Pedrotti and Gafta 1999)
Galio laevigati-Carpinetum betuli lamietosum orvalae subass. nova hoc loco
Luzulo-Quercetum petraeae (Pedrotti and Gafta 1999)
Lamio orvalae-Sambucetum nigrae (Poldini 1980)
Prunetum mahaleb (Nevole 1931)
Robinietum s.l.
Tunico-Koelerietum gracilis phleetosum phleoidis (Braun-Blanquet 1961)
Chelidonio majoris-Alliarietum officinalis (Görs et T. Müller 1969)

On the hills of Gocciadoro there are two vegetation series or sigmeta, the series (sigmetum) of European or common hornbeam (*Carpinus betulus*) and the series (sigmetum) of sessile oak (*Quercus petraea*):

1. Alpic south-central acidophilous series of European or common hornbeam (*Carpinus betulus*) [*Galio laevigati-Carpineto betuli* sigmetum]:
   – forest of *Carpinus betulus* (*Galio laevigati-Carpinetum betuli*)
   – forest of *Carpinus betulus* in a phase of degeneration (*Galio laevigati-Carpinetum betuli lamietosum orvalae*)
   – shrubwood of *Sambucus nigra* (*Lamio orvalae-Sambucetum nigrae*)
   – shrubwood of *Robinia pseudacacia* (*Robinietum* s.l.)
   – nitrophilous herbacious vegetation (*Chelidonio majoris-Alliarietum officinalis*)
2. Alpic south-central acidophilous series of sessile oak (*Quercus petraea*) [*Luzulo niveae-Querceto petraeae* sigmetum]
   – forest of *Quercus petraea* (*Luzulo niveae-Quercetum petraeae*)
   – shrubwood of *Prunus mahaleb* (*Prunetum mahaleb*)
   – shrubwood of *Robinia pseudacacia* (*Robinietum* s.l.)
   – meadow (*Tunico-Koelerietum gracilis*)

In the Valletta del Rio Salé, where the environment is cooler and more moist, on a slope with north-eastern exposure between 200 and 250 m, there is European beech (*Fagus sylvatica*). This site of beech is heterotopic, because the lower limit of the beechwood in Val d'Adige runs to about 800 m with the *Carici albae-Fagetum sylvaticae* association, which belongs to the montane belt (Pedrotti 1981); in the Valletta del Rio Salé, because of the microclimate, *Fagus sylvatica* descends 600 m further below, in the hilly belt.

The soils of the hills of Gocciadoro belong to the type of luvisoils (leached brown soils), as seen in the soil map of Trentino by Ronchetti (1965); consult also Sartori et al. (1997).

The average annual precipitation is 915 mm and the average annual temperature is 12.4 °C; the rainfall chart shows an equinoctial distribution type for the precipitation. The climate is lower mesotemperate subhumid prealpic subcontinental (Gafta and Pedrotti 1998).

8.4 The *Galio laevigati-Carpinetum betuli* Association of Gocciadoro

The *Galio laevigati-Carpinetum betuli* of the hills of Gocciadoro has the appearance of a high forest with very large *Quercus robur* trees as well as *Carpinus betulus, Fraxinus excelsior, Acer pseudoplatanus, Tilia europaea,* and *Tilia platyphyllos* trees of notable dimensions. The shrubby layer is formed by *Carpinus betulus, Tilia cordata, Fraxinus excelsior, Acer campestre, Ulmus minor* and other species. At the time of the 1969 surveys, the herbaceous layer was characterized by *Galium laevigatum, Vinca minor, Lamium orvala, Primula vulgaris, Sanicula europaea, Luzula nivea, Campanula trachelium, Salvia glutinosa, Anemone trifolia, Viola reichenbachiana, Carex digitata, Festuca heterophylla, Melica nutans, Brachypodium sylvaticum, Poa nemoralis, Lathyrus niger* and many other species (Pedrotti and Gafta 1999).

In Valsugana, this association also contains *Anemone nemorosa, Ornithogalum pyrenaicum, Mercurialis perennis, Paris quadrifolia, Lathraea squamaria* and rarely, *Erythronium dens-canis.*

8.5 Phenomena of Degeneration and Regression

In 1969, the woods of *Galio laevigati-Carpinetum betuli* still had good structural and floristic characteristics, as can be noted from surveys 1–3 of Table 8.1, drawn from Pedrotti and Gafta (1999). Today, over 45 years later, the situation has changed and the association has undergone phenomena of degeneration and regression (sensu Falinski and Pedrotti 1990), due to the construction of homes on the edge of the woods and poor forest service management, marked by the use of the area as public parkland with construction of roads and trails in the woods and picnic areas with benches for visitors in the wooded areas.

8.6 Degeneration of the *Galio laevigati-Carpinetum betuli*

In 2015, some surveys were conducted in the *Galio laevigati-Carpinetum betuli* (Table 8.1, survey 1a–5a), which we can compare with the 1969 by Pedrotti and Gafta (1999) (Table 8.1, survey 1–3). One can note:

(a) the disappearance of almost half of the species: the average number of species per survey went from 44 species in 1969 to 23 in 2015;
(b) impoverishment of the species composition. All the categories of species have undergone a reduction: species characteristic of alliance, order and class, and also accompanying species. The differential species of the association and those characteristic of the alliance and suballiance (taking into consideration only the herbaceous species, namely *Galium laevigatum, Luzula forsteri, Luzula nivea, Anemone trifoliata* and *Vinca minor*) have passed from a

8 Degeneration and Regression Processes in *Carpinus betulus* Forests...

Table 8.1 Galio laevigati-Carpinetum betuli (rel 1–3) and subass. Lamietosum orvalae (rel 1a–5a)

Relevé number	1	2	3	1a	2a	3a	4a	5a	P	Pa
Altitude (m a.s.l.)	240	245	230	230	240	240	245	240		
Exposure	O	O	O	N-O	E-S-E	O	O	O		
Slope (%)	35	30	30	30	30	35	35	35		
Coverage (%)	100	100	100	100	100	100	100	100		
Area (m^2)	200	200	200	100	100	100	100	100		
Number of species	42	45	45	26	22	23	25	21		
Diff. species of ass. (*Galio laev.-Carpinet. Betuli*)										
Galium laevigatum	1.2	1.2	+	.	.	.	.	.	3	.
Luzula forsteri	.	+	+	.	.	+	.	.	2	1
Luzula nivea	+	.	+	.	.	.	.	.	2	.
Diff. species of suball. (*Aspar. ten.-Carpinenion*)										
Lamium orvala	2.3	1.2	1.2	5.5	5.5	5.5	5.5	5.5	3	5
Lonicera caprifolium	+	+	+	.	.	.	+	.	3	1
Anemone trifolia	.	.	1.1	.	+	.	.	.	1	1
Char. species of all. (*Carpinion*)										
Carpinus betulus (tree)	1.1	3.1	4.1	3.3	3.3	3.3	3.3	3.3	3	5
Carpinus betulus (shrub)	.	1.1	.	1.1	1.1	1.1	+	1.1	1	5
Tilia cordata (tree)	.	+	2.1	1.1	.	1.1	.	1.1	2	3
Tilia cordata (shrub)	+	+	1.1	+	+	+	.	.	3	3
Vinca minor	2.3	2.3	2.3	.	+	.	1.1	1.2	3	3
Prunus avium (tree)	.	.	.	.	.	+	+	.	.	2
Prunus avium (shrub)	+	+	.	+	.	+	.	.	2	2
Char. species of ord. (*Fagetalia*)										
Fraxinus excelsior (tree)	.	.	.	1.1	.	1.1	.	+	.	3
Fraxinus excelsior (shrub)	1.1	1.1	1.1	.	+	+	+	.	3	3
Primula acaulis	+	+	+	+	.	.	.	.	3	1
Salvia glutinosa	1.2	+	1.1	.	.	.	.	+	3	1
Campanula trachelium	+	+	1.2	.	+	+	.	.	3	2
Carex sylvatica	.	+	.	+	.	.	+	.	1	2
Acer pseudoplatanus (tree)	.	.	.	.	.	+	.	+	.	2
Acer psedoplatanus (shrub)	.	.	+	+	+	.	.	.	1	2
Sanicula europaea	+	+	.	.	.	.	.	.	2	.
Veronica urticifolia	.	.	+	.	.	.	.	.	1	.
Lathyrus vernus	.	.	1.1	.	.	.	.	.	1	.
Euphorbia dulcis	+	.	.	.	.	.	.	.	1	.
Ribes rubrum var. *rubrum*	.	.	+	.	.	.	.	.	1	.
Tilia platyphyllos	.	.	+	.	.	.	.	.	1	.
Fagus sylvatica (tree)	.	.	.	1.1	1.1	.	.	.	.	2
Circaea lutetiana	.	.	.	.	.	.	+	.	.	1

(continued)

Table 8.1 (continued)

Char. species of cl. (*Querco-Fagetea*)										
Quercus robur (tree)	3.1	3.1	1.1	2.2	2.2	1.1	1.1	3.3	3	5
Hedera helix (on the soil)	1.2	1.2	+	2.2	2.2	2.2	2.2	2.2	3	5
Viola reichenbachiana	1.1	1.1	+	+	+	+	1.1	2.3	3	5
Acer campestre (tree)	1.1	.	.	.	+	.	+	.	1	2
Acer campestre (shrub)	2.2	1.1	+	+	+	1.1	1.1	.	3	4
Ulmus campestris (tree)	.	+	.	.	.	.	+	.	1	1
Ulmus campestris (shrub)	+	1.2	.	+	.	+	+	.	2	3
Poa nemoralis	1.2	.	1.2	+	.	.	+	.	2	2
Carex digitata	.	+	+	1.1	+	.	.	.	2	2
Melica nutans	+	.	+	.	.	+	.	.	2	1
Brachypodium sylvaticum	+	+	+	.	.	.	.	+	3	1
Festuca heterophylla	+	+	+	.	.	.	.	.	3	.
Castanea sativa (tree)	+	.	1.1	.	.	1.1	1.1	.	2	2
Hepatica nobilis	+	.	.	+	.	.	.	.	1	1
Lathyrus niger	1.1	1.1	1.1	.	.	.	.	.	3	.
Melittis melissophullum	+	+	+	.	.	.	.	.	3	.
Cornus mas	+	+	.	.	.	.	.	.	2	.
Cephalanthera longifolia	.	+	.	.	.	.	.	.	1	.
Ulmus montana	.	.	.	.	+	+	+	.	.	3
Euphorbia dulcis	.	.	.	+	+	.	.	.	.	2
Viburnum opalus	.	.	.	+	.	.	.	.	.	1
Other species										
Corylus avellana	1.1	+	+	+	+	+	+	.	3	4
Sambucus nigra	+	+	+	+	+	.	+	1.1	3	4
Mycelis muralis	+	+	+	.	.	+	+	+	3	3
Geum urbanum	+	1.1	+	+	.	.	.	+	3	2
Clematis vitalba	+	+	.	.	+	+	+	1.1	2	4
Aegopodium podagraria	.	.	+	1.1	.	.	.	.	1	1
Tamus communis	+	+	.	.	.	.	.	+	2	1
Cruciata glabra	+	+	+	.	+	.	.	.	3	1
Prunella vulgaris	+	+	+	.	.	+	.	.	3	1
Fragaria vesca	+	+	+	.	.	.	.	.	3	.
Hieracium racemosum	1.1	+	+	.	.	.	.	.	3	.
Glechoma hederacea	+	+	.	+	.	.	.	.	2	1
Veronica chamaedryis	.	+	+	.	.	+	.	.	2	1
Ajuga reptans	.	+	+	.	.	.	.	.	2	.
Ligustrum vulgare	.	+	.	.	.	.	.	+	1	1
Mespilus germanica	+	+	.	.	.	.	.	.	2	.
Geranium robertianum	.	+	.	.	.	.	+	.	1	1

(continued)

Table 8.1 (continued)

Carex muricata	.	+	.	.	.	.	.	+	1	1
Crataegus monogyna	+	.	.	.	.	.	.	.	1	.
Evonymus europaea	.	+	.	.	.	.	.	.	1	.
Prunus spinosa	+	.	.	.	.	.	.	.	1	.
Solidago virgaurea	.	.	+	.	.	.	.	.	1	.
Oxalis acetosella	.	.	+	.	.	.	.	.	1	.
Athyrium filix-foemina	.	.	+	.	.	.	.	.	1	.
Melissa officinalis	.	+	.	.	.	.	.	.	1	.
Rumex acetosa	.	.	+	.	.	.	.	.	1	.
Robinia pseudacacia (shrub)	.	.	.	+	1.1	1.1	1.1	+	.	5
Cornus sanguinea	.	.	.	+	.	.	+	.	.	2
Ailanthus altissima	.	.	.	.	.	1.1	1.1	.	.	2
Melissa officinalis	.	.	.	.	.	.	.	+	.	1

P presences (rel, 1–3); Pa presences (rel, 1a–5a)
1—Gocciadoro 1968 (rel. 1 Table 8.3, Pedrotti and Gafta 1999); 2—Gocciadoro 1968 (rel. 2 Table 8.3, Pedrotti and Gafta 1999); 3—Gocciadoro 1968 (rel. 3 Table 8.3, Pedrotti and Gafta 1999); 1a—Valley of Rio Salé, left orographic slope, 9-IV-2015; 2a—Valley of Rio Salé, right orographic slope, 9-IV-2015; 3a—Slope under villa Bernardelli, 9-IV-2015; 4a—Slope under villa Bernardellii, 9-IV-2015; 5a—Slope under the little Church of Gocciadoro, 9-IV-2015

presence of 3.6 species per survey to 0.8; the species characteristic of the order (only herbaceous species) have passed from 6.0 species per survey to 3.6; those of the class (only herbaceous species) from 8.3 to 2.8; thus, the decrease has been very strong. Among them, *Carex sylvatica*, *Poa nemoralis*, *Luzula forsteri*, *Primula acaulis*, *Hepatica nobilis*, *Festuca heterophylla*, *Brachypodium sylvaticum*, and *Anemone trifolia* are present today in the surveys only as isolated plants and with a degree of cover less than 1%. *Galium laevigatum* has completely disappeared, though it has been found in a site with a residual strip of *Galio laevigati-Carpinetum betuli* (Table 8.2). Other species (such as *Sanicula europaea*, *Athyrium filix-foemina*, *Melittis melissophyllum*, and *Lathyrus niger*) are still present in Gocciadoro, albeit rarely, but not in the surveys of *Galio laevigati-Carpnetum betuli*. Among the accompanying species, *Oxalis acetosella*, *Solidago virgaurea* and *Hieracium racemosum* were no longer found, while other accompanying species are deemed to be less significant;

(c) reduction of the degree of cover of some species, such as *Poa nemoralis* (from 20% to less than 1%) and *Vinca minor* (from 40% to less than 1%);
(d) increase in the degree of cover of some species, above all *Lamium orvala*, which has passed from 10–20% to 100%, such that the herbaceous layer is almost monospecific;
(e) appearance of Neophytes: *Robinia pseudacacia* and *Ailanthus altissima*;

Table 8.2 *Galio laevigati-Carpinetum betuli*

Relevé number	1
Altitude (m a.s.l.)	235
Exposure	E
Slope (%)	35
Coverage (%)	100
Area (m^2)	100
Number of species	26
Diff. species of ass. (*Galio laev.-Carpinet. betuli*)	
Galium laevigatum	1.1
Diff. species of suball. (*Aspar. Ten.-Carpinenion*)	
Lamium orvala	1.1
Char. species of all. (*Carpinion*)	
Carpinus betulus (tree)	4.4
Carpinus betulus (shrub)	1.1
Tilia cordata (tree)	1.1
Tilia cordata (shrub)	+
Char. species of ord. (*Fagetalia*)	
Fraxinus excelsior (shrub)	+
Fagus sylvatica (tree)	+
Fagus sylvatica (shrub)	+
Campanula trachelium	+
Primula acaulis	+
Salvia glutinosa	+
Dryopteris filix-mas	+
Euphorbia amygdaloides	+
Char. species of cl. (*Querco-Fagetea*)	
Hedera helix (on the soil)	2.2
Viola reichenbachiana	2.2
Quercus robur (tree)	1.1
Carex digitata	1.1
Acer campestre (shrub)	+
Corylus avellana	+
Ulmus campestris (shrub)	+
Poa nemoralis	+
Euphorbia dulcis	+
Melica nutans	+
Brachypodium sylvaticum	+
Festuca heterophylla	+
Other species	
Mycelis muralis	1.1
Robinia pseudacacia (shrub)	+
Polygonatum officinale	+

1—Valley of Rio Salé, right orographic slope, northern exposition 9-IV-2015

(f) *Cedrus deodara* and *Pinus nigra* were planted in the woods before 1969, and this has provoked the almost complete disappearance of arborescent and shrubby species; in 2015, the presence of *Carpinus betulus* (some trees and shrubs), *Corylus avellana, Sambucus nigra, Robinia pseudacacia* and *Hedera helix* was observed, while in the herbaceous layer, *Geum urbanum, Viola reichenbachiana, Erigeron annuus, Euphorbia peplus* and a few other species were noted.

Surveys 1a–5a of Table 8.1, because of their species composition, have been attributed to a new subassociation of *Galio laevigati-Carpinetum betuli* that is denominated *lamietosum orvalae* subass. nova hoc loco, because of the degree of cover of *Lamium orvala*, which in all the surveys is between 80 and 100%. The current floristic makeup is the consequence of the progressive anthropization of the woods and of the environment in general.

8.7 Regression of *Galio laevigati-Carpinetum betuli*

The process of regression has produced a thinning of the woods, with the formation of clearings invaded by two associations, *Lamio orvalae-Sambucetum nigrae* and *Robinietum*.

In *Galio laevigati-Carpinetum betuli*, clearings have formed in spaces left free by trees of *Castanea sativa* and *Ulmus minor,* which withered due to specific diseases (chestnut cancer and Dutch Elm disease); other clearings are due to the construction of roads and picnic areas. A mesophilous shrubby association, *Lamio orvalae-Sambucetum nigrae*, has settled in these clearings. This association has also grown in some gorges that cut into the western slope of the hills. The European elder (*Sambucus nigra*) was present in the surveys of *Galio laevigati-Carpinetum betuli* of 1969, with a very low degree of cover, less than 1%; in later years this species has prevailed over the others where the woods have thinned, and has expanded to form an autonomous association, the *Lamio orvalae-Sambucetum nigrae*. Table 8.3 reports some surveys of *Lamio orvalae-Sambucetum nigrae*; in this association there are also two woody species, *Robinia pseudacacia* and *Ailanthus glandulosa,* as well as some species of the *Prunetalia spinosae* order such as *Corylus avellana, Clematis vitalba* and *Cornus sanguinea*. The herbaceous layer is dominated by *Lamium orvala*, which reaches degrees of cover of 80 and 100%. In the herbaceous layer, there can also be species of *Galio-laevigati-Carpinetum betuli*, such as *Vinca minor, Viola reichenbachiana, Sanicula europaea, Carex digitata, Festuca heterophylla, Luzula forsteri* and *Carex sylvatica*.

Today *Robinia pseudacacia* is widespread in *Galio laevigati-Carpinetum betuli* subass. *lamietosum* and in the newly formed shrubwoods of *Lamio orvalae-Sambucetum nigrae,* as can be seen in Tables 8.1 and 8.3. However, in some points of Gocciadoro, the species has formed a dense *Robinietum*, a survey of which is reported in Table 8.3, from which one notes that the species composition is similar to that of *Lamio orvalae-Sambucetum nigrae*; it is not possible, at the moment, to specify its phytosociological placement.

Table 8.3 *Lamio orvalae-Sambucetum* (rel. 1–4) end *Robinietum* (rel. 5)

Relevé number	1	2	3	4	5	P
Altitude (m a.s.l.)	230	235	240	220	240	
Exposure	O	O	N-O	O	O	
Slope (%)	30	30	35	35	30	
Coverage (%)	100	100	100	100	100	
Area (m^2)	20	20	30	50	20	
Number of species	17	14	16	17	15	
Shrub layer						
Sambucus nigra	3.3	3.3	3.3	1.1	+	4
Robinia pseudacacia	1.1	+	+	3.3	5.5	4
Corylus avellana	+	+	1.1	1.1	+	4
Acer campestre	+	+	+	1.1	+	4
Prunus avium	.	+	+	+	+	3
Clematis vitalba	+	.	.	+	.	2
Cornus sanguinea	+	.	.	.	.	1
Fraxinus ornus	.	+	.	.	.	1
Ailanthus altissima	.	.	.	1.1	1.1	1
Herbaceous layer						
Lamium orvala	4.4	4.4	5.5	4.4	1.1	4
Hedera helix	2.2	1.1	1.1	2.2	2.2	4
Vinca minor	+	+	+	1.1	+	4
Viola reichenbachiana	+	+	+	+	.	4
Geum urbanum	+	.	+	+	+	3
Euphorbia dulcis	.	+	+	.	.	2
Gum urbanum	.	+	+	.	+	2
Carex digitata	.	.	+	+	.	2
Hepatica nobilis	+	.	.	.	.	1
Tamus communis	+	.	.	.	.	1
Sanicula europaea	+	.	.	.	.	1
Taraxacum officinale	+	.	.	.	.	1
Geranium robertianum	+	.	.	.	+	1
Melittis melissophyllum	+	.	.	.	.	1
Festuca heterophylla	.	+	.	.	.	1
Stellaria media	.	+	.	.	.	1
Luzula forsteri	.	.	+	.	.	1
Agrimonia agrimonioides	.	.	+	.	.	1
Omithogalum umbellatum	.	.	+	.	.	1
Carex sylvatica	.	.	.	1.1	+	1
Anemone trifolia	.	.	.	+	.	1
Duchesnea indica	.	.	.	+	.	1
Aegopodium podagraria	.	.	.	+	.	1
Solanum nigrum	.	.	.	.	+	.
Viola denhardtii	.	.	.	.	+	.

P presences (rel, 1–4)
1—Slope under villa Bernardelli, 9-IV-2015; 2—Slope under villa Bernardelli, 9-IV-2015; 3—Slope under villa Bernardellili, 9-IV-2015; 4—Small valley of the slope under Villa Bernardelli, 9-IV-2015; 5—Slope under villa Bernardelli, 28-VI-2015

Table 8.4 *Chelidonio majoris-Alliarietum petiolatae*

Relevé number	1
Altitude (m a.s.l.)	235
Exposure	0
Slope (%)	20
Coverage (%)	100
Area (m^2)	2
Number of species	10
Chelidonium majus	4.4
Geranium robertianum	1.1
Stellaria media	1.1
Taraxacum officinale	+
Euphorbia peplus	+
Alliaria officinalis	+
Glechoma hederacea	+
Poa annua	+
Geum urbanum	+
Duchesnea indica	+

1—Slope under villa Bernardelli 28-VI-2015

8.8 The Border Vegetation of *Galio laevigati-Carpinetum betuli*

The border vegetation is formed of species of the *Geo urbani-Alliarion petiolatae* alliance, among which the most common are *Alliaria petiolata* and *Geum urbanum*; other species are *Chelidonium majus*, *Veronica chamaedryis*, *Lapsana communis*, *Glechoma hederacea*, *Carex muricata*, *Melissa officinalis*, *Geranium robertianum*, *Aegopodium podagraria*, *Cruciata glabra*, *Prunella vulgaris*, *Ajuga reptans*, *Fragaria vesca*, which form the *Chelidonio majoris-Alliarietum petiolata* association (Table 8.4).

8.9 Degeneration and Regression of *Luzulo niveae-Quercetum petraeae*

Today in Gocciadoro, the sessile oak woods (*Luzulo niveae-Quercetum petraeae*) exist only as coppice. In its area of distribution the following species have been observed (2015): *Carex humilis*, *Silene nutans*, *Peucedanum cervaria*, *Arabis turrita*, *Vincetoxicum officinale*, *Chamaecytisus hirsutus*, *Lathyrus niger*, *Bromus erectus*, *Koeleria gracilis*, *Trifolium medium*. *Inula conyza*, *Phyteuma betonicifolium*, *Melica ciliata*, *Artemisia alba*, *Centaurea maculosa*, *Phleum phleoides*, *Festuca valesiaca*, *Bromus erectus*, *Tunica saxifraga*, *Odontites lutea*, *Asplenium adiantum-nigrum*, *Cytisus nigricans*, and *Viola alba* ssp. *denhardtii*.

Thus the phenomena of degeneration and regression of the *Luzulo niveae-Quercetum petraeae* can be summarized as: (a) transformation of the high wood to coppice, during the last centuries; (b) plantings of *Pinus nigra* as isolated trees or groupings, here and there; (c) thinning of the coppices, which today are very open, with formation of clearings; (d) development in the clearings of *Prunus mahaleb*, of the *Prunetum mahaleb* association; and (e) invasion of *Robinia pseudacacia* in the clearings and formation of the *Robinietum*.

8.10 Conclusion

In 1969, the year in which surveys 1–3 of Table 8.1 were conducted (Pedrotti and Gafta 1999), the general environmental conditions of the hills of Gocciadoro were already fairly compromised, as shown by some very precise species data. In fact, in the past many very interesting species were reported for Gocciadoro (see Dalla Torre and Sarnthein 1900–1913), but later they were no longer found; among them at least the following species should be noted: *Anthericum liliago*, *Dianthus armeria*, *Gypsophila muralis*, *Jasione montana*, *Lathyrus vernus*, *Pyrola chlorantha*, and *Sieglingia decumbens*.

The invasion of Neophytes was already under way, in particular that of *Robinia pseudacacia*, even if in not such a massive and extensive way as in the following years. However, the areas with natural woods were still well represented. Today the situation has changed completely, above all regarding the floristic composition of the herbaceous layer, which has undergone great modification, strong impoverishment of the number of species, and the disappearance of some species.

From the phytocoenotic point of view, the consequence has been the disappearance of the *Galio laevigati-Carpinetum betuli* association, which today remains only in a small area of little more than 100 m^2, in semi-rocky conditions in the Valletta del Rio Salé, where the above-mentioned survey (releve) 1 of Table 8.2 was conducted. The *Galio laevigati-Carpinetum betuli* association has been transformed into a new type of coenosis, which has been called *Galio laevigati-Carpinetum betuli lamietosum orvalae*, and which is very poor in species. Finally, the forest of *Galio laevigati-Carpinetum betuli* has lost its continuity because of the formation of clearings in which the shrubwoods of *Lamio orvalae-Sambucetum nigrae* and *Robinietum* have settled.

References

Dalla Torre KW, von Sarnthein L (1900–1913) Flora der gefürsteten Grafschaft Tirol, des Landes Vorarlberg und des Fürstentums Liechtenstein. Wagner, Innsbruck

Falinski JB, Pedrotti F (1990) The vegetation and dynamical tendencies in the vegetation of Bosco Quarto, Promontorio del Gargano, Italy. Braun-Blanquetia 5:1–31

Gafta D, Pedrotti F (1998) Fitoclima del Trentino-Alto Adige. Stud Trent Sci Nat 73:55–111

Largaiolli T (1971) Fenomeni franosi nel bacino del Rio Salé (Trento). Mem Mus Trident Sci Nat XVIII(3):167–183

Largaiolli T, Sacerdoti M, Sommavilla E (1964) Su un lembo di vulcaniti paleozoiche nei pressi della città di Trento. Stud Trent Sci Nat XLI(2):133–137

Pedrotti F (1963) Nota sulla vegetazione steppica (Stipeto-Poion xerophilae e Diplachnion) dei dintorni di Trento. Stud Trent Sci Nat LX(3):288–301

Pedrotti F (1981) Carta della vegetazione del foglio Trento. Collana programma finalizzato "promozione qualità ambiente", C.N.R., Roma, AQ/1/17:1–38

Pedrotti F (2010) Le serie di vegetazione della Regione Trentino-Alto Adige. In: Blasi C (ed) La vegetazione d'Italia. Palumbi, Roma, pp 83–109

Pedrotti F, Gafta D (1999) Sintassonomia e distribuzione di alcune associazioni di caducifoglie nel Trentino-Alto Adige. Doc Phytosoc XIX:495–508

Poldini L (1980) Übersicht über die Vegetation des Karstes von Triest und Görz (NO-Italien). Stud Geobot 1(1):79–130

Ronchetti G (1965) Nota illustrativa alla carta dei suoli della provincia di Trento. Istituto Sperimentale Studio Difesa Suolo, Firenze, pp 1–74

Sartori G, Corradini F, Mancabelli A (1997) Verso un catalogo dei suoli del Trentino. 1. I suoli bruni lisciviati. Stud Trent Sci Nat 72:55–77

Phytoindicating Comparison of Vegetation of the Polish Tatras, the Ukrainian Carpathians, and the Mountain Crimea

Ya. Didukh, I. Kontar, and A. Boratynski

Abstract

The methods of synphytoindication were used to analyze relevés of syntaxa typical for the Polish Tatra Mountains, the Ukrainian Carpathians, and the Crimean Mountains. Ecological scales of flora species originally developed for Ukraine, were applied to determine the place of the syntaxa in relation to the main ecofactors. The nature of correlation between the change of values of latters, and the estimation of vegetation differentiation, and the place of corresponding regions in the general ecospace. The authors determined the grade values for certain types of syntaxa and mountain ecosystems, also the nature of correlation between the ecofactors' values. These values qualitatively change depending on altitude, and consequently, they are different for the mountain ecosystems of the temperate zone (Tatra Mountains and the Carpathians), and for the submediterranean zone (the Crimean Mountains). We determined that the altitudinal differentiation of vegetation cover is caused by the change in hydrothermal regime. Thus, it has an indirect impact on changes of the other edaphic factors. This cumulative effect is more significant than the direct impact of climate. We further determined the most vulnerable syntaxa, on the base of estimation of the place of syntaxa in accordance to the ecofactors' values. These vulnerable communities are under the threat of extinction if the existing tendency of climate change remains the same.

Y. Didukh (✉) • I. Kontar
M. G. Kholodny Institute of Botany NAS of Ukraine, Kyiv, Ukraine
e-mail: ya.didukh@gmail.com

A. Boratynski
Institute of Dendrology, Polish Academy of Sciences, Kórnik, Poland
e-mail: borata@man.poznan.pl

Keywords

Tatra Mountains • Carpathians • Crimea • Vegetation • Ecological differentiation • Comparative analysis • Synphytoindication

9.1 Introduction

At the present stage of nature-ecological science development, application of quantitative methods analysis gives possibility to provide coenosis comparison, to assess a degree of their differentiation in relation to the effect of different environmental factors, and consequently, to develop predictions of possible changes. Such assessment is possible for the comperance and determination coenosis position among theirselves as well as their response to the external factor changment. In both cases it is necessary to operate with quantitative indicator, in particular valuation scales. In the first case can be used the similarity of the species composition, which is reflected in the classification of syntaxa on the basis of classical relevés. The second case needs to be operated by quantitative indicators of ecofactors that is not easy. As certain ecofactors have no such units of measurement and sometimes it is impossible to calculate the indicator value for the specific type of plant community for a long period of time, in this case it is prospective to use the phytoindication method, where the grade scales are used instead of the direct measuring (Ramensky 1938; Ellenberg 1979; Tsyganov 1983; Landolt 1977; Zólyomi et al. 1966; Zarzycki et al. 2002; Pignatti 2005; Borhidi 1995; Didukh 2011). We analyzed of the number of scales, suggested a method of their comparison, and developed own scales for the species of flora of Ukraine (Didukh 2011). The study of these phytoindicator ranges were verified on the model objects of different types of syntaxa, their territorial-landscape distribution in the different regions of Ukraine and Poland. Such method of evaluation of syntaxa was named as "synphytoindication" (Didukh and Plyuta 1994; Didukh 2012).

This method allows to compare not only certain syntaxa (α—coenodiversity) and/or their change within the landscape (β—coenodiversity), but also compare by a sum of synphytoindicative values between quite distant regions (γ—coenodiversity). It is applicable for the mountain systems, which are characterized by quite heterogenic plant cover, and is important considering the reaction of ecosystems to climatic changes. The ecosystem reaction depends on the different factors, limitations, resistance, etc. For such analysis we have selected mountain systems of the Tatra Mountains, the Ukrainian Eastern Carpathians, and the mountains of the Crimea. The research questions were: (1) how syntaxonomic composition of the Crimean mountains differs from the Eastern Carpathians and/or Tatras, (2) to which extend the reaction of different syntaxa types is the same concerning to the ecological factors in different mountain systems, (3) how much is the nature of such dependence changed, and (4) what position these three compared regions occupy in the system of ecospace of the biosphere of Europe. It is impossible to give comprehensive answers to these questions now, and therefore we are considering our work as the preliminary assessment, made at the example of selected syntaxa options of the three mountain regions that predicts in the future their more comprehensive analysis.

9.2   Materials and Methods

Vegetation of the Polish Tatra, Ukrainian Carpathians and Crimean Mountains were the object of investigation. Maximum altitude of the Tatra in Poland is 2503 m (Rysy), the Ukrainian Carpathians 2061 m (Hoverla) and the Crimean Mountains 1545 m (Roman-Kosh). While all vegetation belts of Ukrainian Carpathians and Crimean Mountains (Fig. 9.1) were covered by analysis, in the Tatra were covered only to the subnival vegetation zone, because above 2250 m the vegetation is represented mostly by cryptogamic communities, which have not been studied. Along with the difference in altitudes these mountain systems are distinguished by altitudinal zonation. For the Tatras and Carpathians it is indicative the

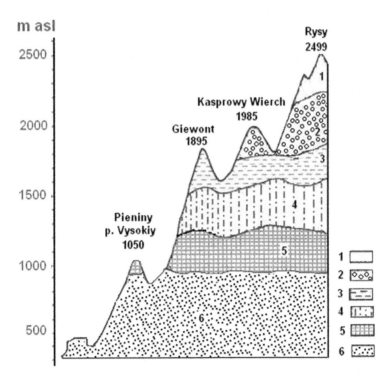

Fig. 9.1 High altitude climate zones of Tatras. 1—subnival belt; 2—alpine belt: *Loiseurio-Vaccinietea, Salicetea herbaceae,* Juncetea trifidi; 3—subalpine belt: *Mulgedio-Aconitetea, Roso pendulinae-Pinetea mugo*; 4—mountane belt: *Vaccinio-Piceetea, Calluno-Ulicetea*; 5—foothill: *Querco-Fagetea, Molinio-Arrhenatheretea*; 6—plain: *Festuco-Brometea*

atlantic or humidic type (Walter 1968; Grebenshchikov 1957), similar to the distribution of vegetation belts of the central Italian Alps (Pedrotti 2013), while the Mountain of Crimea is represented by variants from humidic (northern macroslope) to xerophytic (southern macroslope) of the sub-Mediterranean type (Grebenshchikov 1957; Walter 1968; Didukh 1992; Rivas-Martinez 2005). Such topological distribution of syntaxa in dependence on the change of altitude we interpreted as an analysis β—coenodiversity.

In the work we are limited by an analysis of the typical biotopes that describe altitudinal zonation and geomorphological and/or lithological characteristics of the mountain systems. The hydromorphic and hydrophilous biotopes have not been included in the study. In total, the 37 classes, 65 alliances are represented by 1447 relevés. This material does not cover all coenotic diversity, but sufficiently represents patterns of differences between compared regions.

This analysis predicted a general assessment of gradient (range of amplitudes) by the leading ecofactors. It reflects β-coenodiversity. At that case it is necessary to note that though a material for each mountain systems has different representativity and has been analysed at the level of different syntaxonomical categories it was not reflected on general conclusions.

For the Tatry we are used 234 relevés performed with support of Kasa Mianovskiy in 1998–1999, for the Carpathians—448 relevés, for the Mountain Crimea—765 ones. Besides own relevés covered different biotopes of these mountain systems there were used different literature sources (Derzhypilsky et al. 2011; Klimuk et al. 2006; Kobiv 2014; Malynovski and Kricsfalusy 2002; Solomakha et al. 2004; Chorney et al. 2005; Balcerkiewicz 1984; Pawłowski et al. 1927; Szafer et al. 1923; Szafer and Sokołowski 1925) with the aim to receive more representative data (not less 10 relevés of each syntaxon).

Indicators of 12 ecofactors by the developed scale (Didukh 2011) are totally estimated in the following dimension: soil humidity (Hd = 23 marks), humidity ficklity (Fh = 11 marks), soil airation, content of nitrogens (Nt = 11), soil acid regime (Rc = 15), salt regime (Sl = 19), carbonate content (Ca = 13), thermoregime (Tm = 17), ombroregime (Om = 23), crioregime (Cr = 15), continentality (Kn = 17), lightness in a coenosis (Lc = 9). Unlike scales of Ellenberg, Landolt they reflect not mean quantities but their amplitudes (max-min), though for the further calculations there have been taken mean value of all species, which are present in a coenosis with taking into account of a degree their participation (1 mark—up to 1%; 2—1 to 5%, 3—5 to 20%; 4—21 to 50%; 5—51 to 100%). So, with such an assessment method a coenosis can never have maximum and minimum values, and their amplitude narrows in certain marks from the theoretically possible. The relevés were stored into the TURBOVEG database, to which the database of ecological scales ECODID were integrated (Didukh 2011). This allowed to use a synphytoindication analysis with applying of an arsenal of different mathematic methods. Data processing is performed with the JUICE (Tichy 2002) program with clusters forming by the TWINSPAN program Modified (Roleček et al. 2009). The further analysis of clusters separated with the algorithm TWINSPAN Modified is realized with distinguishing of individual ecological groups. To visualize data there was used 2D protection. Moreover, on the basis of calculation of ecological markers for every relevé performed on the scale base of Ya. P. Didukh (2011), with the STATISTICA 10 program (StatSoft Inc 2005) we created the graphical diagrams "box-and-whiskers plots" for every ecological scale to visualize the distribution of every individual syntaxon relatively ecological optimum, and also fractal images of dependencies between changes of three indicators of climatic and edaphic factors [SurfacePlot].

9.3 Results

9.3.1 Vegetation of Mountain Systems and Its Distribution Regularities

Vegetation of Polish Tatras Tatras are located in the humid climate zone (Fig. 9.1), which defines an Atlantic type of zonality: upper belt subnival (from 2300 to the highest top 2499 m a.s.l. Mt Rysy) *Rhizocarpetea geographici (Rhizocarpion geographicae), Oreochloetum distichae subnivale)*—on lithosols, alpine belt (1800–2300 m a.s.l.—*Oreochloo distichae-Juncetum trifidi*—on acidic rankers, *Festuco versicoloris-Seslerietum tatrae*—on carbonates rendzines), subalpine belt (1550–1800 m a.s.l.—*Pinetum mughi silicosum*—on regosols, *Pinetum mughi calcicosum*—onrendzines, *Adenostylion alliariae)*, mountain-forest (upper-mountane belt) (1200–1500 m a.s.l.—*Plagiothecio-Picetum*—podzols on granite *Polysticho-Piceetum*—on *rendzines*), submountain (lower montane) (500–1200 m a.s.l. *Dentario glandulosae-Fagetum, Luzulo-Fagetum)*—on cambisols (Piękoś-Mirkowa and Mirek 1996; Komornicki and Skiba 1996). Classic premountain (submountain) layer in the Tatra Mts. is absent, therefore for full ecological profile we have conducted research in the Peniny Mountain (Pieniny National Park) (*Fagion sylvaticae, Carpinion betuli, Asplenion trichomanes*) and in the jurasic limestone rocks out of mountain from the Ojców NP (xerophytic grass community *Festuco-Brometea* class, *Cirsio-Brachypodion pinnati*). Great phytocenotic diversity of the Tatras and mentioned adjacent territories is indicated not only by marginal values of the climatic factors (from plains to subnival belt) but also the geology, from acidic granitoids to alkaline jurasic carbonates, and by different moisture of the soils. It is important, because many of rare species of the Eastern Ukrainian Carpathians occur quite frequently there: *Leontopodium alpinum, Biscutella laevigata, Heliosperma quadripetala, Ranunculus thora, Rodiola rosea, Saxifraga caesia, S. androsacea, S. moschata, S. cernua, Selaginella saliginoides, Oreochloa disticha, Veronica alpina* etc.

Different types of vegetation were investigated from the plain to the subnival belt, and also specific petrophytic communities at the outcrops of carbonate and crystal rocks that stipulates for high climatic and edaphic differentiation. There is no representation of azonal communities such as aquatic, riverside-mire, flood forests, which are formed on the hydrogenic substrates. We did not intend to assess in detail all syntaxonomical richness, and selected only certain "rapper" communities at the level leading classes, for that we used more than ten geobotanical relevés made by us and published. It should be noted that the Tatras in that respect are studied very well, by well-known classics of geobotany, such as W. Szafer and B. Pawłowski, who have described (locus classicus) a number of associations, alliances and orders. These relevés represent different types of communities, selection of which is conducted on the base of routs that covered different mounts, as Kasprowy Wierch, Mały Kościelec, Nosal, Giewont (1895 m), Kondratcka Przełęcz, Ciemniak (1900) Małołączniak (2095), Krzesanica (2122), Stoły (1900), Wielki Kopieniec (1328) and others, and valleys (Kościeliska, Chochołowska, Smytna, Jaworzyńska, Strążyska, Kondratowa, Dolina Pieciu stawów Polskich, Dolina Małej Łąki, Koński Żleb, Dolina Olczyskiego Potoku, Dolina Suchej Wody, Dolina Filipki, etc.). The following separate syntaxa that represent the most characteristic patterns of altitudinal and edafic distribution of vegetation were chosen for the analysis.

Cl. *Juncetea trifidi* Hadač in Klika et Hadač 1944

1. All. *Juncion trifidi* Krajina 1934. (*Oreochloetum distichae* Pawl. 1926 *Oreochloo distichae-Juncetum trifidi*)

Cl. *Carici rupestris—Kobresietea ballardii* Ohba 1971

2. All. *Festucion versicoloris* Krajina 1931

Cl. *Loiseurio-Vaccinietea* Eggler ex Schubert 1960

3. All. *Loiseleurio-Vaccinion* Br.-Bl. in Br.-Bl. et Jenny 1926

Cl. *Elyno-Seslerietea* Br.-Bl. 1948

4. All. *Sesleriontatrae* Pawłowski 1935
5. All. *Seslerion caerulae*
6. All. *Caricion firmae* Gams. 1926.

Cl. *Salicetea herbacea* Br.-Bl. 1948

7. All. *Salicion herbaceae* Br.-Bl. in Br.-Bl. et Jenny 1926
8. All. *Arabidion caeruleae* Br.-Bl. in Br.-Bl. et Jenny 1926

Cl. *Tlaspietea rotundifolii* Br.-Bl. 1948

9. All. *Arabidion alpinae* Beguin in Richard 1972
10. All. *Papaverion tatrici* Pawłowski et al. 1928 corr. Valachovič 1995

Cl. *Roso pendulinae-Pinetea mugo* **Theurillat in Theurillat et al. 1995**

11. All. *Pinion mugo* Pawł. 1928

Cl. *Mulgedio-Aconitetea* **Hadač et Klika in Klika et Hadač 1944**

12. All. *Adenostylion alliariae* Br.-Bl. 1926

Cl. *Vaccinio-Piceetea* **Br.-Bl. in Br.-Bl. et al.** 1939

13. All. *Piceion excelsa* Pawłowski et al. 1928

Cl. *Calluno-Ulicetea* **Br.-Bl. EtTx. Ex Klika et Hadac 1944**

14. All. *Nardo-Agrostion tenuis* Sillinger 1933

Cl. *Querco-Fagetea* **Br.-Bl. et Vlieger in Vlieger 1937**

15. All. *Fagion sylvaticae* Pawł. in Pawł., Sikoł. et Wall. (1928)

Cl. *Molinio-Arrhenatheretea* R. Tx. 1937

16. All. *Cynosurion cristati*

Cl. *Festuco-Brometea* **Br.-Bl. et Tx. ex Klika et Hadač 1944**

17. All. *Cirsio-Brachypodion pinnati* Hadač et Klika 1994 in Klika et Hadač ex Klika 1951.
18. All. *Festucion valesiacae* Klika 1931

Cl. *Asplenietea trichomanis* **(Br.-Bl. in Meier et Br.-Bl. 1934) Oberd. 1977**

19. All. *Cystopteridion* Richard 1972.
20. All. *Potentillion caulescentis* Br.-Bl. in Br.-Bl. et Jenny 1926.

Thus, in the Tatra Mts very high climatic and edafic vegetation differentiation, higher than in the Ukrainian Carpathians and in the Crimea.

Vegetation of the Ukrainian Carpathians is characterized by an altitudinal differentiation typical for humidic Atlantic type of zonality (Walter 1968; Grebenshchikov 1974), analogous to those found in the Tatras (Fig. 9.2). Just as the highest top of the Ukrainian Carpathians Hoverla Mt has altitude of 2061 m, so subnival vegetation belt is absent, and Alpine belt is represented fragmentarily, only on the mountain ridges and tops with altitude more than 1750–1800 m (Chornogirsky Ridge, Svydovetsky and Marmarosky massifs). A lower limit of subalpine belt is at altitude of about 1200–1250 m in the north-western part and

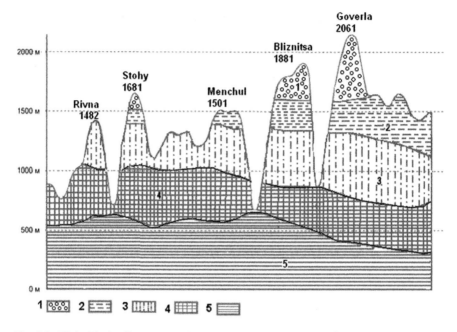

Fig. 9.2 High altitude climate vegetation zones of the Ukrainian East Carpathians. 1—subnival belt; 2—alpine belt: *Juncetea trifidi, Loiseurio-Vaccinietea, Salicetea herbaceae*; 3—subalpine belt: *Mulgedio-Aconitetea, Roso pendulinae-Pinetea mugo*; 4—mountane belt: *Vaccinio-Piceetea, Calluno-Ulicetea*; 5—sub-montane (foothill): *Querco-Fagetea, Molinio-Arrhenatheretea*; 6—plain: *Festuco-Brometea*

reach to 1450–1650 m in the south-eastern part and upper is 1750–1800 m. Alpine and subalpine belts are termed highland, indicator communities of the lower limit of which is expansion of shrub communities *Roso pendulinae-Pinetea mugo* with predominance *Pinus mugo, Duscheckia viridis, Juniperus sibirica*. Forest vegetation belongs to three belts: upper—forests *Vaccinio-Piceetea* with predominance *Picea abies*, mean—*Querco-Fagetea*—with predominance *Fagus sylvatica* and lower—with predominance *Quercus petrea* (Golubets and Milkina 1988). At the place of composite coniferous woods the communities of the *Calluno-Ulicetea* are formed with dominance *Calluna vulgaris, Narduus stricta, Vaccinium myrtillus*, and in more humid conditions—*Mollinio-Arrhenatheretea* with dominance *Arrhenatherium elatior, Cynosurus cristatus, Dactylis glomerata, Deschampsia caespitosa, Molinia caerulea* etc. Petrophytic types of communities are poorer than in the Tatras, but also are represented by carbonates and granitoides *Elyno-Seslerietea, Asplenietea trichomanis, Thlaspietea rotundifolii*). In spite of the fact that the diversity of plant communities is lower in the East Carpathians than in the Tatras, for an analysis were used relevés having more quantities of syntaxa:

Cl. *Juncetea trifidi* Hadač 1946

21. All. *Caricion curvulae* Br.-Bl. 1925
22. All. *Juncion trifidi* Krajina 1933

Cl. *Carici rupestris-Kobresietea bellardii* Ohba 1974

23. All. *Oxytropido-Elynion* Br.-Bl. 1949

Cl. *Salicetea herbaceae* Br.-Bl. 1948

24. All. *Salicion herbaceae* Br.-Bl. in Br.-Bl. et Jenny 1926
25. All. *Festucion pictae* Krajina 1933
26. All. *Arabidion caeruleae* Br.-Bl. in Br.-Bl. et Jenny 1926 (*Salicion retusae* Horv. 1949)

Cl. *Loiseleurio procumbentis-Vaccinietea* Eggler ex Schubert 1960

27. All. *Rhododendro-Vaccinion* Br.-Bl. 1926
28. All. *Loiseleurio-Vaccinion* Br.-Bl. in Br.-Bl. et Jenny 1926
29. All. *Juniperion nanae* Br.-Bl. et al. 1939

Cl. *Elyno-Seslerietea* Br.-Bl. 1948

30. All. *Festuco saxatilis-Seslerion bielzii (Pawł. et Wal. 1949) Coldea 1984*

Cl. *Thlaspietea rotundifolii* Br.-Bl. 1948

31. All. *Papavero-Thymion pulcherrimi* I. Pop 1968—7on,
32. All. *Androsacion alpinae* Br.-Bl. in Br.-Bl. et Jenny 1926

Cl. *Roso pendulinae-Pinetea mugo* Theurillat in Theurillat et al. 1995

33. All. *Pinion mugo* Pawł. 1928

Cl. *Mulgedio-Aconitetea* Hadac et Klika in Klika 1948

34. All. *Adenostylion alliariae* Br.-Bl. 1926
35. All. *Calamagrostion villosae* Pawł. Et al. 1928

Cl. *Oxycocco-Sphagnetea* Br.-Bl. et R. Tx. ex Westhoff et al. 1946

36. All. *Sphagnion magellanici* Kästner et Flössner 1933
37. All. Oxycocco-Empetrion hermaphroditi **Nordh. 1936**

Cl. Vaccinio-Piceetea Br.-Bl. in Br.-Bl. et al. *1939*

38. *All. Piceion excelsae* Pawłowski et al. *1928 (Piceion abietis* Pawł. et al. *1928)*
39. *All. Abieti-Piceion (*Br.-Bl. in Br.-Bl. et al. *1939)* Soó *1964*

Cl. Calluno-Ulicetea Br.-Bl. et R. Tx. ex Westhoff et al. **1946** *(Nardo-Callunetea* Preising *1949, Calluno-Ulicetea* Br.-Bl.et Tx. ex Klika et Hadaé **1944***)*

40. *All. Nardo-Agrostion tenuis* Sillinger *1933*
41. *All. Genistion pilosae* Bocher *1943*
42. *All. Vaccinion vitis-idaeae* Bocher *1943*

Cl. Querco-Fagetea Br.-Bl. et Vlieg. in Vlieg. *1937*

43. *All. Fagion sylvaticae* R. Tx. et Diem. *1936 (Asperulo-Fagion* Tuxen *1955)*
44. *All. Carpinion betuli* Oberd. *1953*
45. *All. Luzulo-Fagion* Lohmeyer et Tx. in Tx. *1954 (Luzulo-Fagenion* Lohm. et R. Tx. *1954)*

Cl. Molinio-Arrhenatheretea R. Tx. *1937*

46. *All. Arrhenatherion elatioris* Luquet *1926*
47. *All. Cynosurion cristati* R. Tx. *1947*

Cl. Quercetea pubescenti-petraeae Jakucs **1960**

48. *All. Quercion petraeae* Issler *1933*

Cl. Asplenietea trichomanis (Br.-Bl. in Meier et Br.-Bl. 1934) Oberd. 1977

49. All. *Hypno-Polypodion vulgaris* Mucina 1993
50. All. *Cystopteridion* Richard 1972

Vegetation of the Crimean Mountains Zonality of the Crimean Mountains differ from the previous mountain systems and is indicative for the mountains of sub-Mediterranean region (Fig. 9.3). The zonality of the southern and northern slope is different; therefore, we distinguished following belts:

Northern macroslope: (1) below an altitude of 400 m—the lower forest-steppe of hemixerophylic forests *Carpino orientalis—Quercion pubescentis* and authentic steppes *Veronici multifidae Stipion ponticae,* (2) between altitudes of 400–450 and 700–800 m—a middle—forests *Paeonio dauricae-Quercion petraeae* and meadow steppes *Adonidi-Stipion tirsae* (from a.s.l.) and (3) upper between 700–800 and 1200–1250 m—nemoral forests *Dentario quinquefoliae-Fagion sylvaticae.* The highest parts of table-like tops of yailasat altitudes 1200–1250 to 1445 m are

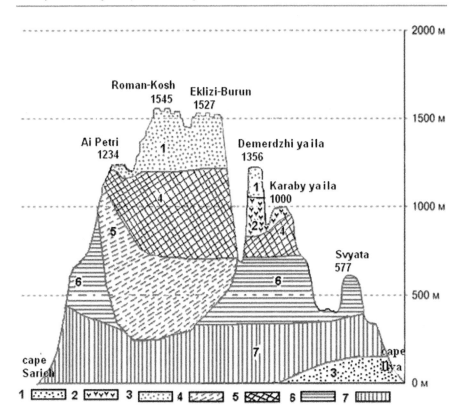

Fig. 9.3 High altitude climate zones of Crimea. 1—high yaila: *Festuco-Brometea (Androsaco-Caricion humilis)*; 2—lower yaila: *Festuco-Brometea (Adonidi-Stipion tirsae)*; 3—foothill: *Festuco-Brometea (Veronici multifidae-Stipion ponticae)*; 4—upper forest belt: *Querco-Fagetea (Dentario-*Fagion *sylvaticae)*; 5—middle and upper forest belt: Erico-Pinetea; 6—middle forest belt: *Paeonio dauricae-Quercion petraeae*; 7—lower forest belt: *Quercetea pubescenti—petraeae*

occupied by mountain meadows and petrophytic steppes *Androsaco—Caricion humilis*, in the karsts *Helictotricho (compressi)-Bistortion officinalis*. **The southern macroslope:** (1) up to altitude about 450 m—the lower belt of hemixerophylic forests and thin forests *Elytrigio nodosae-Quercion pubescentis, Jasmino-Juniperion excelsae* and annual cereals communities *Bromo-Hordeion murine*, ("savannoids"), (2) between 450 and 700–800 m—the middle forest belt of mesoxerophylic and xeromesophylic forests *Paeonio dauricae-Quercion petraeae, Brachypodio rupestris-Pinion pallasianae*, (3) between 600–700 and 1200 m the upper zone of nemoral *Dentario quinquefoliae-Fagion sylvaticae*, and (4) above 900 m—mountain-boreal forests *Carici humilis-Pinion kochianae*. The zonality of the northern macroslope corresponds to the sub-Mediterranean type of humid rank of zonality, where steppes penetrate. A southern macroslope is a transition from humidic to xerophytic rank, and the last with altitude decrease is replaced from the southern aridic to the northern continental type (Didukh 1992).

For the phytoindication analysis there were chosen the relevés of communities of following syntaxa:

Cl. *Querco-Fagetea*

51. All. *Dentario quinquefoliae-Fagionsylvaticae* Didukh 1996
52. All. *Paeonio dauricae-Quercion petraeae* Didukh 1996

Cl. *Quercetea pubescenti—petraeae*

53. All. *Carpino orientalis– Quercion pubescentis* Korzh. et Shelyag 1983 (*Carici michelii—Quercetum pubescentis* Didukh 1996)
54. All. *Elytrigio nodosae—Quercion pubescentis* Didukh 1996
55. All. *Carpino orientalis– Quercion pubescentis* Korzhenevski et Shelyag 1983 (*Physospermo-Carpinetum orientalis* Didukh 1996)
56. All. *Jasmino-Juniperion excelsae* Didukh et al. 1986, Didukh 1996

Cl. *Erico-Pinetea*

57. All. *Carici humilis-Pinion kochianae* Didukh 2003
58. All. *Brachypodio rupestris-Pinion pallasianae* Didukh 2003

Cl. *Molinio-Arrhenatheretea*

59. All. *Trifolio (pratense)-Brizion elatioris* Didukh, Kuzemko 2009 та *Helictotricho (compressi)-Bistortion officinalis* Didukh, Kuzemko 2009

Cl. *Festuco-Brometea*,

60. All. *Androsaco—Caricion humilis* Didukh 2014
61. All. *Adonidi-Stipion tirsae* Didukh 2014
62. All. *Veronici multifidae- Stipion ponticae* Didukh 2014

Cl. *Chenopodietea*

63. All. *Bromo-Hordeion murini* Hejný 1978

Cl. *Alysso-Sedetea*,

64. All. *Drabo cuspidatae-Campanulion tauricae* Ryff 2000.

Cl. *Drypsidetea* (*Onosmato polyphyllae-Ptilostemonetea* Korzhenevsky 1990)

65. All. *Ptilostemonion echinocephali* Korzhenevsky 1990.

Thereby, subnival belt is the above limit of the analyzed biotopes for the Tatras, the Alpine belt for the East Carpathians while meadow-steppe yaila belt for the Crimean Mountains. The oak forest *Quercetea pubescenti–petraeae*—is the lower forest belt of the Tatras and East Carpathians, but forest-steppe of hemixerophylic thin forests (*Quercetea pubescenti–petraeae*) and steppes (*Veronici multifidae-Stipion ponticae*) for the Crimean Mountains.

9.3.2 Assessment of Ecofactor Ranges

On the basis of synphytoindication method there have been calculated amplitudes of ecofactors for each alliance and regularity of distribution of the last has been established. The clearest regularity of distribution is typical for climatic quantities: thermo-, ombroclimate, cryoclimate, but for continentality it is worse.

The Fig. 9.4 demonstrates that the amplitudes of values for syntaxa of every mountain system by some factors are fluctuated within certain limits. Under that considerable difference by the quantities of the most factors for the mountain regions of the central Europe and the Mediterranean, and for the Tatras and East Carpathians they remain quite similar. Thus, a range of amplitudes of humidity of the Tatras is slightly wider than in the Carpathians because of considerable diversity in the Tatras and adjacent premountain communities, *Festuco-Brometea* (8.7–9.5), *Seslerion caerulae* and *Potentillion caulescentis* (9.6–10.2). In the East Carpathians there are well developed associations from *Oxycocco-Sphagnetea*, which are scarce in the West Carpathians. While in the Crimean Mts. polar community types are represented by the most humid forests *Dentario quinquefoliae-Fagion sylvaticae*, 11.7 and driest steppes *Veronici multifidae-Stipion ponticae*, 8.0 marks, and also by annual and saxicolous communities *Bromo-Hordeion murini*, *Drabo-Campanulion tauricae and Ptilostemonion echinocephali* 8.0–8.2 marks.

Fluctuation of humidity because of the absence of hydrophilous coenoses of the flood type is changed in narrower limits; their quantities are more closed and considerably exceeded, although, as it is expected for the Tatras, the highest quantities are peculiar for the typical meadow and steppe communities of the *Molinio-Arrhenatheretea* and *Festuco-Brometea* classes. For the other communities of the Tatras this indicator is oscillated between 4.0 and 5.5. The East Carpathians are characterized by the wider amplitude from 3.5 (*Oxycocco-Sphagnetea, Pinion mugo*) to 7.2 (*Androsacion alpinae, Genistion pilosae*). In the Crimea the amplitude is quite narrowed 5.0–6.0 marks (lowest quantities are typical for nemoral forests *Dentario-Fagion* and driest hazmophytic communities *Ptilostemonion echinocephali*, where moisture are not delayed at skeleton mounds, and highest amplitude is for meadow communities *Molino-Arrhenatheretea*, where during a season humidity values are the most fluctuated.

The amplitudes of values of soil aeration (Ae) and content of mineral nitrogen (Nt) is quite considerably overlapped that is related to moistening regime. Under that aeration values for the Tatras and East Carpathians coincide (except of mire coenoses *Sphagnion magellanici, Oxycocco-Empetrion hermaphroditi*):

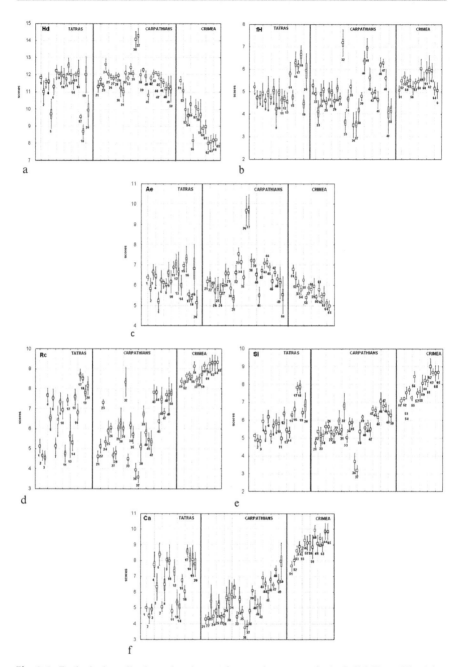

Fig. 9.4 Ecological amplitudes and optimum of vegetation syntaxa in the Polish Tatry, Ukrainian Carpathians and Mountain Crimea by ecological factors: (a) Hd, (b) fH, (c) Ae, (d) Rc, (e) Sl, (f) Ca, (g) Nt, (h) Tm, (i) Om, (j) Kn, (k) Cr, (l) Lc. (syntaxa numbers and values of factors are presented in text)

9 Phytoindicating Comparison of Vegetation of the Polish Tatras, the... 199

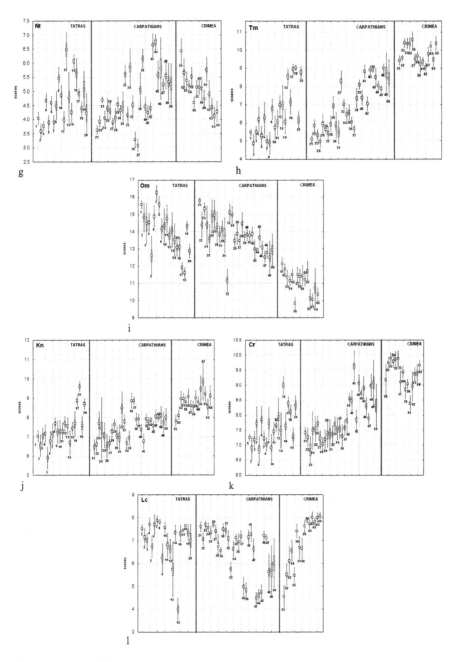

Fig. 9.4 (continued)

Tatras— 5.0 marks *Asplenietea trichomanis*; 7.0–7.5—*Cynosurion cristati, Adenostylion alliariae. Aconietea*; the East Carpathians from 5.5—*Papavero-Thymion pulcherrimi* to 7.3—*Pinion mugo*. For the Crimean Mts, because of drier conditions, these values are replaced to higher aeration from 5.0 (*Ptilostimion*) to 6.9 (*Dentario quinquefoliae-Fagion sylvaticae*: *Ranunculo constantinopolitani--Fraxinetum excelsae*).

By mineral nitrogen forms' content (Nt) these amplitudes are fully overlapped in fact (*Oxycocco-Sphagnetea*). Thus, the *Loiseleurio-Vaccinion* (3.6) communities are the poorest, and *Adenostylion alliariae* (6.7) are the richest in the Tatras; relatively *Juncion trifidi* (3.75)—*Querco-Fagetea* (6.6–6.7) for the East Carpathians. Although for the Crimean Mts an absolute values are close and community types entirely other here. The dry saxicolous are the poorest: *Drabo-Campanulion tauricae* (4.0), and ash forests are the most wealth (*Ranunculo constantinopolitani-Fraxinetum excelsioris, Dentario-Fagion sylvaticae*) (6.5).

Changes of trophic factors are characterized by the considerably sharper gradient, though values of the Tatras and Carpathians are quite overlapped. So, range of soil acidity (Rc) for the Tatras, where there are both deposits of alkaline and acidic rocks, are divided into three levels. The most acidophytic conditions (4.5) belong to *Loiseurio-Vaccinion* communities, most basiphytic (8.6)—to steppe communities of plain part of Poland *Cirsio-Brachypodion* and *Festucion valesiacae*. If the East Carpathians *Sphagnion magellanici* and *Oxycocco-Empetrion hermaphroditi* ($<$4.0) are under the most acidophytic conditions, and *Androsacion alpinae* ($>$8.0) is under the most basiphytic conditions. In the Crimean Mts. minimum values of an acidic regime are close to maximum of the Tatry and Carpathians (*Quercetea pubescentis-petraeae* forests), and maximum reaches the 9.3 level (*Drabo-Campanulion tauricae* and *Ptilostemion echinocephali* saxicolous communities).

In fact, analogous situation is peculiar for values of soil salinisation (Sl), where ranges for the Tatras and East Carpathians match (unless on the one hand *Cirsio-Brachypodion* and *Festucion valesiacae*, and on the other hand—*Sphagnion magellanici* and *Oxycocco-Empetrion hermaphroditi*): minimum—(4.75) *Loiseurio-Vaccinion* (Tatras), (5.1) *Caricon curvulae* (Carpathians), maximum— (6.85) *Potentillion caulescentis* (Tatras), (7.0) *Arrhenatherion elatioris* (Carpathians). For the Crimea as by previous factor the minimum values (7.0) match with maximum of the Tatras and East Carpathians and peculiar for nemoral forests *Dentario-Fagion, Lathyro aurei-Fagetum, Lasero trilobi- Carpinetum betuli*, and maximum (9.1)—for steppes (*Veronici multifidae-Stipion ponticae*).

Under conditions of soil carbonates (Ca) values for different mountain systems a few differences has been found. Thus, the lowest content of carbonates in a soil is fixed in the East Carpatian conenoses, from 3.8–4.2 marks (*Sphagnion magellanici, Oxycocco-Empetrion hermaphroditi* and alpine meadows *Juncion trifidi*) to 8.6 (carbonate outcrops *Cystopteridion*). In the Tatras communities are divided into two groups: carbonate-phobic ($<$6.0) i carbonate-phylic ($>$7.0). These are, in fact, not overlapped. Minimum (4.5 marks) is peculiar for alpine hazmophytic *Festucion versicoloris*, and maximum (8.7) for *Cirsio-Brachypodion* and *Caricion firmae*.

Intermediate position (6–7 marks) belongs to the communities of the *Cynosurion cristati, Seslerion caerulae, Arabidion caerulae* and *Fagion sylvaticae* alliances. For the Crimea the lowest (7.6) carbonate values are peculiar for the broadleaved forests *Dentario quinquefoliae-Fagion sylvaticae*, and the most upper (9.9–10.0) for biotopes of carbonate outcrops *Androsaco-Caricion humilis, Drabo-Campanulion tauricae* and *Ptilostemonion*.

Instead of an izokhora range of cryoclimate in the Crimean Mts is placed considerably higher; the lowest value (8.4 marks −6 to 10 °C) is typical for steppe yailas (*Adonidi-Stipion tirsae*), and the highest—(9.7–10 marks −2 to +2 °C) for communities of southern side *Jasmino-Juniperion excelsae* and forests *Carpino orientalis-Quercion pubescentis, Elytrigio nodosae-Quercion pubescentis*.

Other climatic factors (Om, Kn) depend on both temperature and atmospheric precipitation amount. Therefore, their quantities are characterized by the same change tendencies. In particular, a continentality scale (Kn) reflects ecosystem peculiarities, which is defined by influence of the large sea and land areas to the climate and is changed from extraoceanic (<60%) to ultracontinental (>200%). Values of continentality for communities of the Tatra Mountains carbonate outcrops are the lowest (5.9–6.0 = 100% suboceanic). These are good warmed thoroughly (*Seslerion caerulae*) and differed from other alliances (*Seslerion tatrae, Caricion firmae*) of the *Elyno-Seslerietea* class. But in the East Carpathians the communities of such type are absent. Here the lowest continentality values are typical for communities of Alpine zone *Juncetea trifidi* (6.5 marks—110%). The highest for the Tatra Mountains values (135–145%—hemicontinental) are typical for xerophytic communities of a plain part (*Cirsio-Brachypodion, Festucio valesiacae, Potentillion caulescentis*), and for the East Carpathians (135%—oligotrophic mires (*Oxycocco-Sphagnetea*). The main community number of the Tatras and the East Carpathians belongs to the hemioceanic type. In the Crimean Mts the continentality values are slightly higher: minimum 7.8 marks (121%) for the forests *Dentario quinquefoliae*-Fagion *sylvaticae*, maximum—10.0 to 11.0 (145–155% subcontinental) for the steppes *Veronici multifidae-Stipion ponticae*. The main part of the communities belongs to hemicontinental type, i.e. two marks more than of the Tatra Mountains and East Carpathians.

Ombroregime (Om) still more depends on precipitation amount. Though the same general tendency is observed here, and a gradient of fluctuations is considerably more abrupt on the whole. As for thermoregime it reaches 6 marks changing step by step from the most western Tatra Mountains to the East Carpathians and is quite abrupt relatively to the Crimea. The highest values of an ombroregime (>16 marks—600 to 800 mm) are typical for the Tatra Mountains alpine communities *Salicion herbaceae*, the lowest are for the premountain steppes—11.5 marks (−200 mm). For the East Carpathians maximum quantities (15.5 marks) are typical for *Caricion curvulae* communities, minimum (12.5) for *Androsacio alpinae*, i.e. saxicolous communities occupy polar positions. The main diversity of syntaxa belongs to subombrophytic and mezoombrophytic type with sufficient moistening (0 to 600 mm). However the biotopes of the Crimea are sharply differed from the previous one. Their maximum quantities are lower (12.2 or—100 to 120 mm) of minimum of the Tatra Mountains and Carpathians (unless of steppe communities) and inherent to nemoral mesophytic forests *Dentario-Fagion*, while lowest (9.9 or—500 mm) to dry xerophytic thin forest *Jasmino-Juniperion excelsae*. Majority of syntaxa are characterized by subaridophytic conditions with precipitation deficiency (100–400 mm).

By light (Lc) the communities are set from umbrophytic (4.0) to heliophytic coenoses (8.0) and form the range from shady forests (*Querco-Fagetea, Vaccinio-Piceetea*) to open steppe and rocky (*Festuco-Brometea, Drypsidetea, Alysso-Sedetea, Salicetea herbaceae*).

Assessment of Dependence Between Fluctuation of Ecofactor Values and Plant Communities To assess dependence between fluctuations of different factors and to determine biotope places in the different factor coordinates there was used a method of ordination and DCI-analysis. Results of such analysis showed that between change of factor majority the dependence is quite slight, amplitudes of values are overlapped, and consequently these factors play no differentiation importance. The Fig. 9.3 demonstrates that all graphics can be divided into three groups:

1. Linear (direct or inverse) dependence, i.e. high correlation degree between ecofactors of all three regions: Ca-Hd, Hd-Ae, Nt-Lc, Nt-Ae, Rc-Om, Sl-Kn, Sl-Om, Sl-Kn, Sl-Tm, Om-Kn, Tm-Om, Tm-Kn, Tm-Cr;
2. Correlation is observed only within certain mountain system: of the East Carpathians—Tatra Mountains: Rc-Nt, Nt-Tm, Nt-Om, Om-Cr, Tm-Lc, Om-Lc, Lc-Cr; of the Mountain of Crimea: Hd-Rc, Hd-Sl, Hd-Ca, Hd-Nt, Hd-Om, Ca-Nt, Ca-Om, Ca-Kn, Nt-Om, Nt-Kn, Lc-Hd, Lc-Ae, Lc-Sl, Lc-Ca, Om-Lc, Lc-Kn);
3. Correlation is absent for quantities fH concerning other factors. It happened that in European temperate (the Tatra Mountains and East Carpathians) and sub-Mediterranean (the Crimea) mountain systems there have been observed a contrary correlation (for example, Lc-Om, Nt-Rc, Nt-Sl, Nt-Ca, Tm-Ae for the East Carpathians—Tatra Mountains rectilinear, and for the mountain of Crimea inversely linear or one of factors is changed (Tm-Hd—soil humidity—for the Crimea, thermoregime for the Carpathians-Tatra Mountains).

It should be emphasized that climatic factors not only directly but also indirectly change the impact of edaphicones that is observed basing on their interrelationship. The nitrogen regime of soil (Nt) is the most dependent on climatic factors. Linear relationship between Tm and Nt and inversed linear between Om and Nt means that increase of Tm and decrease of Om fastens the decomposition of organic matter and humus, the level of mineral nitrogen in soil also increases. These processes are the most dangerous for mountain communities *Arabidion caeruleae, Salicion herbaceae, Loiseleurio-Vaccinion, Pinion mughi*. On the other hand, the steppe plain communities (*Cirsio-Brachypodium Festucion valesiacae*) are also in the risk zone. Instead, in the mountains there are more possibilities for the expansion of communities *Adenostylion alliariae*, and on the plains—for *Fagion sylvaticae*.

In order to describe the cumulative impact of 12 leading ecofactors, we used the method of non-metric multidimensional scaling (NMDC), which showed the distribution of unions in relation to the factors and the complex interrelationship between them (Fig. 9.5). Dependence of plant communities of the Tatra, East Carpathians and Crimean Mts. in coordinates of leading ecological factors: Nt-Cf-Tm and Hd-Om-Rc has a complex and nonlinear nature of relationship (Fig. 9.6).

Dendrograms show in details the similarity between ecofactors and syntaxa. As illustrates Fig. 9.7a, b, dendrograms for the Tatras and East Carpathians are very similar according to the correlation level of ecofactors: Hd and Om form a separate cluster; Nt-fH, Ae-Sl-Rc-Ca, and Kn-Cr have high relation. The value of Tm and Lc changes significantly. Instead, in the Crimean Mountains there is observed the following pattern: soil moisture (Hd), aeration (Ae), and nitrogen level (Nt) show straight-line correlation.

In the hydrogenic communities Nt level is higher, and in various types of autogenic and lithogenic communities it is overlapped. However, between Ae and Nt this relation is slightly changing. In the forest coenoses, which have high stability level, it rises linearly, and in hydrogenic conditions it limits at the level of 6.5 points, as the eutrophication processes do not happen in the Crimea. Between Hd and Ca there is observed inversed linear correlation, as even in the hydrogenic coenoses, where accumulates the heavy layer of soil, the carbonate content is high. Between Hd and Sl it is observed more complicated relation. The lowest value of Sl is observed in the optimum moisture conditions *Fagion sylvaticae* (Hd = 12). The

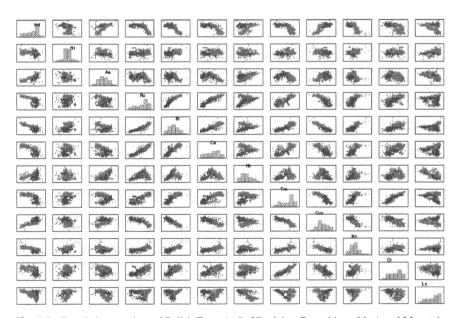

Fig. 9.5 Correlation matrices of Polish Tatry (red), Ukrainian Carpathians (blue) and Mountain Crimea (green), reflecting the dependence of communities in relation to the main ecological factors

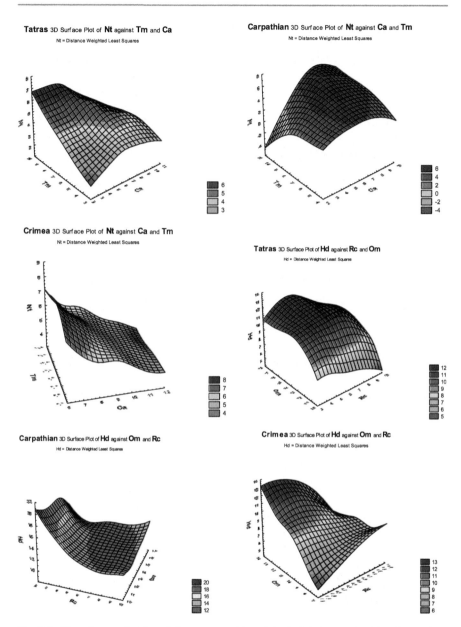

Fig. 9.6 Changing regularities of plant communities of the Polish Tatry, Ukrainian Carpathians and Mountain Crimea in coordinates of leading ecological factors: Nt-Cf-Tm and Hd-Om-Rc

salts content in soil increases towards drier and wetter conditions. Acidity (Rc) decreases with the increase of moisture. Soil moisture highly correlates with the climatic factors (Om, Cr, Tm). It means that the change of latter will cause the change in soil moisture. Between the soil moisture, thermal regime (Tm) and

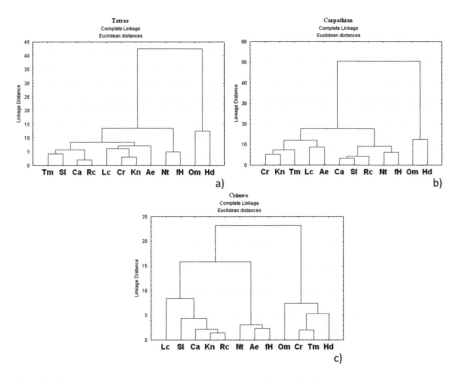

Fig. 9.7 The resemblance dendrogram of ecological factors by the nature of changing their indicators for the Polish Tatry (**a**), Ukrainian Carpathians (**b**) and Mountain Crimea (**c**)

cryoregime (Cr), there is observed inversed linear relation: the wettest conditions—the coldest conditions. Thus, in the forest coenoses from dry (*Jasmino-Juniperionexcelsae*) to wet (*Ornithogalo-Alnetumglutinosae*) these changes are more rapid than within the grass coenoses of autogenic and lithogenic types. It is important that between the soil moisture and ombroregime (Om), which depends on the changes of temperature and precipitation and in this way depicts hydrothermal peculiarities of coenoses, there was determined liner relation. This relation is indicative for autogenic and lithogenic types of communities, while hydrogenic type is out of this predicted pattern. This means that due to the climatic changes it is possible to predict the character of vegetation changes of autogenic and lithogenic types, while hydrogenic coenoses are located on the boundaries, where such changes may cause their destruction or even extinction. The chemical conditions of soil are closely connected with each other and with climate continentality. The increase of continentality causes the rise of salt content (Sl) and pH, which is highly undesirable for the south.

After analyzing the general trends and distribution of habitats in relation to changes of specific ecofactors, we consider the character of the relationship between the latter on a comprehensive assessment of these factors. The distribution of habitats (Fig. 9.8) on the x-axis is observed from the warmest and the driest

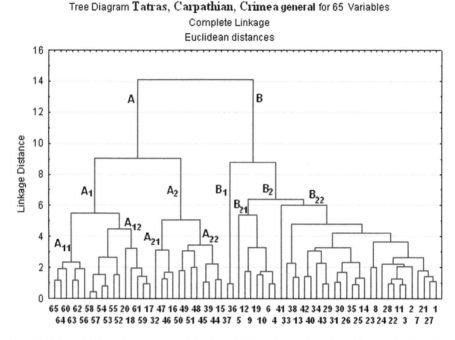

Fig. 9.8 Distribution of syntaxons of the Polish Tatry, Ukrainian Carpathians and Mountain Crimea according to cumulative assessment of leading ecological factors (syntaxa numbers is given in text)

Ptilostemonion echinocephali, which is spread on the rock and gravel slides on the southern coast of the Crimea, to the alpine cryophytic *Loiseleurio-Vaccinietea* in the Tatra Mountains. Consequently, this distribution is determined by the changes of climatic factors that specify the altitudinal zonation of mountains.

The dendrogram (Fig. 9.8) is divided into two big categories on the level $D > 10$. The first one (A) is formed by the Crimean sub-Mediterranean habitats, which include the forest of the lower zone (*Quercetea pubescenti-petraeae*), coniferous forest (*Erico-Pinetea*), and steppes and rock communities. The second category (B) includes all the habitats of the Tatra Mountains and the East Carpathians, and also the Crimean nemoral forests *Querco-Fagetea*. That makes sense and indicates that the impact of environmental factors and not the territorial delimitation is crucial for the syntaxonomical differentiation of vegetation.

The category B is divided into two groups on the level $D > 8$. The first one (B_1) includes nemoral forests, subalpine meadows, and rock communities of the forest belt. The second one (B_2) includes conifer forests and communities of the alpine Carpathians and Tatra Mountains. On the level $D > 5$ each of the groups is divided into two: A_{11}—the Crimean steppe, savannah, and rock communities. A_{12}—sub-Mediterranean deciduous and conifer forests, and also meadows and meadow steppes. It is remarkable that grass habitats due to the ecological factors are closer to the forests of lower zones than that of the same belt. Group A_{11} is divided into two: the first one includes petrophytic rock communities *Ptilostemonion, Bromopsido tauricae-Asphodelinetum tauricae, Drabo-Campanulion tauricae, Androsacio-Caricion humilis* and juniper woodlands *Jasmino-Juniperion excelsae;* and the second one includes the steppe communities of the lower zone *Veronico multifidae-Stipion ponticae* and savannah communities *Bromo-Hordeionmurini.* In particular, the communities of juniper woodlands *Jasmino-Juniperionexcelsae* are included to that group, and not to the neighbor one, where are the forests of class *Quercetea pubescenti-petraeae.* This confirms the view of those phytocenologists, who consider this group as a part of the class *Junipero sabinae-Pinetea sylvestris*, not the *Quercetea pubescenti-petraeae*, and indicates that the syntaxonomy of these types of communities needs critical revision, which is already carried out within the preparation of "Prodromus Vegetation Europeae" (Mucina et al. 2016).

It is possible that the communities of alliance *Androsace-Carici humilis* due to their environmental peculiarities are closer to the rock communities, than the typical steppe ones, as they are considered as part of the group *Stipo pulcherrimae-Festucetalia pallentis*, not *Festucetalia valesiacae.*

On the same level (D > 8) the group B_1 produces a cluster B_{11}—nemoral forests of the Crimea, East Carpathians, and Tatra Mountains (*Querco-Fagetea, Quercetea robori-petraeae, Quercion petraeae* (despite the relation of the latter to class *Quercetea pubescenti-petraeae*, still it is a debatable question), and also rock habitats of class *Asplenietea trichomanis.* To the second cluster B_{12} are included meadows and rock communities of the Tatra Mountains. The group B_2 is divided into two: B_{21}—conifer forests of the East Carpathians and Tatra Mountains, and subalpine communities of crooked forests *Mulgedio-Aconitetea*; and B_{22}, which includes rock communities of the East Carpathians and alpine habitats of classes *Juncetea trifidi* and *Loiseleurio-Vaccinietea* of the Tatra Mountains and the East Carpathians.

Such division is quite logical, although, certain rock type habitats brake this logic. This could be explained by their specificity, poor floristic composition, and lack of representativeness of the sample. The revealed patterns indicate that between the alpine nemoral-mountain and sub-Mediterranean vegetation zones there is a significant qualitative difference in the interrelation of ecofactors values, which is important to take into consideration while predicting vegetation changes.

The possibilities of made analysis are not depleted by mentioned above. As can be noticed from the received data, such dependencies are nonlinear, and have complex character. Their determination reveals new aspects of the organization, structure, differentiation, and changes of vegetation cover.

9.3.2.1 Findings

Analysed data based on synphytoindication let us assess in global ecospace the place of crucial ecofactors of the three mountain ranges, to make their comparison, and identify their specific characters. In particular, there was determined that the mountain ranges of temperate zone (the Tatra Mountains and East Carpathians) are more similar to each other and vastly differ from the Crimean Mountains, located in the sub-Mediterranean zone. Based on the assessment of crucial ecofactors gradient, there was determined that the main factors are climatic (hydrothermal) factors, which form the specificity of edaphic factors and differentiation of vegetation cover. The latter appears in three qualitatively different macrobelts for these mountain ranges: alpine-boreal (Junceteatrifidi-Vaccinio-Piceetea), temperate (Querco-Fagetea), and hemi-thermophilic (Querceteapubescenti-petreae-Festuco-Brometea).

The results showed a complex, nonlinear character of interdependencies between changes of the most important ecofactors, specificity of distribution and differentiation of syntaxa. Overall, between nemoral and sub-Mediterranean areas there is a significant difference, which appears in qualitative changes of interdependencies between ecofactors and their limiting function. It determines the specificity of directions of successions and processes, which is especially important for their prediction.

It was determined that among the climatic factors the highest differentiating value has the hydrothermal regime, which affects the change of edaphic factors values (acidity, salt mode, nitrogen level). Unlike the East Carpathians and Tatra Mountains, in the Crimea the significant limiting value has continentality, which rise causes the increase of salt level and pH.

Current tendencies of climate changes cause displacement of vegetation belts in the mountains. Considering the limited altitude of the East Carpathians by 2000 m, a number of communities of alpine type may be lost. The most susceptible to changing conditions are communities of high-alpine belt (*Salicionherbaceae, Arabidionalpinae, Junciontrifidi, Papaveriontatrae*). On the plains and submountain zones the biggest threats exist for grassxerophytic communities (*Festuco-Brometea*), where appeared sylvatization because of the intensification of soil nitrification and lack of traditional rural economy. In certain communities of hydrogenous type, there are possible significant qualitative changes that may lead to their loss.

Acknowledgements We wish to express appreciation to Zygmund Denisiuk for the financial support of the forest research in Tatra Mountains. We thank Kateryna Norenko for the provided help in the translation of this article.

References

Balcerkiewicz S (1984) Roślinność wysokogórska Doliny Pięciu Stawów Polskich w Tatrach i jej przemiany antropogeniczne. Wyd Naukowe UAM, Ser Biol 25:1–191

Borhidi A (1995) Social behaviour types, the naturalness and relative ecological indicator values of the higher plants in the Hungarian Flora. Act Bot Hung 39(1–2):97–181

Chorney II, Budzhak VV, Yakushenko DM, Korzhyk VP, Solomakha VA, Sorokan YI, Tokariuk AI, Solomakha TD (2005) National nature park "Vyzhnytsky". Plant World. – Naturale reserve territories of Ukraine. Plant world. Iss., 4. Phytosociocentre, Kyiv, p 248

Derzhypilsky LM, Tomych MV, Yusyp SV, Losyuk VP, Yakushenko DM, Danylyk IM, Chorney II, Budzhak VV, Kondratyuk SY, Nyporko VO, Virchenko VM, MikhaiLyuk TI, Darienko TM, Solomakha VA, Prorochuk VV, Stefurak YP, Fokshey SI, Solomakha TD, Tokariuk AI

(2011) National nature park "Hutzulshchyna". Plant World. – Naturale reserve territories of Ukraine. Plant world. Iss 9. Phytosociocentre, Kyiv, 360 p

Didukh YaP (1992) The vegetation cover of the Mountain Crimea (structure, dynamics, evolution andc onservation). Nauk, Kyiv, 256 p

Didukh YaP (2011) The ecological scales for the species of Ukrainian flora and their use in synphytoindication. Phytosociocentre, Kyiv, 176 p

Didukh YaP (2012) Fundamentals of bioindication. Nauk, Kyiv, 342 p

Didukh YaP, Plyuta PG (1994) The phytoindication of ecological factors. Nauk, Kyiv, 280 p

Didukh YP, Chetvertnykh IS, Boratynski A (2014) The symphytoindication estimation of vegetation of northen macroslope Tatras and surrounding areas. Visn Lviv Univ Ser Biol 67:35–47

Ellenberg H (1979) Zelgerwerte der Gefässpflanzen Mitteleuropas. Scr Geobot 9:1–122

Golubets MA, Milkina LI (1988) Ukrainian carpathians. Nature Nauk Kiev 208:51–63

Grebenshchikov OS (1957) Vertical zonation of vegetation in the mountains of the eastern part of Western Europe. Bot J 42(6):834–854

Grebenshchikov OS (1974) On the zones of vegetation cover in the mountains of the Mediterranean in the latitude band 35–40° N. Probl Bot 12(12):128–134

Klimuk YV, Miskevych UD, Yakushenko DM, Chorney II, Budzhak VV, Nyporko VO, Shpilchak MB, Chernyavsky MV, Tokaryuk AI, Oleksiv TM, Tymchuk YY, Solomakha VA, Solomakha TD, Mayor RV (2006) Nature park reserve "Gorgany". Plant world. Naturale reserve teritories of Ukraine. Plant world. Iss. 6. Phytosociocentre, Kyiv, 400 p

Kobiv Y (2014) Populations of rare plant of Ukrainian Carpathians: structure, dynamics, conservation. Doctoral dissertation of Biological Sciences, Kyiv, 457 p

Komornicki T, Skiba S (1996) Gleby. In: Mirek Z (red) Przyroda Tatrzańskiego Parku Narodowego, Tatrzański Park Narodowy, Kraków-Zakopane, 787 p

Landolt E (1977) Ökologische Zeigerwerte zur schweizer Flora – Veröff. Geobot. Inst. der Eidgen Techn. Hochschule in Zürich – H. 64, S. 1–208

Malynovski K, Kricsfalusy V (2002) Plant communities of the Ukrainian Carpathian highlands. Karpatska Vezha, Uzhgorod, 244 p

Mucina L, Bültmann H, Dierßen K, Theurillat J-P, Raus T, Čarni A, Šumberová K, Willner W, Dengler J, García RG, Chytrý M, Hájek M, Di Pietro R, Iakushenko D, Pallas J, Daniëls FJA, Bergmeier E, Guerra AS, i Ermakov N, Valachovič M, Schaminée JHJ, Lysenko T, Didukh YP, Pignatti S, Rodwell JS, Capelo J, Weber HE, Solomeshch A, Dimopoulos P, Aguiar C, Hennekens SM, Tichý L (2016) Vegetation of Europe: hierarchical floristic classification system of vascular plant,bryophyte, lichen, and algal communities. Appl Veg Sci 19(Suppl 1):3–264

Pawłowski B, Sokołowski M, Wallisch K (1927) Zespoły roślin w Tatrach,. Cz. VII. Zespoły roślinne i flora Doliny Morskiego Oka. Rozpr. Wydz. Mat.-Przyr. PAU, T. 67, Ser. A/B, pp 171–311

Pedrotti F (2013) Plant and vegetation mapping. Springer, Heidelberg

Piękoś-Mirkowa H, Mirek Z (1996) Szata roślinna Tatr Polskich – stan poznania, potrzeby i perspektywy badań. Przyroda Tatrzańskiego Parku Narodowego a Człowiek. T.2. Biologia, Krakow-Zakopane, s. 9–23

Pignatti S (2005) Valori di bioindicazione delle piante vascolari delle flora D'Italia. Braun-Blanquetia 39:97

Ramensky LG (1938) Introduction in complex soil – geobotanical in vegetation of lands. Selkhoziz, Moskva, 620 p

Rivas-Martinez S (2005) Notions on dynamic-catenal phytosociology as a basis of landscape science. Plant Biosyst 139(2):135–144

Roleček J, Tichý L, Zelený D, Chytrý M (2009) Modified TWINSPAN classification in which the hierarchy respects cluster heterogeneity. J Veg Sci 20:596–602

Solomakha VA, Yakushenko DM, Kramarets VO, Milkina LI, Vorontsov DP, Vorobyov ЄO, Voytyuk BY, Vinnichenko TS, Kohanets MI, Solomakha IV, Solomakha TD (2004) National nature park "Skolivski Beskydy". Naturale reserve territories of Ukraine. Plant World. Iss. 2. Phytosociocentre, Kyiv, 240 p

StatSoft, Inc (2005) STATISTICA for Windows. Version 7.0. http://www.statsoft.com
Szafer W, Sokołowski M (1925) Zespoły roślin w Tatrach. Cz.V. Zespoły roślin w dolinach położonych na północ od Giewontu. Rozpr. Wydz. Matem.-Przyr. PAU, T. 30, Ser. B., tab. 6, pp 123–140
Szafer W, Pawłowski B, Kulczyńński S (1923) Zespoły roślin w Tatrach. Cz.I. Zespoły roślin w dolinie Chochołowskiej. Rozpr. Wydz. Matem.-Przyr. PAU. T.63, Ser. B, pp 1–66
Tichy L (2002) JUICE, software for vegetation classification. J Veg Sci 13:451–453
Tsyganov DN (1983) Phytoindication of ecological regimes in the mixed coniferous-broad-leaved forest subzone. Nauka, Moskva, 198 p (Rus.)
Walter H (1968) Die vegetation der Erde in öko-physiologischer Betrachtung. Band II: Die gemäßigten und arktischen Zonen. VEB Gustav Fischer Verlag, Jena, 1001 S
Zarzycki K, Trzcicńska-Tacik H, Różański W, Szeląg Z, Wołek J, Korzeniak U, Rylacski W et al (2002) Ekologiczne liczby wskaźnikowe roślin naczyniowych Polski. Różnorodność Biologiczna Polski. Cz. 2. W. Szafer Institute of Botany PAS, Kraków, pp 1–183
Zólyomi B, Baráth Z, Fekete G et al (1966) Einreihung von 1400 Arten der ungarischen Flora in ökologische Gruppen nach TBR-Zahlen. Fragm Bot Musei Hist Natur Hung 4(1–4):101–142

Beech (*Fagus grandifolia*) in the Plant Communities of Long Island, New York, and Adjacent Locations, and Some Comparisons Across Eastern North America

10

Andrew M. Greller

Abstract

A review was undertaken of the distribution of *Fagus grandifolia* (American beech) in ligneous communities on Long Island (Oak-Chestnut Association of the Eastern Deciduous Forest). There are seven forest types described in which beech is a dominant or co-dominant; and three microclimatic variants of those types. On Long Island, beech occupies moist well-drained to overly well-drained sites, either in pure stands, in combination with a few other tree species of moist soils, or in combination with tree species of well-drained uplands. Dwarfed beech forest is represented on Long Island ("Grandifolia Sandhills"), as it is in coastal New England and in the "Beech Gaps" of the southern Appalachian Mountains. Finally, the phytosociological role of *Fagus grandifolia* in all the forest regions of eastern North America and Mexico is examined.

Keywords

Fagus · Long Island · North America

10.1 Introduction

American Beech (*Fagus grandifolia*) is widely distributed in eastern North America from approximately 29 to 47° N latitude (at which northern latitude it has been stable for 8000 years (Maycock 1994); and from the Atlantic Coast at 60° W longitude to 95° W longitude (Little 1971). According to Maycock (1994), it achieves an upper elevational limit of 1875 m in the Appalachian Mountains and reaches 1000 m in the Adirondack Mountains. It occurs in 16 USDA "Forest Cover

A.M. Greller (✉)
Department of Biology, Queens College, CUNY, Flushing, New York, USA
e-mail: andrew.greller@qc.cuny.edu

© Springer International Publishing AG, part of Springer Nature 2018
A.M. Greller et al. (eds.), *Geographical Changes in Vegetation and Plant Functional Types*, Geobotany Studies,
https://doi.org/10.1007/978-3-319-68738-4_10

Types." It is associated with 18 dominant trees in those communities. The only forest region of eastern North America in which it does not occur is the Maple-Basswood Association. Carpenter (1974) summarized the distribution, description, growth characteristics and properties, as well as commercial aspects of the beech. Maycock (1994) reported that beech reaches 125 ft (38.1 m) tall and averages 60–80 ft (18.3–24.4 m) tall in southern Ontario, near its northern limits. It reaches 400 years of age in Pennsylvania. Camp (1951) argued for the recognition of three basic types of Beech in eastern North America, 'White', 'Red' and 'Gray'. He outlined their distributions: White in the southeastern coastal plain, Gray at higher elevations and latitudes, and Red widely in the deciduous forest region. He said the beech of the Beech-Maple climax forest of this region is made up of genetic elements of the three basic types. Maycock (1994) summarized the putative distributions of each of the three races. Subsequent students of beech genetics have sometimes recognized the White Beech as a separate subspecies, *F. grandifolia* var. *caroliniana* (viz., A.J. Rehder, cf. Flora of North America 2012). More generally, Camp's segregation of the three races has not been supported by recent investigations (Hardin and Johnson 1985). In Mexico, *Fagus* has a restricted distribution in montane forests that contain a number of Holarctic genera that are similarly deciduous (Miranda and Sharp 1950). It has been treated as a variety of *Fagus grandifolia* (*F. g.* var. *mexicana*) by Little (1965) and by Williams-Linera et al. (2003).

From its extensive range, it is clear that beech can tolerate a wide range of average winter temperatures, down to January Normals of -4 °C (Huntley et al. 1989). Maycock (1994) reported that beech tolerates extreme low temperatures of between -30 and -40 °F (-34.4 to -40 °C). He further stated that it reproduces by seeds and root sprouts. It is shade tolerant and co-occurs with *Acer saccharum*, which is also shade tolerant. It is fire tolerant. It can occur on a wide range of soils, and on two major soil types, Gray-Brown Podzolic in the north, and 'Lateritic" in the south. Soils range in pH from 4.1 to 6.0, seldom exceeding pH 7.0.

Ecologically, beech occupy soils that are well-drained, mesic to hydro-mesic and not depleted of nutrients (Carpenter 1974). Thus, it often dominates stands on coarse soils at the bases of hills. Yim (1983) states that North American beech-dominated forests occur on mesic sites, on moderately acidic soils of moderate to high fertility, in a variety of topographic sites.

10.2 Classification of Forest Zones Where Beech Is Present in Eastern North America

Greller (1988) recognized three Forest Zones or Ecotones in eastern North America, the Conifer-Hardwoods Zonoecotone (CD, or Hemlock-White Pine-Northern Hardwoods of Braun 1950), the Deciduous Forest Zonobiome (DF), and the Deciduous Dicot-Evergreen Dicot-Conifer Zonoecotone (DEC). The latter includes most of the Southern Mixed Hardwood Forest of Quarterman and Keever (1962), excluding only those regions where evergreen dicots do not reach the canopy

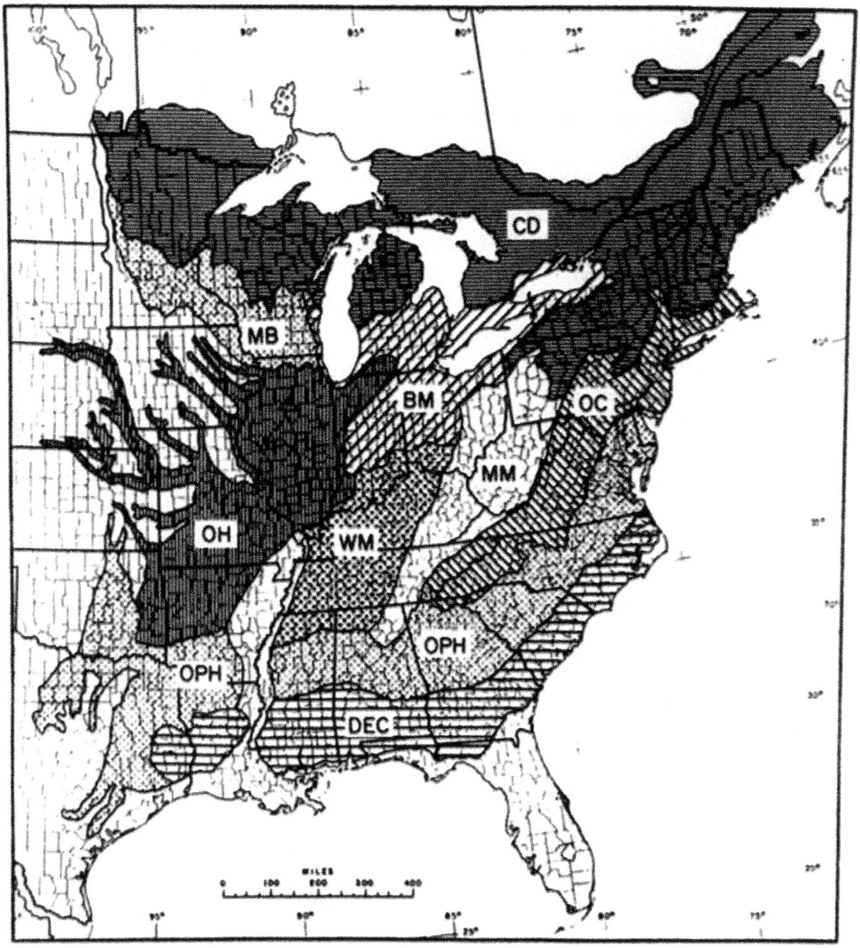

Fig. 10.1 Map of Eastern North America to show the forest associations according to Greller (1988), following Braun (1950) and Quarterman and Keever (1962). Code: CD = Conifers-Deciduous Dicots (northern hardwoods); [Decidous Forest Associations:] B-M = Beech-Maple [-northern hardwoods], MB = Maple-Basswood [-northern hardwoods], MM = Mixed Mesophytic, OH = Oak-Hickory, OPH = Oak-Pine [-Hickory-southern hardwoods]; DEC = [Southern] Deciduous Dicots–Evergreen Dicots-Conifers [in part, the Southern Mixed Hardwood Forest of Quarterman and Keever (1962)]

because of cold climate. Figure 10.1 shows all the forest regions of eastern North America, with CD located at the north and DEC at the southern edge and the associations of the Deciduous Forest between them. Figure 10.2 is a H.P. Bailey nomograph that shows the three zones can be distinguished climatically: Conifer-Hardwood (CD) from Deciduous Forest (DF) by warmth, specifically the radian of $W = 12.5\ °C$. Evergreen Broadleaved-Deciduous Broadleaved-Evergreen Conifer

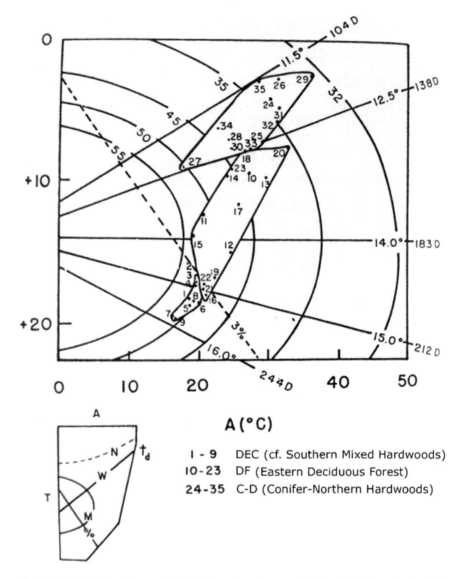

Fig. 10.2 H.P. Bailey Nomograph of the three forest zones of Eastern North America in which deciduous trees dominate the canopy (from Greller 1988). Code for Nomograph: A = Annual range of mean monthly temperatures; T = Mean annual temperature, W = Warmth in °C, % = Percentage hours of the year with temperatures ≤ 0 °C, td = number of days with temperatures above W, N = Water need [not analyzed]. Code for forest zones: 1–9 = DEC (Greller 1988); 10–23 = Deciduous forest (sensu Greller 1988); 24–35 = CD (Hemlock-White Pine Northern Hardwoods of Braun 1950). Refer to Greller (1988) for names of numbered stations

Forest (DEC or Southern Mixed Hardwoods, in part) can be distinguished approximately from Deciduous Forest by an annual freeze frequency of 3% hour/year.

10.2.1 Conifer-Hardwoods Zonoecotone (CD)

This is the [Needleleaved] Coniferous/Deciduous [Broadleaved] Dicotyledonous Forest [Greller's (1989) CD]. The following is a summary of the bioclimatology of the Conifer-Hardwoods Zonoecotone. It is abstracted from Greller (1989, 1990).

Walter's (1979) zone: ZE VI–VIII; W = 11.5–12.5 °C (Cool); M = 32–51 (Intemperate-Subtemperate); f ≤ 0 °C > 10% hour/year; Growing season (td) = 104.1–137.9 days; S = 4.4–15.1 inches (Semiarid-Humid)

In the present study this type is considered to be separate from the Deciduous Forest, a practice that Braun (1950) acknowledges as common. According to Braun this region ranges from "Minnesota to Atlantic Coast, occupying a position between the Deciduous Forest and the Boreal (Conifer) Forest." The CD, which Braun calls Hemlock-White Pine-Northern Hardwood Forest (HWPNH), is characterized by the pronounced alternation of deciduous dicot, coniferous, and mixed communities. The name Conifer/Deciduous Dicot Forest (CD), gives more emphasis to the physiognomy, and allows one to include Küchler's (1964) Spruce-Hardwood forest type as a subdivision of equal rank to the HWPNH. "[*Acer saccharum, Fagus grandifolia, Tilia americana, Betula lutea, Tsuga canadensis,* and *Acer saccharum, Fagus grandifolia, Tilia americana, Betula lutea, Tsuga canadensis,* and *Pinus strobus*] are the climax dominants of the region as a whole. . . . [*Pinus resinosa*] and, in the east, [*Picea rubens*], are characteristic species of the region, the former in dominantly coniferous communities, the latter in dominantly deciduous communities" (Braun 1950). Referring to Little (1971), it is clear that the arborescent dominants of the CD are broadleaved deciduous taxa and evergreen conifers of northern distribution in eastern and central North America. In Wisconsin, Curtis (1959: 535) listed *Dirca palustris* and *Sambucus pubens* as characteristic shrubs of CD mesic forests; other common shrubs are *Corylus cornuta* and *Lonicera oblongifolia*. The diversity of shrub and herb layers is lower in CD than in DF mesic forests, and decreases with an increase of coniferous trees (Braun 1950; Rogers 1982). Curtis (1959) lists 21 common angiospermous herbs.

10.2.2 Deciduous Forest Zonobiome (DF)

This forest zone (Zonobiome) has, as arborescent dominants, taxa that are cold-tolerant and deciduous. Deciduous taxa also dominate a well-developed subcanopy tree layer, a discontinuous shrub layer, and a dense, rich, seasonally varied herb layer. Evergreen dicotyledons, when present, are confined to the subcanopy tree, shrub, and herb layers; they comprise little of the total basal area and belong to only a few taxa. Dominant deciduous trees have wide distributions in the eastern United States, for example: *Quercus, Fagus, Acer, Tilia, Carya, Fraxinus, Ulmus, Betula,*

Liriodendron, and (formerly) *Castanea. Tsuga canadensis*, an evergreen conifer, is a dominant on some mesic sites. The following is a summary of the bioclimatology of the Deciduous Forest Zonobiome (DF) of eastern North America. It is abstracted from Greller (1989, 1990). Greller (2013) treated the Oak-Pine[-Hickory] Association of DF as a separate ZE for purposes of climatic analysis.

Deciduous [Broadleaved] Forest [Greller's (1989) DF].

Walter's (1979) zone: ZB VI: W = 12.5–14.9 °C (Cool/Mild); M = 36–53 (Subtemperate-Temperate); f ≤ 0 °C = 3 to 10% hours/year; S = 6.8–13.3 inches (Subhumid-Humid)

Growing season (td) = 137.9–209.0 days.

10.2.3 Deciduous Dicot-Evergreen Dicot-Conifer Zonoecotone (DEC)

This is the zonal type of the Gulf and Atlantic Coastal Plains. Here the canopy comprises three physiognomic types: deciduous angiosperms, evergreen broadleaved angiosperms, and evergreen conifers. Among the dominant deciduous angiosperms are *Fagus grandifolia, Liquidambar styraciflua, Carya glabra, Quercus* spp., and often, *Acer* spp. Among the dominant evergreen angiosperms are: *Magnolia grandiflora, Quercus hemisphaerica, Ilex opaca* and, locally, *Q. virginiana*. Dominant conifers are: *Pinus glabra, P. taeda* and *P. echinata*. The following is a summary of the bioclimatology of the Deciduous Dicot-Evergreen Dicot-Conifer Zonoecotone (DEC) of southeastern North America. It is abstracted from Greller (1989, 1990).

Deciduous Dicot—Evergreen Dicot—Coniferous Forest (Greller's [1989] DEC). It is equivalent to Quarterman and Keever's (1962) Southern Mixed Hardwoods forest, excluding the stands where evergreen trees do not reach the canopy.

Walter's zone = VI–V; W = 14.7–15.7 °C (Mild/to Warm); M = 49–53 (Temperate); f ≤ 0 °C = 0.5 to 3.0% hours/year; Growing season (td) = 203.1–233.7 day; S = 8.8–13.0 inches (Humid).

10.2.4 Mexican Holarctic Dicotyledonous Forest (MHDF, cf., Greller 1990)

Another region of North America possesses small and scattered populations of *Fagus* (*F. mexicana* or *F. grandifolia* var. *mexicana*). That is the montane zone of the Sierra Madre Oriental, from Tamulipas to Veracruz States. When analyzed using the H.P. Bailey nomograph, Mexican Holarctic Dicotyledonous Forest (MHDF, cf., Greller 1990) weather stations in that forest type are congruent in Warmth with the DEC of Greller 1990 (in part, Southern Mixed Hardwoods Forest of Quarterman and Keever 1962) and with the northern coastal stations of the Temperate Broadleaved Evergreen Forest (TBEF, sensu Greller 1980, 1990).

10.3 Classification of Forests Containing Beech on Long Island

Long Island beech trees usually have a single trunk, and mature trees reach a height of 70–100 ft. Sexual reproduction resulting in the production of beech nuts is relatively uncommon; asexual reproduction via clone production is common.

Greller (1977, 2001, 2002, cf. Reschke 1990) summarized a classification of Long Island forests. Those types that include *Fagus grandifolia* in the canopy are given in an extracted and slightly modified form below.
(MACROCLIMATIC TYPES)

I. Upland Forests, Woodlands, and Shrublands (Well-Drained Sites)
  i. Non-Oak Hardwoods Dominant
   A. Tuliptree Series
      1. Mixed Mesophytic Association (tuliptree, beech, red oak, red maple, black birch, sweetgum, white ash. ex., Nassau Co.: Grace Forest, North Hills (largely extirpated), Glen Cove, Welwyn Preserve; Queens Co.: Tuliptree Trail, Alley Park.
   B. Beech Series
      2. Beech-Mixed Hardwoods (incl. red oak, red maple) Association; ex., Nassau Co.: Meadowbrook Park (north), William Cullen Bryant Preserve, Roslyn.
      3. Beech Consociation; ex., Nassau Co.: southern edge of Hempstead Plains (Harper 1918), Fox Hollow Preserve, Syosset; Suffolk Co.: Cold Spring Harbor vic., Coastal Sand Dunes on the Bluffs Above Long Island Sound, Riverhead.
  ii. Oaks Dominant
   D. Black oak Series
      7. Black oak-beech-black birch Association; ex., Queens Co.: Oakland Lake; Nassau Co.: Whitney Est., Manhasset; Glen Cove; Suffolk Co.: Jayne's Hill, West Hills Co. Pk.; Lloyd Neck, Huntington; Riverhead Bluffs, Riverhead.
II. Swamp Forests and Related Shrubland (Edaphic Forests of Flooded Soil)
  i. Hardwood Swamps
   B. Red Maple Series (red maple dominant or co-dominant)
      8. Red Maple-American Beech Association; ex., Suffolk Co.: Point Woods, Camp Hero SP, Montauk (now rare; Taylor 1923).
   F. Beech Series (Beech dominant or co-dominant)
      19. Beech-sweetgum Association; ex., Nassau Co.: Wantagh Creek, North Bellmore
      20. Beech-Sourgum-Bitternut-Red Maple Association; ex., Suffolk Co.: Point Woods, Montauk Point SP (common).

(MICROCLIMATIC TYPES)

III. Coastal Variants of Forests on Well Drained Sites
    A. Black Oak Series
        1. Black Oak-Beech [-Black Birch]/American Holly Association ex., Suffolk Co.: Montauk Pt. St. Pk.
    B. Tuliptree Series
        2. Mixed Mesophytic/*Ilex opaca* var. *opaca*/*Polystichum acrostichoides;* Nassau Co.: Cove Neck (Edinger et al. 2008)
IV. Maritime Forests and Shrublands
    A. Beech Series
        1. Dwarfed Beech Low Woodland Association ex., Suffolk Co. : Friar's Head (on "Grandifolia Sandhills"), Riverhead

10.4 Results and Discussion

10.4.1 Beech in the Deciduous Forest Zonobiome (DF) (All Photos by A.M. Greller Unless Otherwise Indicated)

10.4.1.1 Beech in the Oak-Chestnut Forest Association of the Eastern Deciduous Forest (DF)

Long Island

Mixed Mesophytic Community (Fig. 10.3)
I. Upland Forests, Woodlands, and Shrublands (Well-Drained Sites); Non-Oak Hardwoods Dominant; (A). Tuliptree Series; (1). Mixed Mesophytic Association (tuliptree, beech, red oak, red maple, black birch, sweetgum, white ash); ex., Nassau Co.: Grace Forest, North Hills (largely extirpated), Glen Cove, Welwyn Preserve; Matinecock, Coffin Woods; Queens Co.: Tuliptree Trail, Alley Park

Greller et al. (1978) described in some detail a Mixed Mesophytic stand at North Hills, Nassau County, N.Y. The dominant trees were: *Liriodendron tulipifera* (51.3% rel. dom.), *Quercus [rubra]* 11.9%), *Fagus grandifolia* (7.5%), *Acer rubrum* (4.5%), *Quercus velutina* (4.1%), and *Quercus coccinea* (3.5%). Other sub-dominants include: *Acer saccharum, Betula lenta, Carya glabra, Cornus florida, Fraxinus americana, Liquidambar styraciflua,* and *Quercus palustris,* among others. At 35–54 cm below the surface, the soil was mottled. Thus, the water table was high and it fluctuated. This explains the relative paucity of *Quercus* and *Carya* species on the site.

Beech-Black Birch-Black Oak Forest (Fig. 10.4)
Upland Forests, Woodlands, and Shrublands (Well-Drained Sites); Oaks Dominant; (D) Black oak Series; (7). Black oak-beech-black birch Association; ex., Queens:

Fig. 10.3 *Fagus grandifolia* with *Liriodendron tulipifera*, Coffin Woods, Matinecock, Nassau County, NY

Oakland Lake; Nassau Co.: Greentree (fmr. Whitney Est.), Manhasset; Glen Cove; Suffolk Co.: Jayne's Hill, West Hills Co. Pk.; Lloyd Neck, Huntington; Riverhead Bluffs, Riverhead.

Good and Good (1970) presented quantitative data on a forest that included oaks, beeches, birches and other hardwood trees on the bluffs above Long Island Sound, north of Riverhead (Friar's Head). This type of forest is perhaps the most common one in which beech occurs on Long Island. *Fagus, Quercus velutina* and *Betula lenta* appear to be regularly scattered throughout the forest. *Acer saccharum* occurs here with more frequency than in other forest types of Long Island, as does *Ostrya virginiana*.

Beech-Mixed Hardwoods (Fig. 10.5)
Upland Forests, Woodlands, and Shrublands (Well-Drained Sites); Non-Oak Hardwoods Dominant; (B). Beech Series; (2). Beech-Mixed Hardwoods (incl. red oak, red maple) Association; ex., Nassau: Meadowbrook Park (north); William Cullen Bryant preserve, Roslyn.

In the north-facing, forested ravine in William Cullen Bryant Preserve, Roslyn, Nassau County, beech occurs from the upper rim of the ravine down nearly to the coast of Long Island Sound. At the rim, *Fagus* occurs scattered with *Quercus* spp., *Carya* spp., *Liriodendron, Acer* spp., *Ostrya virginiana*, and *Betula lenta*. On

Fig. 10.4 Beech-Black Birch-Black Oak forest at "Greentree," Manhasset, Nassau County, NY

Fig. 10.5 Beech-Mixed Hardwoods, William Cullen Bryant Preserve, Roslyn, New York

Fig. 10.6 Stand of pure beech (consociation) in Point Woods, Montauk, Suffolk Co., NY

steeper slopes in the ravine, beech occurs with *Quercus* species such as *Q. rubra*, *Q. velutina* and *Q. montana*.

Beech Consociation (Fig. 10.6)
Upland Forests, Woodlands, and Shrublands (Well-Drained Sites); (i). Non-Oak Hardwoods Dominant; (B). Beech Series; (3). Beech Consociation: Nassau Co.: Southern edge of Hempstead Plains (Harper 1918); Suffolk Co.: Coastal sand dunes on the bluffs above Long Island Sound; scattered throughout moist sites in Point Woods, Montauk.

Lamont (1998) gave a detailed description of the sand dunes that cap the remnant of the terminal moraine north of Riverhead, Long Island. These dunes stretch about 1.25 miles and have a maximum width of about 0.5 mile; they have been known to science for over 120 years. "Myron L. Fuller (1914), in his geological survey of Long Island, extensively discussed the sand dunes, devot[ing] a full page plate to photographs of sand dunes." Elias Lewis, Jr. recorded the dunes as rising 150 ft above tide, but standing on the bank (moraine) which is half that height, so that 75 ft of that elevation is drifting sand. At least five distinct but interrelated plant communities occur in the sandhills: (1) Coastal Beech Forest, (2) Maritime Dwarf Beech Forest, (3) Maritime Dunes, (4) Pitch Pine-Oak Duneland, and (5) Maritime Shrubland. The first four communities are rare occurrences in New York.

Fig. 10.7 Beech with *Carya cordiformis* near Big Reed Pond, Montauk

One site for Beech Consociation occurs at Wildwood State Park, approximately 3 miles west of the Riverhead sandhills; another site is located approximately 2.5 miles east of the sandhills. Frederick C. Schlauch included approximately 250 acres of beech-dominated vegetation on coastal dunes in that area. According to Lamont (1998), coastal beech forest occurs throughout the protected depressions of the undulating duneland south of the bluffs. Nowhere else in New York does such a mature climax forest occur on sand dunes. Sugar maple (*Acer saccharum*) occurs scattered throughout the coastal beech forest; and a there is a grove of Canada hemlock (*Tsuga canadensis*). Both species occur much more widely further north, but are uncommon south, in forests of the Oak-Chestnut Association. Good and Good (1970) published a detailed phytosociological description of the coastal beech forest of Friars Head, north of Riverhead. Beech can be seen in pure stands on old stabilized dunes in the Point Woods, Montauk.

Beech-Mixed Hardwoods Swamp (Figs. 10.7 and 10.8)
II. Swamp Forests and Related Shrubland (Edaphic Forests of Flooded Soil); (i). Hardwood Swamps; (B). Red Maple Series (Red Maple dominant or co-dominant); (8). Red Maple-American Beech Association; ex., Suffolk: Point Woods, Montauk Point SP (Taylor 1923, p. 55); F. Beech Series (Beech dominant or co-dominant); (20). Beech-Sourgum-Bitternut Hickory-Red Maple Association; ex., Suffolk Co.: Big Reed Pond Trail, Montauk Point.

Fig. 10.8 Beech and red maple swamp. Camp Hero, Point Woods, Montauk

At Big Reed Pond, Montauk, beech occurs in a continuum of soils from well drained to very moist. It is seen on wetter soils where *Symplocarpus foetidus* is present in springs or seeps, intermixed with *Nyssa sylvatica* or *Carya cordiformis,* or above *Acer rubrum* on slightly better drained sites (pers observ., Mar 27–30, 2012). Taylor (1923) described extensive Beech-Red Maple stands at Point Woods, Montauk. Today, Beech-Red Maple is a fragmentary association and is rarely encountered in the many swamps of the Montauk Point Woods.

Fig. 10.9 *Fagus grandifolia* with *Quercus* spp., *Ilex opaca* and *Kalmia latifolia* on morainal till in Point Woods, Montauk

Oak (*Quercus*)-Beech (*Fagus*)/*Ilex opaca*/*Kalmia latifolia* (Fig. 10.9)
III. Coastal and Highland Variants of Forests on Well Drained Sites; (A). Black Oak Series; (1). Black Oak-Beech-Black Birch/American Holly Association ex. Suffolk Co.: Montauk Pt. St. Pk.

In the Point Woods, Montauk, *Fagus* can be seen on well-drained morainal uplands in combination with *Quercus* spp., and an occasional *Betula lenta*, with an understory of *Ilex opaca* and a shrub layer of tall *Kalmia latifolia*.

Maritime Dwarfed Beech Forest (Fig. 10.10)
IV. Maritime Forests and Shrublands; (A). Beech Series; (1). Dwarfed Beech Low Woodland Association ex., Suffolk: Friar's Head [on "Grandifolia Sandhills"], Riverhead).

Lamont (1998) summarized information on the dwarfed beech forests of the Riverhead, Long Island area as follows: they occur on the bluff face, on the bluff top, and on dune ridges. He stated that this unique plant community is considered to be globally rare. Good and Good (1970) published a detailed phytosociological description of the maritime dwarfed beech-forest of Friars Head, north of Riverhead. In addition to those on the 140 acres of sandhills at Friars Head, Riverhead ("Grandifolia Sandhills"), several small, disjunct occurrences of the

Fig. 10.10 Dwarfed beech emerging from dune sand near a large *Quercus velutina*, McQuaid Property, Riverhead, NY

dwarf beech forest have been located and described by Mary Laura Lamont. The dominant species is a dwarf form of American beech; mature individuals are stunted [usually less than 12 ft (3.66 m) tall], often multi-stemmed from the base, and extremely gnarled and contorted. Century old trees often grow horizontally before sending twisted limbs skyward, only to be pruned back by constant exposure to salt spray, sand blowup, cold wind, and winter ice. Asexual reproduction by root crown sprouts may play a significant role in the success of the species at the "Grandifolia Sandhills," the major stand of L.I. Maritime Dwarfed Beech Forest. Colonies of dwarf beech trees on the seaside bluffs probably constitute a single persistently, organically connected clone (Lamont 1998). Other species associated with the dwarf beech include red maple (*Acer rubrum*), hickory (*Carya glabra* and *C. tomentosa*), black oak (*Quercus velutina*), and shadbush (*Amelanchier canadensis*). Lamont used an increment borer to determine the age of one of the old dwarf beech trees as 138 years, in 1998.

Other Areas Near Long Island

Westchester and Putnam Cos North of New York City, at Ice Ponds Conservation Area, Putnam County, New York, Beech-Maple mesic forest occurs where the water table is high (but below the surface) and the soil is slightly acidic (pH 5–6). Native tree, shrub and herb diversity is highest here. The dominant trees are *Liriodendron, Fagus, Betula lenta* and *B. allegheniensis*, *Quercus rubra* and

Q. velutina, Acer rubrum, Prunus serotina, Ulmus rubra, and *Carya* spp. *Lindera benzoin, Viburnum acerifolium* and *Hamamelis virginiana* are the common shrubs. The herb layer contains *Trillium, Asarum canadense, Polystichum, Caulophyllum, Maianthemum canadense* and *Sanguinaria*. Lianas include much *Vitis aestivalis* and some *Celastrus scandens*.

New Jersey West of New York City, Collins and Anderson (1994) report that *Fagus grandifolia* is present in both physiographic regions—the coastal plain of "South Jersey" and the piedmont and highlands of "North Jersey."
In Southern New Jersey, only a small part of the Inner and Outer Coastal Plains can be classified as mesic uplands. Two major forest types occur here, Mixed Oak and Beech-Oak.

A. Mixed Oak. In the western and southwestern part of the coastal plains, upland forests comprise the following trees: *Quercus alba, Q. velutina, Fagus grandifolia, Carya glabra* and *C. alba, Juglans nigra, Liriodendron*, and *Acer rubrum*; less common are: *Quercus falcata, Diospyros virginiana, Liquidambar, Pinus virginiana* and *Ilex opaca*. Understory trees are: *Cornus florida, Carpinus caroliniana, Prunus serotina* and *Sassafras*. Vines are *Lonicera japonica* (exotic and invasive), *Parthenocissus quinquefolia, Vitis* spp., *Toxicodendron radicans. Fagus* is a relatively less important constituent of this forest, after *Quercus* spp., *Ilex* and *Sassafras*.

B. Beech-Oak. *Fagus* is a more important constituent of the Beech-Oak Forest. Near Camden, N.J., pure stands of *Fagus* were reported by W. Stone in 1910 [*Fagus* Consociation]. In general, beech intergrades into the Mixed Oaks Forest. In the Mount Holly area, Chesterfield and Burlington Cos., *Fagus* occurs in large numbers in all size classes, with *Liriodendron, Quercus velutina, Carya ovata, C. glabra* and *Acer rubrum; Cornus florida* and *Carpinus caroliniana* are scattered throughout. At Mill Creek Park in Willingboro, Burlington Co., *Fagus* composes more than 50% of the forest, associated with *Quercus alba, Q. montana, Q. velutina, Liriodendron, Ilex opaca*; other trees are *Prunus serotina, Quercus falcata, Sassafras, Liquidambar, Acer rubrum, Pinus virginiana, Diospyros virginiana, Carya [alba]* and *C. glabra*. In Monmouth Co., at the junction of the coastal plain and piedmont regions, beech occurs with black birch (*Betula lenta*) and tuliptree (*Liriodendron*). Other constituents of this mixed mesophytic community are *Quercus rubra* and *Q. velutina, Liquidambar* and *Sassafras*; understory comprises *Cornus florida, Viburnum* acerifolium, *Viburnum dentatum, Lindera benzoin, Prunus serotina* and *Smilax glauca*.

In Northern New Jersey, on Kittatiny Limestone, in High Point State Park, Collins and Anderson (1994) recognize a Sugar Maple-Mixed Hardwood Forest. *Fagus* is a minor component of this forest. Dominant species are: *Acer saccharum*, with *Quercus alba, Q. velutina, Q. rubra*; also *Fraxinus americana, Liriodendron tulipifera, Betula lenta, B. allegheniensis, Acer rubrum, Tilia americana, Fagus grandifolia, Carya* spp., scattered *Tsuga canadensis, Pinus strobus, Ulmus americana, Juglans nigra.*

Coastal New England
Busby and Matzkin (2009) described a Dwarf Beech Forest in coastal New England "on morainal knobs and ridges characterized by excessively well-drained, sandy soils and well-developed E horizons; soil organic horizon...absent or minimal" due to high winds that remove it. They state that tall beech appears in adjacent depressions on fine-texture soils, with greater soil fertility and protection from the wind. In summary they said, "dwarf growth is a response to harsh edaphic conditions including chronic nutrient depletion, drought stress and wind exposure."

Great Smoky Mountains
The Oak-Chestnut Association is centered on the southern Appalachian Mountains. Above about 1100 m, beech occurs in forests of all ranges of "mesic" (Yim 1983). The reader is referred to Whitaker (1956) for a thorough discussion of forest composition in the Great Smoky Mountains. The only significant associates of *Fagus grandifolia* in the Oak-Chestnut region are: *Acer saccharum, Acer rubrum, Quercus rubra, Betula lenta, Liriodendron tulipifera,* and *Quercus alba.* The following dominance types, sampled by Braun (1950), are listed by Maycock (1994): *Fagus grandifolia-Acer saccharum, Fagus grandifolia, Fagus grandifolia-Quercus alba, Fagus-Castanea dentata.* The first listed type was more commonly sampled than all the other types combined.

10.4.1.2 Beech in the Oak-Pine[-Hickory] Forest Association of the Eastern Deciduous Forest (DF)

In this association (for detailed tree composition see Greller 2013) on the Atlantic and Gulf Coastal Plains, beech is confined to the narrow zone of mesic soils of higher nutrient content. It is absent from most of the upland sites, which are sandy, dry and nutrient-poor (Yim 1983). The mixed mesophytic community in this forest association differs from that in the Oak-Chestnut Association in having a greater, often leading dominance by *Fagus grandifolia,* greater dominance by *Quercus,* especially *Q. alba,* and an understory that includes *Ilex opaca.* Nesom and Treiber (1977) listed the following taxa and their percentage dominance in three stands: *Fagus grandifolia* (49, 47, 82% dom.), *Quercus alba* (9, 10, 8), *Liquidambar styraciflua* (20, 5, 1), *Liriodendron tulipifera* (12, 8, 1), *Quercus rubra* (6, 5, 4), *Carya [alba]* (1, 10, 1), *Nyssa sylvatica* (0, 5, 2), *Quercus falcata* (0, 5, 1), as well as others. Ware (1970), working on the coastal plain in Williamsburg, Virginia, described a mesic forest dominated by *Querus alba* (21.9% dominance), *Pinus taeda* (16.5), *Fagus grandifolia* (13.1), *Quercus falcata* (13.1), *Liriodendron* (10.8), *Pinus virginiana* (5.1), *Quercus rubra* (4.6), as well as lesser percentages of *Acer rubrum* var. *trilobum, Carya* spp., *Liquidambar, Cornus florida,* and *Quercus nigra.* Yim (1983) summarized previous work on an Oak-Beech Forest near Washington, D.C. Here, *Quercus alba* accounted for 41.0% basal area and *Fagus* 24.2%; (MAT = 14.1 °C; MAP = 1029. Habitat was a slope of 10–25%).

10.4.1.3 Beech in the Beech-Maple Forest Association of the Eastern Deciduous Forest (DF)

According to Maycock (1994) extreme southern Ontario between three Great Lakes, below 43° N latitude, has a summergreen deciduous forest of 85 species of trees. This region is the northern limit of the Beech-Maple Association of the Eastern Deciduous Forest. It is a limestone region of low elevation 125–400 m a.s.l; Clay Plains and Sand Plains account for 90% of site types, but Till Plains and morainal hills are present. *Acer saccharum, Fagus grandifolia, Quercus rubra* and *Ulmus americana* are the leading dominants. Other important taxa are *Acer rubrum, Acer saccharinum, Quercus velutina* and *Q. alba*; other "significant" species are: *Acer negundo, Betula papyrifera, Carya cordiformis, C. ovata, Celtis occidentalis, Populus deltoides, P. tremuloides, Quercus macrocarpa, Q. palustris, Salix amygdaloides, S. nigra, Tsuga canadensis* and *Ulmus rubra*. Maycock (1994) gave detailed composition and structure data for some of these stands.

Ohio comprises three of the Associations of the Eastern Deciduous Forest, Beech-Maple, Mixed Mesophytic, and Western Mesophytic in that order of areal extent. Gordon (cf. Yim 1983) recognized five beech-dominated forest types in Ohio: Beech-Sugar Maple, Wet Beech, Hemlock-Beech, Beech-Maple-Tuliptree, and Beech-White Oak. "Wet beech" forests occur on mounds above floodplains, but may endure occasional inundation. Beech forests do not occur on well-drained or "droughty" soils in Ohio. In Cincinnati, Ohio, a till plain mesophytic forest has 50% basal area for *Fagus*; the MAT= 9.9 °C and MAP = 1208 mm (cf. Yim 1983). In the Ontario lowlands of western New York State, where the substrate is lowland glacial drift of dolomitic limestone, Miller in 1973 (cf. Yim 1983) describes a Maple-Beech Forest where *Acer saccharum* accounts for 61% of basal area; here the MAT = 8.4 °C. In the "Beech Border" of Wisconsin, Ward in 1961 (cf. Yim 1983) describes a Beech-Maple Forest where *Fagus* accounted for 9.4% of basal area. Soil pH ranges from 5.5 in the north to 6.5 in the south.

For Warren Woods, Michigan, Yim (1983) summarizes previous work on a Beech-Maple Forest in a till plain, in which *Fagus* accounts for 50% of basal area, and *Acer saccharum* 12%; where MAT = 9.9 °C and MAP = 1020 mm.

10.4.1.4 Beech in the Mixed Mesophytic Forest Association of the Eastern Deciduous Forest (DF)

Braun (1950) compiled extensive data on forest composition in this association. Maycock (1994) summarized the stand data as follows: *Fagus* is the sole dominant in 9 of Braun's stands, *Fagus-Liriodendron* co-dominated in 4 stands, *Fagus-Tsuga canadensis* co-dominated in 3 stands, *Fagus-Acer saccharum* co-dominated in 2 stands, *Fagus-Tilia heterophylla* co-dominated in 2 stands. The following co-dominated stands occurred once in Braun's data tables: *Liriodendron-Fagus, Liriodendron-Tsuga canadensis-Fagus, Castanea dentata-Fagus* (sic). In addition, Maycock reported the following *Fagus* co-dominated types in the Mixed Mesophytic Association region: *Fagus-Quercus alba, Quercus alba-Fagus, Fagus-Magnolia acuminata, Fagus—Picea rubens*, and *Tsuga canadensis-Fagus*.

10.4.1.5 Beech in the Western Mesophytic Association of the Eastern Deciduous Forest (DF)

Maycock (1994) reported that in this association, beech domination occurs with trees of southern distribution such as *Tilia neglecta, Quercus schumardii, Carya glabra, Tilia heterophylla, Aesculus octandra, Carya [alba]* and *Magnolia acuminata*. The following tree species had Constancy greater than 70%: *Acer saccharum, Fraxinus americana, Carya ovata, Quercus alba* and *Liriodendron tulipifera*. *Fagus grandifolia* accounts for 42% of all trees in the stands in which it dominates, accompanied by *Acer saccharum* (11%), *Liriodendron* (8), *Quercus alba* (5), *Fraxinus americana* (4), *Carya ovata* (3), *Tilia heterophylla* (3), and *Aesculus octandra* (3). Maycock listed the following forest types for this association: *Fagus, Fagus-Acer saccharum, Fagus-Liriodendron, Fagus-Quercus alba, Fagus-Aesculus octandra, Fagus-Tilia heterophylla*. The most commonly sampled type was *Fagus-Acer*, followed by *Fagus* and *Fagus-Liriodendron*. Greller (2013) assigned the southern portion of this Association, in the Mississippi Valley, to the Oak-Pine Forest Association.

10.4.1.6 Beech in the Oak-Hickory Forest Association of the Eastern Deciduous Forest (DF)

This is the western Association of the Deciduous Forest. Here the natural uplands are often vegetated by savanna and prairie. Beech-dominated forests occur on middle or lower slopes. In northwestern Arkansas and the Boston Mountains south of the Ozark Mountains, in a Beech-Oak forest, *Fagus grandifolia* occurs with *Nyssa sylvatica* and *Quercus rubra, Q. alba, Ulmus americana, Tilia floridana, Acer saccharum, Carya ovata* and *C. cordiformis* (Maycock 1994).

10.4.1.7 Beech in the Beech-Birch-Maple-Basswood Forest Association of the DF in the Southern Appalachian Mountains of the Eastern United States

Greller (1988) recognized a combination of non-oak deciduous trees at elevations of approximately 1000–1400 m in the Southern Appalachian Mountains, as constituting a new forest Association in eastern North America, (BBMB), based on the Mesophytic Forest of Rheinhardt and Ware (1984). Those authors described a few types of forest in this zone: Beech-Maple, Mesophytic and Dwarf Beech. They gave composition data for each type. It is their Mesophytic type that distinguishes the Association.

Mesophytic Forest (Fig. 10.11) is the most variable type of forest in the Balsam Mountains of southwestern Virginia. Rheinhardt and Ware (1984) distinguish it from the Mixed Mesophyic forest and cove forests of the Oak-Chestnut Association because BBMB-Mesophytic type lacks *Liriodendron, Tsuga, Quercus alba, Nyssa sylvatica, Juglans nigra* and *Carya* spp. It does however include species that are lacking or uncommon in those lower elevation Mixed Mesophytic communities, such as *Tilia heterophylla, Acer saccharum, Fagus grandifolia, Fraxinus americana, Betula lutea, Aesculus octandra, Quercus rubra* and *Acer rubrum*. Mesophytic forest at high elevations is dominated by *Acer saccharum* (28%), *Tilia*

Fig. 10.11 Mesophytic type of BBMB Association, with prominent *Aesculus octandra* trunks and abundant leaves of *Fagus grandifolia* from smaller trees; on Blue Ridge Parkway, North Carolina (2005)

heterophylla (22%), *Fagus grandifolia* (14%), *Fraxinus americana* (12%) and *Betula lutea* (6%). On south-facing slopes the leading dominants are *Quercus rubra*, *Acer* spp. and *Fagus grandifolia*. On north-facing slopes *Tilia heterophylla* and *Fraxinus americana* co-dominate, with lesser percentages of *Acer saccharum* and *Fagus grandifolia*.

The Beech-Maple type occurs on sheltered slopes and flat land at higher elevations. Here the beech is said to be the 'Red' race, and is usually the leading dominant, although *Acer saccharum* can be the leading dominant. Rheinhardt and

Fig. 10.12 Dwarf beech forest on the Blue Ridge Parkway in North Carolina (2005)

Ware (1984) gave the following composition data: *Fagus grandifolia* (38% dominance), *Acer saccharum* (35%) and *Betula lutea* (10%).

The Dwarf Beech (Fig. 10.12) type has the following composition: (*Fagus grandifolia* (69% dominance), *Aesculus octandra* (13%) and *Betula lutea* (7%). This type occurs on and immediately below exposed summits above 1400 m a.s.l. Rheinhardt and Ware (1984) considered the *Fagus grandifolia* to be of the 'Gray' race. Accompanying the beech are small numbers of *Aesculus octandra*, *Betula lutea* and *Acer saccharum*. *Fagus* and *Aesculus* are stunted, with few stems greater than 6 m tall and 40 cm in circumference.

Fig. 10.13 Beech Gap near the Blue Ridge Parkway, North Carolina

"Beech Gaps" (Fig. 10.13) are areas where beeches dominate in low shrub-like communities between peaks in the southern Appalachian Mountains. Whitaker (1956) gave some environmental and compositional data for that type in the Great Smoky Mountains: they occur at 1586–1708 m; MAT = 7–9 °C; MAP = 1250 mm; beech composes 50–80% of stems $\geq$2.5 cm, on mesic, rich soils.

10.4.2 Beech in the Conifer-Hardwoods Zonoecotone (CD)

For central Ontario, Maycock (1994) gave some data on forest composition for a Dry-Mesic Forest and for a Mesic Forest. In the Dry-Mesic Forest the leading trees were: *Fagus grandifolia* (100 Constancy/129 Importance Value), *Acer saccharum* (100/70), *Fraxinus americana* (100/6), and *Prunus serotina* (100/1); other trees of high Constancy and IV are: *Quercus rubra, Ostrya virginiana, Pinus strobus, Acer rubrum* and *Tsuga canadensis*. For the Mesic Forest, the following taxa reached 100% Constancy: *Fagus grandifolia, Acer saccharum, Tilia americana, Acer rubrum, Prunus serotina Tsuga canadensis* and *Betula lutea*. Similar compositions occur in dry-mesic Beech-Maple forests in central Quebec, where *Betula lutea, Acer pensylvanica* and *Abies balsamea* show 100% Constancy values. In southern Quebec, in the Monteregian Hills (including Mt. Royal, Mt St. Bruno, etc.) composition of Dry-Mesic and Mesic stands of Beech-Maple Forest is similar to central Ontario, with the addition of *Betula papyrifera, Acer pensylvanicum*, and *Prunus pennsylvanica*, which also achieve high Constancy ratings.

10.4.3 Beech in the Deciduous Dicot-Evergreen Dicot-Conifer Zonoecotone (DEC)

Beech-dominated forests occur on floodplain terraces of small streams, in addition to lower slopes adjacent to stream floodplains. In those sites, beech is secondary to *Magnolia grandiflora, Quercus* spp., and others. Rarely, it is dominant on those sites. Those sites are mesic and of relatively high fertility compared to the surrounding sandy or loess-derived soils (Yim 1983). Ware et al. (1993) did a thorough review of the vegetation of this phytogeographic region, which is centered on the Gulf and Atlantic coastal plains. Beech (*Fagus grandifolia* 'White' race or *Fagus grandifolia* var. *caroliniana*) occurs in the following types of forest:

1. Southern Mixed Hardwoods (SMHF), where it occurs with a wide variety of deciduous trees, evergreen trees and evergreen conifers. Varying combinations of dominants have been described as: Lower Slope Hardwood Pine Forest, Magnolia-Beech-Holly (Fig. 10.14), Beech-Magnolia[-Maple], Beech-Spruce Pine-Magnolia, Oak-Hickory-Pine, Magnolia-Oak-Loblolly Pine. Beech is a canopy dominant in those so indicated.
2. Pine-Mixed Hardwoods, a mesic forest on nutritionally poorer sites than the SMHF, where *Fagus grandifolia* occurs as a sub-dominant in a forest dominated by *Pinus palustris, P. taeda, P. elliottii* var. *elliottii*, and *Magnolia grandiflora, Liquidambar styraciflua, Quercus hemisphaerica, Quercus alba, Q. nigra, Q. stellata, Liriodendron tulipifera, Nyssa sylvatica*, with *Cornus florida* in the understory.

10.4.4 Beech in Forests (MDHF) Dominated by Holarctic Taxa in the Sierra Madre Oriental of Mexico

Holarctic Dicotyledonous Forest (MHDF) or Bosque Mexicano de Dicotyledoneas Holarcticos, is the name Greller (1990) proposed for the mainly dicotyledonous mountain forests of eastern Mexico. MHDF has a canopy dominated by taxa of

Fig. 10.14 *Fagus grandifolia*, with *Magnolia grandiflora* and *Ilex opaca* in a lower slope forest, Indian Mounds Wilderness, Sabine Co., Texas, by J.T. Williams

Holarctic floristic affinities, both deciduous and evergreen, as detailed by Miranda and Sharp (1950). It is endemic to Mexico and northern Central America, for nowhere else in North America do taxa such as *Oreomunnea (Engelhardia*; Juglandaceae), *Clethra* (Clethraceae), *Dendropanax* (Araliaceae), *Podocarpus* (Podocarpaceae), and *Weinmannia* (Cunoniaceae) dominate in the same stands as *Quercus, Carpinus, Liquidambar, Juglans, Carya*, and *Pinus. Podocarpus* and *Weinmannia* belong to the Austral or Southern Hemisphere Temperate Flora, and it is the admixture of Southern Hemisphere taxa into a matrix of Holarctic dominants that makes this ZE unique. Miranda and Sharp (1950) strongly emphasized this conglomeration of disparate floristic groups in arguing for the uniqueness of their predominantly Holarctic forests. MHDF is distributed from approximately 1000 to 2000 m elev., and occurs above the Orizaban Rain Forest (e.g., Gómez-Pompa 1973: Figs. 20, and 21). Miranda and Sharp (1950) listed six types of forest dominated by dicotyledons: "oak-mixed hardwoods; *Liquidambar; Platanus; Fagus; Weinmannia*; and *Engelhardia [Oreomunnea].*" The first type is most widely distributed. The canopy averages 20 m tall, but reaches 40 m, and

Fig. 10.15 Beech forest in the Sierra Madre Oriental (from Adele Rowden, "Conservation genetics of a threatened Mexican tree species, *Fagus grandifolia* var. *mexicana,*" Ph.D. Thesis, University of Edinburgh, UK)

Fig. 10.16 Cattle grazing in a *Fagus grandifolia* var. *mexicana* forest in the Sierra Madre Oriental (from Adele Rowden, "Conservation genetics of a threatened Mexican tree species, *Fagus grandifolia* var. *mexicana*," Ph.D. Thesis, University of Edinburgh, UK)

includes: *Quercus* spp., *Clethra quercifolia, Liquidambar [mexicana], Meliosma alba, Carpinus [tropicalis], Ostrya virginiana, Engelhardia [Oreomunnea] mexicana,* and *Cornus* spp. There is a small-tree layer reaching 10 m tall, with *Cyathea* (a tree fern), *Inga* (Leguminosae), *Oreopanax* (Araliaceae), *Xylosoma* (Flacourtiaceae), *Turpinia* (Staphyleaceae), *Symplocos* (Symplocaceae), *Podocarpus [matudae],* some other tropical taxa (*Beilschmiedia* and *Rapanea*) and other Holarctic taxa (*Vaccinium, Sambucus, Leucothöe,* and *Cornus*). Epiphytes of Orchidaceae, Bromeliaceae, Cactaceae, and Araliaceae are present. A shrub layer is present, containing a number of Holarctic taxa and some tropical taxa (*Drypetes, Randia*). Lianas and vines are present, as well as some dicotyledonous herbs.

Williams-Linera et al. (2003) examined ten stands of Mexican Beech (*Fagus grandifolia* var. *mexicana*; Figs. 10.15 and 10.16). They stated that

> species richness varied between 3 and 27 tree species in the canopy, and from 9 to 29 species in the understorey. Basal area of trees ≥ 5 cm dbh varied between 27.87 and 70.98 $m^2\,ha^{-1}$, density from 370 to 1290 individuals ha^{-1}. Beech represented 22–99.6% of total basal area and 6.8–83.3% of total density. Beech dominance varied from monodominance to codominance with *Carpinus caroliniana, Quercus* spp., *Liquidambar styraciflua, Magnolia schiedeana* and *Podocarpus* spp. Beech population ranged from 180 to 8300 trees with a total of fewer than 1300 individuals on four of the sites. Anthropogenic disturbance remains a major threat to these forests. It is uncertain whether Mexican beech will be able to survive without conservation efforts.

References

Busby PE, Matzkin G (2009) Dwarf beech forests in coastal New England: topographic and edaphic controls on variation in forest structure. Am Midl Nat 162:180–194

Braun EL (1950) Deciduous forests of eastern North America. Blakiston, Philadelphia, p 596

Camp WH (1951) A biogeographic and paragenetic analysis of the American Beech (*Fagus*). Yearbook, American Philosophical Society, Philadelphia 1950, pp 166–169

Carpenter RD (1974) American Beech. American Woods FS-220. U.S. Department of Agriculture, Forest Service. 8 p + map

Collins BR, Anderson KH (1994) Plant communities of New Jersey: a study in landscape diversity. Rutgers University Press, New Brunswick, xvi + 287 p

Curtis JT (1959) The vegetation of Wisconsin. The University of Wisconsin Press, Madison, xiv + 657 p

Edinger GJ et al (2008) Vegetation classification and mapping at Sagamore Hill national historic site, New York. National Park Service U.S. Department of the Interior Northeast Region Philadelphia, Pennsylvania. Technical Report NPS/NER/NRTR—2008/124

Flora of North America (2012) *Fagus grandifolia* Ehrhart, vol 3. http://efloras.org/florataxon.aspx?_id=1&taxon_id=233500645. Accessed 21 Mar 2012

Fuller ML (1914) The geology of long Island, New York. U S Geol Surv Prof Pap 82:1–231

Gómez-Pompa A (1973) Ecology of the vegetation of Veracruz. In: Graham A (ed) Vegetation and vegetational history of northern Latin America. Elsevier, Amsterdam, pp 73–148

Good RE, Good NF (1970) Vegetation of the sea cliffs and adjacent uplands on the North Shore of Long Island, New York. Bull Torrey Bot Club 97:204–208

Greller A (1977) A classification of mature forests on long Island, New York. Bull Torrey Bot Club 104:376–382

Greller AM (1980) Correlation of some climate statistics with broadleaved forest zones in Florida, USA. Bull Torrey Bot Club 107:189–219

Greller AM (1988) Deciduous forest, Chapt. 10. In: Barbour MG, Billings DW (eds) North American terrestrial vegetation. Cambridge University Press, New York, pp 287–326

Greller AM (1989) Correlation of warmth and temperateness with the distributional limits of zonal forests in eastern North America. Bull Torrey Bot Club 116:145–163

Greller AM (1990) Comparison of humid forest zones in eastern Mexico and southeastern United States. Bull Torrey Bot Club 117:382–396

Greller AM (2001) Outline of a revised classification for mature forests and related vegetation on Long Island, NY. Q Newsl Long I Bot Soc 11(2):16–17; 11(3):33–34

Greller AM (2002) Outline of a revised classification for mature forests and related vegetation on Long Island, NY. Q Newsl Long I Bot Soc 12(1):10

Greller AM (2013) Climate and regional composition of deciduous forest in eastern North America and comparisons with some Asian forests. Bot Pac 2(1):3–18

Greller AM, Calhoon RE, Mansky JM (1978) Grace forest, a mixed mesophytic stand on Long Island, New York. Bot Gaz 139(4):482–489

Hardin JW, Johnson GP (1985) Atlas of foliar surface features in woody plants, VIII. *Fagus* and *Castanea* (Fagaceae) of eastern North America. Bull Torrey Bot Club 112:11–20

Harper RW (1918) The vegetation of the Hempstead Plains. Bull Torrey Bot Club 17:262–286

Huntley B, Bartlein PJ, Prentice IC (1989) Climatic control of the distribution and abundance of beech (*Fagus* L.) in Europe and North America. J Biogeogr 16:551–560

Küchler AW (1964) Potential natural vegetation of the conterminous United States. Am Geogr Soc Spec Publ 36:116

Lamont E (1998) The Grandifolia Sandhills: one of long island's great natural wonders. Long I Botl Soc Newsl 8(3):13–19

Little EL Jr (1965) Mexican Beech, a variety of *Fagus grandifolia*. Castanea 30(3):167–170

Little EL Jr (1971) Atlas of United States trees, volume 1, Conifers and important hardwoods. US Department of Agriculture Miscellaneous Publication 1146, Washington, DC, 9 p

Maycock PF (1994) The ecology of beech (*Fagus grandifolia* Ehrh.) forests of the deciduous forests of southeastern North America, and a comparison with the beech (*Fagus crenata*) forests of Japan. In: Miyawaki A, Iwatsuki K, Grandtner MM (eds) Vegetation in Eastern North America. University of Tokyo Press, Tokyo, pp 351–407

Miranda F, Sharp AJ (1950) Characteristics of the vegetation in certain temperate regions of eastern Mexico. Ecology 31:313–333

Nesom GL, Treiber M (1977) Beech-mixed hardwoods communities: a topo-edaphic climax on the North Carolina coastal plain. Castanea 42:119–140

Quarterman E, Keever C (1962) Southern mixed hardwood forest: climax in the southeastern coastal plain, U.S.A. Ecol Monogr 32:167–185

Reschke C (1990) Ecological communities of New York. New York Natural Heritage Program, New York State Department of Environmental Conservation, Latham, NY

Rheinhardt RD, Ware SA (1984) The vegetation of the Balsam Mountains of Southwest Virginia: a phytosociological study. Bull Torrey Bot Club 111(3):287–300

Rogers RS (1982) Early spring herb communities in mesophytic forests of the Great Lakes region. Ecology 63:1050–1063

Taylor N (1923) The vegetation of Long Island, Part I. The vegetation of Montauk: a study of grassland and forest. Memoirs of the Brooklyn Botanic Garden, vol II, 107 p

Walter H (1979) Vegetation of the earth and ecological systems of the geo-biosphere. Springer, Berlin, 274 p

Ware SA (1970) Southern mixed hardwood forest in the Virginia coastal plain. Ecology 51:921–924

Ware SA, Frost C, Doerr PD (1993) Southern mixed hardwood forest: the former Longleaf Pine Forest. In: Martin WH, Boyce SG, Echternacht AC (eds) Biodiversity in the Southeastern United States: lowland terrestrial communities. Wiley, New York, pp 447–493

Whitaker RH (1956) Vegetation of the Great Smoky Mountains. Ecol Monogr 26(1):1–80

Williams-Linera G, Rowden A, Newton AC (2003) Distribution and stand characteristics of relict populations of Mexican beech (*Fagus grandifolia* var. *mexicana*). Biol Conserv 109(1):27–36

Yim Y-J (1983) On the distribution of beech (Fagus, Fagaceae) and beech-dominated forests in the Northern Hemisphere. Korean J Bot 6(3):153–166

11

Phytosociological Study of *Pteroceltis tatarinowii* Forest in the Deciduous-Forest Zone of Eastern China

Hai-Mei You, Kazue Fujiwara, and Qian Tang

Abstract

Pteroceltis tatarinowii is a Chinese endemic tree species belonging to the family Ulmaceae. This species likes high-calcium soil, can tolerate drought and barren environmental conditions, has strong rooting ability, and is regarded as one of the important tree species for vegetation restoration on limestone mountains in the warm-temperate zone (*sensu sinico*) of eastern China. But existing *Pteroceltis tatarinowii* forests have survived mostly in small areas, and so far there have been only rare phytosociological studies of *Pteroceltis tatarinowii* forest reported. To clarify the phytosociological characteristics of *Pteroceltis tatarinowii* forest in eastern China, this study selected five sites in the main distribution areas of *Pteroceltis tatarinowii* forest and recorded 28 vegetation samples, using Braun-Blanquet methodology. Field data were also collected for *Quercus* forests in the same region (including *Quercus variabilis* forest and *Quercus acutissima* forest), in order to determine the position of *Pteroceltis tatarinowii* forest in the hierarchical syntaxonomic system of deciduous broad-leaved forests. Data on *Pteroceltis tatarinowii* forest were classified by tabular comparison and summarized in a synoptic table, along with typical *Quercus* forests occurring in the same region. The different geographical distributions of *Pteroceltis tatarinowii* forest and deciduous *Quercus* forests were demonstrated by de-trended canonical correspondence analysis (DCCA). The compared

---

H.-M. You (✉)
Department of City and Environment, Jiangsu Normal University, Xuzhou, China
e-mail: 6020030138@jsnu.edu.cn

K. Fujiwara
Graduate School in Nanobioscience, Yokohama City University, Yokohama, Japan

Q. Tang
Key Laboratory of Geospatial Technology for the Middle and Lower Yellow River Regions, Ministry of Education, Kaifeng, China

© Springer International Publishing AG, part of Springer Nature 2018
A.M. Greller et al. (eds.), *Geographical Changes in Vegetation and Plant Functional Types*, Geobotany Studies,
https://doi.org/10.1007/978-3-319-68738-4_11

results show that *Pteroceltis tatarinowii* forest in eastern China can be recognized as a new alliance in the Quercetea variabilis class, named Moro australis-Pteroceltidion tatarinowii. This new alliance can be subdivided into two associations (i.e. Isodo rubescens-Pteroceltido tatarinowii ass. nov. and Pteroceltido tatarinowii-Quercetum variabilis) and one community (i.e. *Platycladus orientalis-Pteroceltis tatarinowii* community). The results of DCCA show that suitable environmental ranges of *Pteroceltis tatarinowii* forest are narrower than for *Quercus* forests and the different geographical distributions can be explained by the geographical and climate conditions. Compared with *Quercus* forests, *Pteroceltis tatarinowii* forest is more likely to occur in relatively harsh habitats at low altitudes, on steep cliffs or hillsides, on dry slopes with much exposed rock, and on calcium-rich limy soils. The constituent species of *Pteroceltis tatarinowii* communities are not many, the average number being about 34, because most species are unable to tolerate unfavorable dry, rocky, or barren environments. The structure of *Pteroceltis tatarinowii* communities is relatively simple, but many seedlings and clonal ramets of *Pteroceltis tatarinowii* appeared in the understory, so natural regeneration of *Pteroceltis tatarinowii* populations is better in forests. The structure and species composition of *Pteroceltis tatarinowii* forests showed significant differences along with environmental temperature and humidity. The community types of *Pteroceltis tatarinowii* forest in the study area change from Isodo rubescens-Pteroceltido tatarinowii to a *Platycladus orientalis-Pteroceltis tatarinowii* community, and then to Pteroceltido tatarinowii-Quercetum variabilis, with increasing environmental temperature and decreasing humidity (from north to south).

Keywords
Vegetation classification • Community characteristics • *Pteroceltis tatarinowii* • Vegetation ecology • Eastern China

11.1 Introduction

Pteroceltis tatarinowii is a deciduous broad-leaved tree species in the monotypic genus *Pteroceltis* (Ulmaceae), a Chinese endemic relict belonging to the tropical flora of the Tertiary period (Tan et al. 2004; Zhang et al. 2008). This species has various values and is commonly utilized for timber, medicine, feed, landscapes, ecological protection, etc. In particular, *Pteroceltis tatarinowii* is a major raw material for paper making, its bark fiber usually being used for making the famous Xuan paper (Fig. 11.1). However, the distribution area of *Pteroceltis tatarinowii* forests has decreased quickly because of human disturbance, and it has been listed as a third-level protected plant in China.

Pteroceltis tatarinowii occurs widely in northeast, north, northwest and south-central China (Fu and committee 2001). It usually lives on limestone or granite mountains and on the slopes of valley streams, but it occurs most often on limestone mountains. *Pteroceltis tatarinowii* occurs mostly as scattered individuals and

Fig. 11.1 *Pteroceltis tatarinowii* is a deciduous broad-leaved tree, growing to 20 m in height. Its bark is grayish white to dark gray (upper left). Leaf blades are broadly ovate to oblong, with base oblique and leaf apex acuminate (upper right). Seeds are yellowish green to yellowish brown, globose to oblong, with a notched apex (lower left). The species can usually produce many clonal sprouts from the stem base (lower right)

dominates forests only in part of its area (Yang et al. 1995). Because *Pteroceltis tatarinowii* has strong rooting ability and can grow in bare rock crevices, it is recognized as an appropriate tree species for vegetation restoration on limestone mountains. Therefore, it is important for conservation and utilization of *Pteroceltis tatarinowii* to understand the phytosociological characteristics of its forests.

Vegetation study has been more focused on the widespread elements in the woody flora of the northern hemisphere, such as species of *Quercus* and *Fagus*, and there has been less study of other species (e.g., *Pteroceltis*). Vegetation-ecological study in China on rare endangered plants (e.g. *Davidia involucrata*, *Tetraena mongolica*, etc.) has been conducted since the 1980s (Ge et al. 1998; Zhi et al. 2008). In recent years, phytosociological studies on *Pteroceltis tatarinowii* forest have also been reported (Fu and Li 1997; Tan et al. 2004; Wang and He 2001; Wang et al. 2011; Zhang et al. 2008; Chen et al. 2011; You et al. 2012; Zhou et al. 2012), but most of the studies have only involved local areas in Hunan, Hubei, Henan, Anhui and Guangxi. Therefore, comprehensive, synthetic and systematic study on *Pteroceltis tatarinowii* forest is still needed.

In this study, we focus on *Pteroceltis tatarinowii* forest in the deciduous-forest zone of eastern China (*sensu* Zhu 1984). We collected field data on *Pteroceltis tatarinowii* forest, *Quercus variabilis* forest and *Quercus acutissima* forest in

Anhui, Shandong, Beijing and Henan, according to Braun-Blanquet methology (1964). Data on *Pteroceltis tatarinowii* forest were classified by the tabular comparison method (Ellenberg 1956) and summarized in a synoptic table along with data from deciduous *Quercus* forests (mainly *Quercus variabilis* and *Q. acutissima* forests) occurring in the same region, because deciduous *Quercus* forest is thought to be the representative, typical forest type of this region (Wu and Committee 1980; Tang et al. 2009). Our goals are: (1) to classify community types and to clarify the phytosociological characteristics of *Pteroceltis tatarinowii* forest in the deciduous-forest zone of eastern China; (2) to compare with deciduous *Quercus* forests in the same region and to determine the position of *Pteroceltis tatarinowii* forest in the syntaxonomic hierarchical system of deciduous broad-leaved forests in this region; and (3) to analyze the different geographical distributions of *Pteroceltis tatarinowii* forest and deciduous *Quercus* forests.

11.2 Materials and Methods

11.2.1 Study Area

The study area is in eastern China, ranging from 113°9′E to 122°42′E and 33°45′N to 39°45′N. The area lies in the North China Plain, north of the Yangtze River (Chang Jiang), east of the Taihang Mountains and extending east to the Yellow Sea and Bohai Sea and north to the Inner Mongolia Plateau. This area includes parts of Beijing, Shandong, Henan, Anhui, Jiangsu and Hebei (see Fig. 11.2).

The study area is situated in the warm-temperate monsoon climate zone, with hot, rainy summers but cold, dry winters. Average annual temperature is 10–15 °C,

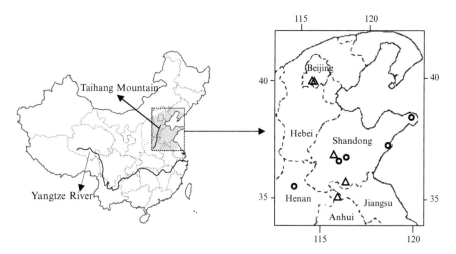

Fig. 11.2 Locations of study areas. Triangles represent sites of *Pteroceltis tatarinowii* forest, and circles represent sites of *Quercus variabilis* forest and *Quercus acutissima* forest

the mean temperature of the coldest month (January) is −5 to 0 °C, and the mean temperature of the warmest month (July) is 26–28 °C. Average annual precipitation varies from 600 mm to 1100 mm/year, with more than 70% of the annual precipitation concentrated mainly in July and August. Duration of sunshine in the study area is long, varying from 2300 h to 2800 h. Mean annual temperature and average annual precipitation of the study area decrease from south to north.

The topography of the study area is mainly plains (North China Plain) and hills, with high mountains situated mostly at the west and north in the study area. The average altitude of the plain is less than 50 m, and hills in the east have mean elevation below 500 m; high mountains usually can reach 1500 m, and the highest mountain, Xiaowutai-Shan, in Hebei, reaches 2882 m. Rock types in the study area are complex and diverse, but the main rock type is granite, followed by gneiss, limestone and volcanic rocks. The predominant soil types in the study regions are brown and yellow-brown mountain soils, while limy soil occurs only in areas over limestone substrate. The main vegetation types of the study area are deciduous *Quercus* forests and conifer plantations. The deciduous *Quercus* forests are mainly *Q. variabilis* forest, *Q. acutissima* forest, *Q. wutaishanensis* forest, etc.; the conifer plantations involve *Platycladus orientalis* and *Pinus tabulaeformis*.

11.2.2 Methods

The field data from *Pteroceltis tatarinowii* forest were collected from 2011 to 2013, by phytosociological methodology (Braun-Blanquet 1964; cf. Fujiwara 1987). Five study sites were surveyed and 28 relevés were recorded, within Beijing, Shandong and Anhui. Environmental factors for each relevé were also recorded, including latitude and longitude, altitude, slope and aspect, percentage of exposed rock, etc. (Table 11.1).

Table 11.1 Environmental information for *Pteroceltis tatarinowii* forest at the study sites

Site	LO (°E)	LA (°N)	AL (m)	AS	SL (°)	Tm (°C)	Pm (mm)	SH (h)	PER (%)
BF	115.52	39.67	260–533	NW, NE, SE, SW	20–85	10–12	600–705	2800	5–98
BP	115.54	39.71	376–400	SW, NE	2–43	10	610	2063	85–90
SL	116.98	36.36	312–367	NW	3–10	12.8	715	2121	10–40
SZ	117.54	34.77	135–177	NW, SW, SE	60–78	13	800	2500	98
AH	117.05	34.02	150–201	SE, NE, NW	3–30	14.3	875	2400	2–85

Note: *LO* longitude, *LA* latitude, *AL* altitude, *AS* aspect, *SL* slope, *Tm* annual mean temperature, *Pm* annual mean precipitation, *SH* sunshine duration, *PER* percentage of exposed rock

In addition, in order to determine the position of *Pteroceltis tatarinowii* forest in the syntaxonomic system of deciduous broad-leaved forests, we also collected field data from typical deciduous *Quercus* forests of this region, mainly *Q. variabilis* and *Q. acutissima* forest, because *Pteroceltis tatarinowii* could often be found occurring together with these two *Quercus* species.

All data were compiled into a matrix and compared with Excel software. A synoptic table was constructed according to species abundance, by the tabular comparison method (Ellenberg 1956). The constancy values of the constituent species are represented in the synoptic table by frequency classes for the samples in which the species was present, as follows: r = 0–5%, + = 6–10%, I = 11–20%, II = 21–40%, III = 41–60%, IV = 61–80%, V = 81–100%). For communities with fewer than five samples, the constancy values of constituent species are represented by the number of samples in which each species was present.

De-trended canonical correspondence analysis (DCCA, Canoco for Windows 4.5) was used to analyze the relationship between communities and environmental factors. The values of nine environmental factors were determined for all communities: latitude and longitude, altitude, slope and aspect (on an eight-point scale: 1 = N; 2 = NE; 3 = NW; 4 = E; 5 = W; 6 = SE; 7 = SW; 8 = S), mean annual temperature, average annual precipitation, percentage of exposed rock and soil pH (see Table 11.1). The importance values for constituent species were calculated using the following formula (Qian et al. 1999):

$$IV_{ijk} = \frac{O_{ijk} - L_{jk}}{\sum_{i=1}^{n} O_{ij}}$$

where IV_{ijk} is the importance value of species i, O_{ijk} is percent cover of species i in layer k of community j with a total of n species, and L_{jk} is the projected cover (in percent) of layer k of community j.

11.3 Results

11.3.1 Community Classification and Description

Data from *Pteroceltis tatarinowii* forests collected in the deciduous-forest zone of eastern China were summarized in a synoptic table and were compared with forests of *Quercus variabilis* and *Quercus acutissima*. As a result, it was possible to recognize *Pteroceltis tatarinowii* forests in the deciduous-forest zone of eastern China as a new alliance, Moro australis-Pteroceltidion tatarinowii, in the class Quercetea variabilis (see Table 11.2). This new Moro australis-Pteroceltidion tatarinowii alliance includes two associations and one community. According to Braun-Blanquet terms, we named the community composed of more than 10 plots as an "association", and that of less than 10 plots as a "community".

Table 11.2 Synthetic table of *Pteroceltis tatarinowii* forest and *Quercus* forests in the deciduous-forest zone of eastern China

Relevé reference number	1	2	3	4	5	6	7	8
Number of relevés	16	7	5	3	3	5	2	6
Number of species	34	31	40	42	46	47	37	25
Diagnostic species of Isodo rubescens-Pteroceltido tatarinowii ass. nov.								
Flueggea suffruticosa	V	.	.	.	.	.	.	.
Ampelopsis aconitifolia	IV	.	.	.	.	.	.	.
Isodon rubescens	IV	.	.	.	.	.	.	.
Myripnois dioica	IV	.	.	.	.	.	.	.
Thalictrum aquilegifolium var. *sibiricum*	IV	.	.	.	.	.	.	.
Lespedeza floribunda	IV	.	.	.	.	.	.	I
Gleditsia heterophylla	IV	.	.	.	2	.	.	.
Fraxinus chinensis	III	.	.	.	.	.	.	.
Clematis heracleifolia	III	.	.	.	.	.	.	.
Lonicera pekinensis	III	.	.	.	.	.	.	.
Sageretia paucicostata	III	.	.	.	.	.	.	.
Differential species of *Platycladus orientalis-Pteroceltis tatarinowii* community								
Platycladus orientalis	.	V	.	.	.	II	.	.
Ziziphus jujuba	II	IV	.	.	.	.	.	.
Acalypha australis	I	IV	.	.	.	.	.	.
Cocculus orbiculatus	.	III	.	.	.	.	.	.
Aconitum jeholense var. *angustius*	.	III	.	.	.	.	.	.
Arisaema erubescens	+	III	.	.	.	.	.	.
Punica granatum	.	III	.	.	.	.	.	.
Cudrania tricuspidata	.	III	I	.	.	.	.	.
Diagnostic species of Pteroceltido tatarinowii-Quercetum variabilis Assn. Tang, Fujiwara et You in Box et Fujiwara 2014								
Acer mono	.	.	V	.	.	II	.	.
Ligustrum quihoui	.	.	V	.	.	.	.	.
Aristolochia mollissima	.	.	IV	.	.	.	.	I
Firmiana simplex	.	.	IV	.	.	.	.	.
Cornus walteri	.	.	III	.	.	.	.	.
Dalbergia hupeana	.	.	III	.	.	.	.	.
Diagnostic species of Moro australis-Pteroceltidion tatarinowii all. nov.								
Pteroceltis tatarinowii	V	V	V	.	.	.	1	.
Morus australis	V	I	II	.	.	.	.	.
Cynodon dactylon	V	III	III	.	.	.	.	.
Liriope spicata	IV	I	IV	.	.	.	.	.
Aristolochia debilis	II	III	II	.	.	.	.	.
Chrysanthemum indicum	I	IV	IV	.	.	.	.	IV
Viola selkirkii	III	I	II	.	.	.	.	.
Paederia scandens	II	III	I	.	.	.	.	.
Rubia cordifolia	II	II	I	.	.	.	.	.
Angelica dahurica	I	I	I	.	.	.	.	.

(continued)

Table 11.2 (continued)

Pinellia ternata	V	III	·	·	·	·	1	I
Clematis brevicaudata	V	II	·	·	·	·	·	·
Arthraxon hispidus	·	V	V	·	·	·	·	I
Vitex negundo var. cannabifolia	·	III	II	·	·	·	·	·
Differential species of *Stephanandra incisa-Quercus variabilis* community								
Prunus japonica	·	·	·	3	·	·	·	·
Quercus serrata	·	·	·	3	·	·	·	·
Sorbus alnifolia	·	·	·	3	·	·	·	·
Carex ciliata-warginata	·	·	·	2	·	·	·	·
Cirsium japonicum	·	·	·	2	·	·	·	·
Lilium chindaoensis	·	·	·	2	·	·	·	·
Lysimachia barystachys	·	·	·	2	·	·	·	·
Meliosma oldhamii	·	·	·	2	·	·	·	·
Pinus thunbegii	·	·	·	2	·	·	·	·
Prunus serralata	·	·	·	2	·	·	·	·
Sapium japonicum	·	·	·	2	·	·	·	·
Stephanandra incisa	·	·	·	2	·	·	·	·
Weigera florida	·	·	·	2	·	·	·	·
Symplocos paniculata	·	·	·	2	·	·	·	·
Lindera glauca	·	·	·	2	·	·	·	·
Differential species of *Vitex negundo-Quercus variabilis* community								
Aconitum henryi	·	·	·	·	3	·	·	·
Artemisia shangnanensis	·	·	·	·	3	·	·	·
Parthenocissus himalayana	·	·	·	·	3	·	·	·
Thalictrum przewalskii	·	·	·	·	3	·	·	·
Isodon inflexa	·	·	I	·	3	·	·	·
Akebia trifoliata	·	·	·	·	2	·	·	·
Amorpha fruticosa	·	·	·	·	2	·	·	·
Celtis biondii	·	·	·	·	2	·	·	·
Liriope minor	·	·	·	·	2	·	·	·
Stephania tetrandra	·	·	·	·	2	·	·	·
Juglans regia	+	·	·	·	2	·	·	·
Diagnostic species of Quercetum alienae-variabilis								
Allium anisopodium	·	·	·	·	·	V	·	·
Atractylodes macrocephala	·	·	·	·	·	V	·	·
Viburnum mongolicum	+	·	·	·	·	V	·	·
Quercus aliena	·	·	·	·	·	V	·	·
Cotinus coggygria var. pubescens	·	·	·	·	1	V	·	·
Phyllanthus glaucus	·	·	·	·	1	IV	·	·
Carex subpediformis	·	·	·	·	·	III	·	·
Clematis obscura	·	·	·	·	·	III	·	·
Fraxinus bungeana	·	·	·	·	·	III	·	·
Lilium concolor	·	·	·	·	·	III	·	·
Lonicera ferdinandii	·	·	·	·	·	III	·	·

(continued)

Table 11.2 (continued)

Sporobolus indicus	·	·	·	·	·	III	·	·
Zelkova sinica	·	·	·	·	·	III	·	·
Rhododendron molle	·	·	·	·	·	III	·	·
Differential species of *Elaeagnus umbellata-Quercus variabilis* community								
Agrostis clavata	·	·	·	·	·	·	2	·
Elaeagnus umbellata	·	·	·	·	1	·	2	·
Youngia japonica	·	·	·	·	·	·	2	I
Crataegus pinnatifida	·	·	·	·	·	·	1	·
Quercus mongolica	·	·	·	·	·	·	1	·
Lonicera japonica	·	·	·	·	·	·	1	·
Phlomis umbrosa	·	·	·	·	·	·	1	·
Phytolacca acinosa	·	·	·	·	·	·	1	·
Amaranthus viridis	·	·	·	·	·	·	1	·
Artemisia sylvatica	·	·	·	·	·	·	1	·
Aster lavandulaefolius	·	·	·	·	·	·	1	·
Capsella bursa-pastoris	·	·	·	·	·	·	1	·
Celastrus angulatus	·	·	·	·	·	·	1	·
Convallaria majalis	·	·	·	·	·	·	1	·
Vicia unijuga	·	·	·	·	·	·	1	·
Differential species of *Albizzia julibrissin-Quercus acutissima* community								
Quercus acutissima	·	·	·	2	·	·	·	V
Albizzia julibrissin	·	·	·	·	·	·	·	IV
Melandrium apricum	·	·	·	·	·	·	·	IV
Allium tuberosum	·	·	·	·	·	·	1	IV
Ampelopsis brevipedunculata	·	·	·	·	·	·	1	IV
Ampelopsis humulifolia	·	·	·	·	·	·	1	III
Cleistogenes hancei	·	·	·	·	·	·	·	III
Diagnostic species of order and class								
Pistacia chinensis	I	III	IV	2	3	I	1	IV
Koelreuteria paniculata	+	V	V	·	2	II	2	·
Quercus variabilis	·	·	IV	3	3	V	2	·
Grewia biloba var. *parviflora*	V	V	III	·	3	I	·	·
Broussonetia papyrifera	·	III	I	1	1	·	·	I
Diospyros lotus	II	·	I	·	3	·	2	II
Trachelospermum jasminoides	·	III	V	2	·	·	·	II
Toxicodendron succedaneum	II	II	I	·	1	·	·	·
Speranskia tuberculata	·	III	I	·	2	·	·	·
Forsythia suspensa	·	·	·	·	1	V	·	II
Vitex negundo var. *heterophylla*	V	·	·	·	·	·	1	V
Carpinus turczaninowii	III	·	·	·	·	V	·	·
Other species								
Deyeuxia pyramidalis	III	III	III	1	1	IV	·	II
Aster ageratoides	III	·	I	2	1	IV	1	I
Dioscorea nipponica	V	III	II	1	·	III	·	I

(continued)

Table 11.2 (continued)

Carex lanceolata	V	II	III	3	.	.	1	IV
Vitis amurensis	IV	II	.	.	3	II	1	I
Rubia cordifolia	.	.	I	2	3	V	1	IV
Fraxinus chinensis var. rhynchophylla	II	.	.	1	1	I	.	I
Polygonatum odoratum	II	.	III	.	1	V	1	.
Smilax china	+	.	I	2	1	III	.	.
Spodiopogon sibiricus	I	II	I	3	.	.	.	V
Artemisia annua	I	II	II	.	.	.	2	II
Lespedeza pilosa	.	I	I	.	1	IV	.	II
Oplismenus undulatifolius	V	.	.	.	3	I	2	II
Lespedeza bicolor	V	IV	.	1	.	I	.	II
Prunus davidiana	I	I	.	.	2	II	1	.
Viola collina	III	I	I	.	.	.	2	I
Humulus scandens	I	I	V	.	2	.	2	.
Adenophora trachelioides	.	I	.	2	2	V	.	.
Leibnitzia anandria	I	V	.	.	.	III	.	II
Commelina communis	I	.	.	1	.	.	2	IV
Polygonatum sibiricum	II	.	I	.	2	III	.	.
Ailanthus altissima	II	III	I	.	.	.	.	II
Carpesium cernuum	II	III	I	.	.	.	.	I
Rubus parviflorus	.	II	I	.	3	.	.	II
Indigofera kirilowii	+	II	II	1	.	.	.	.
Ixeris sonchifolia	.	III	I	1	.	.	.	I
Setaria viridis	+	II	I	.	.	.	.	II
Celtis koraiensis	.	I	IV	.	.	.	2	I
Clematis kirilowii	.	III	III	.	.	.	2	II
Rhamnus parvifolia	IV	.	.	.	2	.	1	III
Indigofera bungeana	III	.	III	.	3	III	.	.
Pueraria lobata	I	.	.	.	1	.	1	I
Sanguisorba officinalis	.	.	I	3	.	II	.	I
Dioscorea opposita	.	.	I	.	2	.	1	II
Solanum lyratum	.	I	IV	.	.	.	2	I
Smilax stans	I	.	II	.	1	I	.	.
Adenophora wawreana	III	I	.	.	.	I	.	.
Euonymus alatus	.	.	III	3	.	.	.	I
Ixeris denticulata	.	I	III	.	.	.	.	IV
Robinia pseudoacacia	.	IV	.	.	.	.	2	I
Viola variegata	.	.	I	.	2	IV	.	.
Bupleurum scorzonerifolium	I	.	II	.	.	II	.	.
Cynanchum versicolor	.	.	I	.	2	II	.	.
Ophiopogon japonicus	.	.	III	.	1	I	.	.
Achyranthes bidentata	.	II	I	.	.	.	1	.
Andrachne chinensis	I	.	I	.	.	.	.	I
Oxalis corniculata	.	I	I	.	.	.	.	II

(continued)

Table 11.2 (continued)

Vitex negundo	.	IV	.	.	3	II	.	.
Spiraea pubescens	II	.	.	.	1	III	.	.
Bidens parviflora	+	II	.	.	.	.	.	V
Sedum bulbiferum	.	I	I	1	.	.	.	.
Isodon japonicus var. *glaucocalyx*	.	I	I	1	.	.	.	.
Adenophora tetraphylla	.	.	II	2	.	II	.	.
Spiraea trilobata	II	.	.	.	1	I	.	.
Chenopodium album	+	I	.	.	.	.	.	I
Geum aleppicum	I	I	.	.	.	.	1	.
Armeniaca sibirica	.	.	.	.	2	II	1	.
Peucedanum wawrii	.	.	I	2	.	.	.	I
Artemisia gmelinii	.	I	.	1	.	.	.	III
Cynanchum atratum	.	.	.	1	.	.	2	IV
Cocculus trilobus	.	.	.	1	.	.	1	I
Isodon inflexus	.	.	.	1	.	.	1	I
Arundinella hirta	.	.	III	1	.	.	.	I
Crataegus cuneata	.	II	.	.	.	V	.	.
Artemisia eriopoda	.	.	.	2	.	.	.	I
Rhamnus utilis	.	.	.	2	.	.	2	.
Taraxacum mongolicum	+	.	.	.	.	.	.	I
Viola betonicifolia	.	.	II	.	.	.	.	II
Allium macrostemon	+	.	II	.	.	.	.	.
Carex humilis var. *nana*	.	.	I	.	.	.	.	I
Microstegium nudum	.	.	I	.	.	.	1	.
Ulmus parvifolia	.	.	I	.	.	.	.	I
Cotinus coggygria	II	I	.	.	.	.	.	.
Amphicarpaea trisperma	I	I	.	.	.	.	.	.
Artemisia apiacea	I	I	.	.	.	.	.	.
Periploca sepium	I	I	.	.	.	.	.	.
Tilia amurensis	I	I	.	.	.	.	.	.
Metaplexis japonica	+	II	.	.	.	.	.	.
Ipomoea purpurea	+	I	.	.	.	.	.	.
Chenopodium glaucum	.	II	I	.	.	.	.	.
Lespedeza cuneata	.	II	I	.	.	.	.	.
Tilia miqueliana	.	I	II	.	.	.	.	.
Acer buergerianum	.	I	I	.	.	.	.	.
Deyeuxia sylvatica	.	.	II	1	.	.	.	.
Lysimachia clethroides	.	.	I	2	.	.	.	.
Smilax sieboldii	.	.	.	.	3	III	.	.
Clematis florida	.	.	.	.	3	IV	.	.
Lespedeza formosa	.	.	.	.	2	II	.	.
Rhamnus davurica	.	.	.	.	2	IV	.	.
Schisandra sphenanthera	.	.	.	.	2	IV	.	.
Clematis obtusidentata	.	.	.	.	1	IV	.	.

(continued)

Table 11.2 (continued)

Species								
Asparagus schoberioides	·	·	·	·	3	I	·	·
Salvia miltiorrhiza	·	·	·	·	1	III	·	·
Berchemia giraldiana	·	·	·	·	1	II	·	·
Muhlenbergia hugelii	·	·	·	·	1	II	·	·
Cephalanthera longifolia	·	·	·	·	1	II	·	·
Rhus potaninii	·	·	·	·	1	II	·	·
Styrax obassia	·	·	·	·	1	I	·	·
Galium bungei	·	·	·	·	1	I	·	·
Grewia biloba	·	·	·	·	·	·	2	IV
Chenopodium ambrosioides	·	·	·	·	·	·	1	II
Thalictrum minus var. hypoleucum	·	·	·	·	·	·	1	I
Toxicodendron vernicifluu	·	·	·	·	·	·	1	I
Dryopteris goeringiana	·	·	·	·	·	·	1	I
Dianthus chinensis	·	·	·	1	·	·	·	III
Themeda triandra var. japonica	·	·	IV	·	·	·	·	III
Zanthoxylum bungeanum	I	·	·	·	1	·	·	·
Pinus densiflora	·	·	·	2	·	·	·	I
Carex siderosticata	·	·	·	2	·	·	·	I
Cynanchum chinense	·	·	·	2	·	·	·	I
Euphorbia pekinensis	·	·	·	·	·	II	·	I
Lespedeza cyrtobotrya	·	·	·	·	·	I	·	II
Patrinia scabiosaefolia	·	II	·	1	·	·	·	·
Celastrus orbiculatus	·	I	·	1	·	·	·	·
Rhus chinensis	I	·	·	·	·	I	·	·
Synurus deltoides	·	·	·	2	·	I	·	·
Viola philippica ssp. munda	I	·	·	·	·	·	·	I
Albizia julibrissin	·	·	I	·	1	·	·	·
Artemisia argyi	·	I	·	·	·	·	·	I
Galium aparine var. tenerum	+	·	·	·	·	·	·	I
Gypsophila odlnamiana	·	·	·	1	·	·	·	I
Lespedeza chinensis	·	·	II	·	·	I	·	·
Rubus crataegifolius	·	·	·	1	·	·	·	I
Rhamnus globosa	·	·	I	·	·	I	·	·
Ulmus pumila	·	·	II	·	·	·	·	I
Deutzia grandiflora	·	·	·	·	·	I	·	I
Viola mandshurica	·	·	·	·	·	I	·	I

1. Moro australis-Pteroceltidion tatarinowii all. nov.

Diagnostic species: *Pteroceltis tatarinowii, Morus australis, Cynodon dactylon, Liriope spicata, Aristolochia debilis, Chrysanthemum indicum, Viola selkirkii, Paederia scandens, Pinellia ternata, Arthraxon hispidus, Vitex negundo* var. *cannabifolia* and *Cudrania tricuspidata*.

Holotype: Isodo rubescens-Pteroceltido tatarinowii ass. nov., Relevé reference number 1 in Table 11.1.

This new alliance can be found in limestone areas in Beijing, Shandong and Anhui (eastern China). The *Pteroceltis tatarinowii* forests generally remain on relatively low hills, at altitudes from 135 m to 550 m. They occur usually on poor soil, with exposed bare rock reaching an areal average of 61.6%. The alliance includes both natural and disturbed forests, but some larger individuals of *Pteroceltis tatarinowii* still remain in these forests. The forests are commonly short and can be divided into three layers. The height of the canopy layer is mostly less than 11 m but can attain 25 m (in Huangcangyu nature reserve, in Anhui). The cover of *Pteroceltis tatarinowii* forests reaches 75–90%, but the species composition is not diverse, with an average of about 34 species per relevé.

The *Pteroceltis tatarinowii* forest of this alliance shows different floristic composition under different environmental conditions. For example, in the stands with more exposed rock (e.g. Beijing and Shandong), *Pteroceltis tatarinowii* usually is the single dominant species of the canopy layer, and *Quercus* species do not occur in the forest; but in other stands, with less exposed rock (in Huangcangyu nature reserve), *Pteroceltis tatarinowii* is less dominant and often shares tree-layer dominance with *Quercus variabilis*.

The constituent tree species of this alliance are mostly deciduous, but a few evergreen species (e.g. *Ligustrum quihoui*) occur in the forest of Huangcangyu nature reserve (Xiaoxian County, Anhui), located in the transition between the warm-temperate and subtropical zones.

Syntaxonomy: This new alliance can be divided into two associations and one community, i.e. Isodo rubescens-Pteroceltido tatarinowii, Pteroceltido tatarinowii-Quercetum variabilis and the *Platycladus orientalis-Pteroceltis tatarinowii* community.

1) Isodo rubescens-Pteroceltido tatarinowii ass. nov.

Diagnostic species: *Flueggea suffruticosa, Ampelopsis aconitifolia, Isodon rubescens, Myripnois dioica, Thalictrum aquilegifolium* var. *sibiricum, Lespedeza floribunda, Gleditsia heterophylla, Fraxinus chinensis, Clematis heracleifolia, Lonicera pekinensis* and *Sageretia paucicostata*.

Type relevé: Relevé reference number 12 in Table 11.3.

Stands of this new association are situated in the Shidu scenic area and Puwa nature reserve of the Fangshan area of Beijing. Most *Pteroceltis tatarinowii* forests of this area had been destroyed by human activities and are secondary forests. The natural forests remain only on steep slopes, with much exposed rock. The soil is usually poor, with pH value ranging from 8.13 to 8.25.

This association can be found at altitudes from 260 m to 533 m, on steep south-facing slopes (averaging 43°), with exposed rock covering an average of 69.3% of the total area. Stratification of this association is not complicated and usually involves only three layers. The canopy layer reaches 5–10 m, with an average cover of 80.3%, and is dominated by *Pteroceltis tatarinowii*, mixed with *Gleditsia heterophylla, Morus australis, Ailanthus altissima, Diospyros lotus, Pistacia chinensis* and *Carpinus turczaninowii*; companion species, however, are relatively

Table 11.3 Isodo rubescens-Pteroceltido tatarinowii assn. nov.

Relevé reference no.	1	2	3	4	5	6	7	8
Original relevé number	BF1	BF2	BF3	BF4	BF5	BF6	BF7	BF8
Relevé date	2013.08.02	2013.08.02	2013.08.02	2013.08.02	2013.08.02	2013.08.02	2013.08.02	2013.08.02
Relevé size (m^2)	120	120	150	100	225	225	120	150
Altitude (m)	260	271	284	329	330	363	533	280
Aspect	NW	SE	NE	NW	SW	NW	SE	SE
Slope (°)	30	28	85	75	80	38	53	70
Percentage of exposed rock (%)	30	50	95	70	75	10	5	75
Height(m) of Tree layer 1 (T1)	–	–	–	–	–	–	–	–
Cover(%) of Tree layer 1	–	–	–	–	–	–	–	–
Height(m) of Tree layer 2 (T2)	8	9	10	11	8	8	7	10
Cover(%) of Tree layer 2	85	75	80	80	75	85	85	75
Height(m) of Shrub layer (S)	2.5	3	3.5	4	2	2.5	3.5	3
Cover(%) of Shrub layer	35	30	15	20	10	15	25	25
Height(m) of Herb layer (H)	0.4	0.5	0.3	0.3	0.4	0.5	0.5	0.4
Cover(%) of Herb layer	20	30	8	15	15	20	20	15
Number of species	37	43	30	28	34	36	24	35
Diagnostic species of Isodo rubescens-Pteroceltido tatarinowii assn. nov.								
Flueggea suffruticosa	1·2	+	+	+	+·2	1·1	1·1	2·2
Isodon rubescens	+·2	+·2	+	.	+	+	.	1·1
Ampelopsis aconitifolia	+	+	+	+	.	.	+	+
Lespedeza floribunda	+	+·2	+	.	+	+	+	+
Myripnois dioica	1·1	+·2	+	+	+	.	.	+
Thalictrum aquilegifolium var. *sibiricum*	+	+	.	+	.	+	.	.
Gleditsia heterophylla	+	+	+	.	+	.	.	+
Fraxinus chinensis	.	+	+	+	.	+	+	+

11 Phytosociological Study of *Pteroceltis tatarinowii* Forest in the...

Species										
Lonicera pekinensis	.	+	+	.	.	+	.	.	+	+
Clematis heracleifolia	+	.	.	.	.	.	.	.	.	.
Sageretia paucicostata	.	1·1	.	.	+	.	.	.	+	.
Diagnostic species of Moro australis-Pteroceltidion tatarinowii all. nov.										
Pteroceltis tatarinowii	5·5	4·5	5·5	5·5	5·5	4·4	.	5·5	5·5	5·5
Morus australis	+	+	+·2	+·2	+·2	+	.	+	+	+
Cynodon dactylon	1·1	+·2	.	.	+·2	.	.	+·2	+·2	+·2
Pinellia ternata	+	+	+	+	.	+	.	+	.	+
Liriope spicata	.	.	.	+	.	+	.	.	+	+
Viola selkirkii	+	.	+	.	.	+	.	.	+	+
Paederia scandens	.	.	+·2	+·2	+	+	.	+	.	.
Aristolochia debilis	+	+	+	.	.	+	.	.	.	.
Chrysanthemum indicum	.	.	.	.	.	.	.	.	.	.
Diagnostic species of higher unit of Quercetea variabilis Class Tang, Fujiwara et You in Box et Fujiwara 2014										
Vitex negundo var. *heterophylla*	2·3	1·1	.	.	+·2	+	+	.	1·1	1·1
Grewia biloba var. *parviflora*	+	1·1	+	+	+	+·2	+·2	+	1·1	+·2
Carpinus turczaninowii	.	.	.	.	.	.	+	.	.	+
Diospyros lotus	+	.	.	.	.	1·1	.	.	.	.
Toxicodendron succedaneum	+	+	+	+	+	.	+·2	.	.	.
Pistacia chinensis	.	.	.	.	.	+	.	.	+	.
Other species										
Clematis brevicaudata	+	+	+	+	.	+	.	+	+	+
Dioscorea nipponica	+	+	+	+	+	+	+	+	+	+
Carex lanceolata	+	2·2	+	2·2	1·1	+·2	2·2	+	2·2	+·2
Oplismenus undulatifolius	2·2	+	+·2	.	1·1	1·1	.	.	.	1·1
Lespedeza bicolor	.	+	+	+	+	.	+	+	+	+

(continued)

Table 11.3 (continued)

Relevé reference no.	1	2	3	4	5	6	7	8
Vitis amurensis	+	.	+	.	+	+	.	.
Rhamnus parvifolia	.	1·1	+	+	.	.	+	+
Viola collina	.	+	.	.	+	+	.	+
Aster ageratoides	.	.	.	.	+	+	+	+
Adenophora wawreana	+	+	.	+	+	.	.	+
Deyeuxia arundinacea	.	+	.	+	.	.	+·2	+
Indigofera bungeana	+·2	+	.	.	.	.	.	+
Rubia cordifolia	.	+	.	.	.	.	.	+
Polygonatum odoratum	.	.	.	.	+	+	+	.
Polygonatum sibiricum	.	.	.	+	.	.	+	+
Carpesium cernuum	.	+	.	.	+	.	.	+
Spiraea pubescens	.	.	+	+	.	.	.	.
Ailanthus altissima	.	.	.	.	.	.	.	.
Zizyphus jujuba	.	+	.	.	.	.	.	+
Spiraea trilobata	1·1	+	+	+·2	.	.	.	.
Cotinus coggygria	+·2	+	.	.	.	.	.	.
Adiantum capillus-veneris	.	1·1	+·2	.	+	+	.	+
Pueraria lobata	.	+	.	.	.	.	.	.
Zanthoxylum bungeanum	.	+	.	.	.	.	.	.
Humulus scandens	.	+	.	.	.	.	.	.
Leibnitzia anandria	.	.	.	.	.	.	.	.
Cypripedium macranthum	.	+	.	.	.	.	.	.
Commelina communis	.	+	.	.	.	.	.	.
Caragana rosea	.	.	.	.	.	.	.	+
Morus alba	.	.	+	.	+	.	.	.
Ligularia xanthotricha	+·2	.	+	+	.	.	.	.

Species	9	10	11	12	13	14	15	16
Viola dissecta	.	.	.	.	.	.	.	.
Spodiopogon sibiricus	+	+	.	.	.	.	.	.
Saussurea bodinieri	+	+	.	.	.	.	.	.
Aquilegia yabeana	.	.	.	+	.	.	.	.
Amphicarpaea trisperma	+	.	.	.	.	.	.	.
Allium schoenoprasum	+	.	.	+	.	.	.	.
Moss sp.	+-2	.	.	.	.	.	.	.
Lygodium japonicum	+	.	.	.	+	.	.	.
Tilia amurensis	.	.	+	.	.	.	.	+
Smilax stans	.	.	.	.	.	.	.	+
Rhus chinensis	.	.	.	.	+	.	.	+
Geum aleppicum	.	.	.	.	.	.	.	+
Leptopus chinensis	.	.	.	.	.	.	.	.
Prunus davidiana	.	.	.	.	.	.	.	.
Periploca sepium	.	.	.	.	.	.	.	.
Viola philippica ssp. *munda*	+	.	.	.	.	.	.	.
Boea hygrometrica	.	.	.	.	.	.	.	.
Angelica dahurica	.	.	.	.	.	.	.	.
Acalypha australis	.	.	.	.	.	.	.	.
Euphorbia fischeriana	.	.	.	.	+	.	.	.
Deutzia grandiflora	.	.	.	+	.	.	.	+
Dennstaedtia wilfordii	.	.	.	+	.	.	.	.
Bupleurum scorzonerifolium	.	.	.	.	.	.	+	.
Asparagus cochinchinensis	.	.	.	.	.	.	+	.
Artemisia apiacea	+	.	.	.	.	.	.	.
Artemisia annua	.	+	.	.	.	.	.	.
Relevé reference no.	9	10	11	12	13	14	15	16

(continued)

Table 11.3 (continued)

Relevé reference no.	9	10	11	12	13	14	15	16
Original relevé number	BF9	BF10	BF11	BF12	BF13	BP2	BP3	BP4
Relevé date	2013.08.03	2013.08.03	2013.08.03	2013.08.03	2013.08.03	2013.08.04	2013.08.04	2013.08.04
Relevé size (m^2)	225	225	100	400	300	300	150	200
Altitude (m)	271	275	282	301	335	400	383	376
Aspect	SE	SE	SW	SE	SE	SW	SW	NE
Slope (°)	26	30	85	20	24	2	43	2
Percentage of exposed rock (%)	95	95	98	85	60	85	90	90
Height(m) of Tree layer 1 (T1)	–	10	–	10	–	–	–	–
Cover(%) of Tree layer 1	–	85	–	85	–	–	–	–
Height(m) of Tree layer 2 (T2)	10	5	8	6	8	5	5	6.5
Cover(%) of Tree layer 2	80	10	65	15	85	75	85	85
Height(m) of Shrub layer (S)	2	2.5	3	2	3	2	2	3.5
Cover(%) of Shrub layer	20	15	35	25	30	20	15	10
Height(m) of Herb layer (H)	0.6	0.6	0.3	0.6	0.6	0.6	0.6	0.6
Cover(%) of Herb layer	20	5	5	20	15	30	25	15
Number of species	49	27	20	29	27	46	38	36
Diagnostic species of Isodo rubescens-Pteroceltido tatarinowii assn. nov.								
Flueggea suffruticosa	1·1	1·1	+	+·2	1·1	+	+	+
Isodon rubescens	1·1	1·1	+	.	+·2	1·1	1·1	1·1
Ampelopsis aconitifolia	+	+	+	.	+	1·1	+·2	.
Lespedeza floribunda	+	+	+	+·2	+·2	.	.	.
Myripnois dioica	+	+	.	+	.	.	+·2	+·2
Thalictrum aquilegifolium var. sibiricum	+·2	+·2	+	.	+	+	+	+
Gleditsia heterophylla	+	+	+	+·2	1·1	.	.	.
Fraxinus chinensis	+	+	.	.	+	.	.	.
Lonicera pekinensis	.	.	.	.	1·1	.	.	.

Species	1	2	3	4	5	6	7	8	9	10
Clematis heracleifolia	+	+	.	.	.	.	.	.	1·2	+
Sageretia paucicostata	+	+	.	.	+·2	.	1·1	.	.	.
Diagnostic species of Moro australis-Pteroceltidion tatarinowii all. nov.										
Pteroceltis tatarinowii	4·4	5·5	4·4	.	5·5	.	5·5	4·5	5·5	5·5
Morus australis	+·2	+	.	+·2	+·2	+·2	+	+	+	+
Cynodon dactylon	+	+	+	+·2	+·2	+·2	+·2	+·2	+·2	+
Pinellia ternata	+	+	+	+·2	+·2	.	+	+	+	+
Liriope spicata	+	+	.	+·2	+·2	.	+	+	.	.
Viola selkirkii	+	.	+	.	.	.	.	.	.	.
Paederia scandens	.	.	.	+	.	.	.	.	.	+
Aristolochia debilis	.	.	.	.	.	.	+	.	.	.
Chrysanthemum indicum	.	.	.	.	.	.	+	+	.	+
Diagnostic species of higher unit of Quercetea variabilis Class Tang, Fujiwara et You in Box et Fujiwara 2014										
Vitex negundo var. heterophylla	1·1	+·2	2·2	+·2	+	+·2	+·2	1·2	+·2	+·2
Grewia biloba var. parviflora	+·2	+	1·1	.	.	1·1	1·1	+	.	.
Carpinus turczaninowii	+·2	+	.	.	+·2	+	+	+	+	+
Diospyros lotus	+·2	+	+·2	.	+·2	.	.	.	+	.
Toxicodendron succedaneum	+	.	.	.	.	.	.	.	.	.
Pistacia chinensis	.	.	.	.	.	.	.	.	.	.
Other species										
Clematis brevicaudata	+	+	+	.	+	+·2	+·2	+·2	+	+·2
Dioscorea nipponica	+·2	+	+	.	.	.	+	+·2	+	+
Carex lanceolata	2·2	+·2	+·2	1·1	1·1	.	.	+	+	+
Oplismenus undulatifolius	+	+	.	+·2	+·2	.	+	+·2	+·2	1·1
Lespedeza bicolor	+·2	+·2	+·2	.	.	+·2	+·2	+	+	+
Vitis amurensis	+	.	.	+	+	.	.	+	+	+
Rhamnus parvifolia	+	+·2	+·2	+	+·2	.	+·2	.	.	.

(continued)

Table 11.3 (continued)

Relevé reference no.	9	10	11	12	13	14	15	16
Viola collina		+	+	.	+	.	.	+
Aster ageratoides		+	+	.	.	.	+	+·2
Adenophora wawreana		+	.	.	.	.	+	+
Deyeuxia arundinacea		+	.	.	.	+·2	.	+·2
Indigofera bungeana		+	.	.	+	.	+	+
Rubia cordifolia		+	.	.	.	+·2	+·2	+
Polygonatum odoratum		+	+	+	.	.	.	.
Polygonatum sibiricum		+	.	.	+	.	.	.
Carpesium cernuum		+	.	.	.	+	.	.
Spiraea pubescens		+	.	.	+	.	+	.
Ailanthus altissima		+	.	.	+	+	+	.
Zizyphus jujuba		+	.	.	+	.	.	.
Spiraea trilobata		.	.	.	.	.	.	.
Cotinus coggygria		1·1	.	.	+	.	.	.
Adiantum capillus-veneris		.	.	.	.	.	.	.
Pueraria lobata		+	.	.	.	.	.	.
Zanthoxylum bungeanum		.	.	.	.	+·2	+	.
Humulus scandens		.	.	.	.	1·1	+	.
Leibnitzia anandria		+	.	.	.	+	+	.
Cypripedium macranthum		.	.	.	.	.	+	+
Commelina communis		.	.	.	.	+	.	+
Caragana rosea		.	.	+	.	.	.	.
Morus alba		.	.	.	.	+	.	.
Ligularia xanthotricha		.	.	.	.	.	.	.
Viola dissecta		+	.	.	.	.	.	.
Spodiopogon sibiricus		.	+	.	.	.	.	.

Species	1	2	3	4	5	6	7	8	9	10	11	12	13	14	15	16
Saussurea bodinieri	.	.	.	.	.	.	.	.	.	.	.	.	.	.	.	.
Aquilegia yabeana	.	+	.	.	.	.	.	.	.	.	.	.	.	.	.	+
Amphicarpaea trisperma	.	.	.	.	.	.	.	.	.	.	.	.	.	.	.	+
Allium schoenoprasum	.	1·1	.	.	.	.	.	.	.	.	+	.	.	+	.	.
Moss sp.	.	.	.	.	.	.	.	.	.	.	.	.	.	.	.	.
Lygodium japonicum	.	.	.	.	.	.	.	.	.	.	.	.	.	.	.	.
Tilia amurensis	.	.	.	.	.	.	.	.	.	.	.	.	.	.	.	.
Smilax stans	.	.	.	.	.	.	+	.	.	.	.	.	.	.	.	.
Rhus chinensis	.	.	.	.	.	.	.	.	.	.	.	.	.	.	.	.
Geum aleppicum	.	.	.	.	.	.	.	.	.	+	.	.	.	.	.	+
Leptopus chinensis	.	.	.	.	.	+	.	.	.	.	.	.	.	.	.	.
Prunus davidiana	.	+	.	.	.	.	.	.	.	.	.	.	.	.	.	+
Periploca sepium	.	+	.	.	.	.	.	.	.	+	.	.	.	.	.	.
Viola philippica ssp. munda	.	.	.	.	.	.	.	.	.	+	.	.	.	.	.	.
Boea hygrometrica	.	.	.	.	.	.	.	.	.	+	.	+	.	.	.	.
Angelica dahurica	.	.	.	.	.	.	.	.	.	+	.	+	.	.	.	.
Acalypha australis	.	.	.	.	.	.	.	.	.	+	.	+	.	.	.	.
Euphorbia fischeriana	.	.	.	.	.	.	.	.	.	.	.	.	.	+	.	.
Deutzia grandiflora	.	.	.	.	.	.	.	.	.	.	.	.	.	.	.	.
Dennstaedtia wilfordii	.	.	.	.	.	.	.	.	.	.	.	.	.	+	.	.
Bupleurum scorzonerifolium	.	.	.	.	.	.	.	.	.	+	.	.	.	.	.	.
Asparagus cochinchinensis	.	.	.	.	.	.	+	.	.	.	.	.	.	.	.	.
Artemisia apiacea	.	.	.	.	.	.	.	.	.	.	.	.	.	.	.	+
Artemisia annua	.	.	.	.	.	.	.	.	.	+	.	.	.	.	.	.

Location: running numbers 1–13 are from the Shidu scenic area and numbers 14–16 from the Puwa nature reserve (both in the Fangshan area of Beijing)
Additional species occurring once in relevé reference no.2: *Taraxacum mongolicum* H-+; no.3: *Abelia biflora* S-+; no.5: *Rheum officinale* H-+·2, *Viburnum mongolicum* H-+; no.6: *Smilax china* H-+; no.7: *Indigofera kirilowii* H-+; no.9: *Koelreuteria bipinnata* S-+, H-+; no.14 *Chenopodium album* H-+·2, *Thalictrum thunbergii* H-+, *Stellaria media* H-+, *Armeniaca sibirica* H-+, *Ipomoea purpurea* S-+, H-+, *Juglans regia* H-+, *Galium aparine* var. *tenerus* H-+, *Artemisia* sp.1 H-+; no.15: *Davallia mariesii* H-+, *Sedum aizoon* H-+, *Bidens parviflora* H-+, *Metaplexis japonica* H-+; no.16: *Begonia evansiana* H-+, *Arisaema erubescens* H-+

few. The shrub layer attains an average height of 2.8 m with a mean cover of 21.6%, and the predominant species include *Pteroceltis tatarinowii*, *Vitex negundo* var. *heterophylla*, *Flueggea suffruticosa*, *Grewia biloba* var. *parviflora*, *Morus australis*, *Lespedeza bicolor*, *Rhamnus parvifolia*, etc. The height of the herb layer is 0.3–0.6 m, and the cover ranges from 10% to 35%. The main species of the herb layer are *Carex lanceolata*, *Clematis brevicaudata*, seedlings of *Pteroceltis tatarinowii*, *Pinellia ternata*, *Oplismenus undulatifolius*, *Dioscorea nipponica*, *Cynodon dactylon*, *Lespedeza bicolor*, *Flueggea suffruticosa*, *Isodon rubescens*, *Vitex negundo* var. *heterophylla*, *Liriope spicata*, *Grewia biloba* var. *parviflora*, *Ampelopsis aconitifolia*, *Lespedeza floribunda*, *Thalictrum aquilegifolium* var. *sibiricum*, *Viola collina*, *Vitis amurensis*, *Aster ageratoides* and *Carpinus turczaninowii*. The average number of species is 34.

2) *Platycladus orientalis-Pteroceltis tatarinowii* community

The stands of this community are located at the Lingyan Temple in Jinan city and Qingtan Temple in Zaozhuang city in Shandong (Table 11.4). This community survives mainly in the limestone area around temples, and large *Pteroceltis tatarinowii* individuals (with DBH more than 50 cm) can often be found in the forest. The differential species of this community are *Platycladus orientalis*, *Ziziphus jujuba*, *Robinia pseudoacacia*, *Cocculus orbiculatus*, *Clematis pogonandra*, *Acalypha australis*, *Arisaema erubescens* and *Punica granatum*. This community was disturbed strongly by human activities, and *Platycladus orientalis* and *Robinia pseudoacacia* can usually be found planted in the forests. The *Platycladus orientalis-Pteroceltis tatarinowii* community is distinguished mainly by the species *Acer mono*, *Ligustrum quihoui*, *Aristolochia mollissima*, *Firmiana simplex*, *Cornus walteri* and *Dalbergia hupeana*. This community can be subdivided into two subunits, i.e. the *Clematis pogonandra* subunit (running numbers 1–3) and *Ziziphus jujuba* subunit (running numbers 4–6). The common species of the tree layer are *Pteroceltis tatarinowii*, *Koelreuteria bipinnata*, *Platycladus orientalis* and *Pistacia chinensis*. The main species of the shrub layer include *Pteroceltis tatarinowii*, *Grewia biloba* var. *parviflora*, *Vitex negundo*, and *Robinia pseudoacacia*. The common herbaceous species include *Arthraxon hispidus*, *Leibnitzia anandria*, *Chrysanthemum indicum*, *Acalypha australis*, *Carpesium cernuum*, *Trachelospermum jasminoides*, and *Ixeris sonchifolia*.

(1) *Clematis pogonandra* subunit

The stands of this subunit are located on gentle northwest-facing slopes (3°–10°) at elevation on 312–367 m at the Lingyan Temple of Jinan city. This subunit occurs in valleys or lowlands, with mean soil pH of 7.3 and exposed rocks covering 10–40% of the surface area. The forest was affected significantly by human activities, and trees of *Platycladus orientalis* and *Robinia pseudoacacia* have been planted in the forest. However, some natural trees of *Pteroceltis tatarinowii*, *Koelreuteria bipinnata*, *Tilia amurensis* and *Morus australis* were retained. The *Clematis pogonandra* subunit is differentiated by herbaceous species *Pinellia ternata*, *Chenopodium glaucum*, *Achyranthes bidentata*, *Viola grypoceras*, *Bidens parviflora*, *Potentilla kleiniana*, *Spodiopogon sibiricus* and *Leonurus japonicas*, and the lianas *Clematis pogonandra*, *Clematis brevicaudata*, *Metaplexis japonica*,

Table 11.4 *Platycladus orientalis-Pteroceltis tatarinowii* community

Relevé reference number		1	2	3	4	5	6	7
Original relevé number (in field)		SL1	SL2	SL3	SZ1	SZ2	SZ3	SZ4
Relevé data		2013	2013	2013	2013	2013	2013	2013
		8	8	8	7	7	7	7
		14	14	14	13	13	13	13
Relevé size (m²)		160	200	200	160	500	400	300
Altitude (m)		367	346	312	164	137	135	177
Aspect		NW	NW	NW	NW	SW	SE	SE
Slope (°)		3	5	10	78	75	75	60
Percentage of exposed rock (%)		10	40	40	98	98	98	98
Height(m)/cover(%) of tree layer 1 (T1)		11/50	–	10/75	–	10/15	–	–
Height(m)/cover(%) of tree layer 2 (T2)		7/75	10/80	6/10	6/80	7/90	10/80	8/75
Height(m)/cover(%) of shrub layer (S)		2.5/25	4/40	2/10	1.5/30	2.5/50	2/15	2.5/40
Height(m)/cover(%) of herb layer (H)		0.6/70	0.8/60	0.6/60	0.3/20	0.4/8	0.3/25	0.3/20
Number of Species		43	36	16	30	31	33	26
Differential species of *Platycladus orientalis-Pteroceltis tatarinowii* community								
Platycladus orientalis	T,S,H	3·3	1·1	3·3	+	+	+	+
Robinia pseudoacacia	T,S,H	·	·	1·1	+	1·1	+	+
Acalypha australis	H	+	+	·	·	+	+·2	+
Arisaema erubescens	H	1·1	+·2	·	·	+	·	·
Differential species of *Clematis pogonandra* subunit								
Clematis pogonandra	S,H	1·1	1·1	1·1	·	·	·	·
Pinellia ternata	H	+·2	+	+·2	·	·	·	·
Chenopodium glaucum	H	·	+	2·2	·	·	·	·
Achyranthes bidentata	H	+	·	+·2	·	·	·	·
Dioscorea nipponica	H	+·2	·	+	·	·	·	·
Clematis brevicaudata	H	2·2	1·1	·	·	·	·	·
Spodiopogon sibiricus	H	+	+·2	·	·	·	·	·
Leonurus japonicus	H	+	+	·	·	·	·	·
Vitis amurensis	T,S,H	+	+	·	·	·	·	·
Metaplexis japonica	S,H	+	+	·	·	·	·	·
Viola grypoceras	H	+	+	·	·	·	·	·
Bidens parviflora	H	+	+	·	·	·	·	·
Potentilla kleiniana	H	+	+	·	·	·	·	·
Differential species of *Zizyphus jujuba* subunit								
Trachelospermum jasminoides	T,H	·	·	·	1·2	1·2	2·3	+·2

(continued)

Table 11.4 (continued)

Zizyphus jujuba	T,S,H	+	·	·	1·1	+·2	+·2	1·1
Broussonetia papyrifera	T,S,H	·	·	·	+·2	+·2	+	+
Cocculus orbiculatus	T,S,H	·	·	·	+	+·2	+	+
Vitex negundo var. cannabifolia	S,H	·	·	·	+·2	·	+	+·2
Cudrania tricuspidata	S,H	·	·	·	+	·	+·2	+
Paederia scandens	S,H	·	·	·	·	+	+	+
Clematis kirilowii	S,H	·	·	·	+	+	+	+
Ixeris sonchifolia	H	·	·	·	+·2	+·2	+	+
Deyeuxia arundinacea	H	·	·	·	+	+	·	+
Pistacia chinensis	T,S,H	·	·	·	+	·	+	+·2
Punica granatum	T,S	·	·	·	·	+	+	+
Crataegus cuneata	T,S	·	·	·	+	·	+	·
Rubus parviflorus	H	·	·	·	+·2	·	+	·
Patrinia scabiosaefolia	H	·	·	·	+	·	+	·
Lespedeza cuneata	H	·	·	·	+	·	+	·
Rubia cordifolia	H	·	·	·	·	+	+	·
Diagnostic species of Moro australis-Pteroceltidion tatarinowii all. nov.								
Pteroceltis tatarinowii	T,S,H	3·3	3·3	2·2	5·5	5·5	5·5	4·4
Chrysanthemum indicum	H	+	1·1	·	+	+	·	+
Aristolochia debilis	S,H	+	·	·	+	·	+	·
Diagnostic species of higher unit of Quercetea variabilis Tang, Fujiwara et You in Box et Fujiwara 2014								
Grewia biloba var. parviflora	S,H	1·2	1·1	·	1·1	+·2	+·2	1·1
Koelreuteria bipinnata	T,S,H	1·2	2·2	+	+	+	+·2	·
Speranskia tuberculata	H	+	+	·	·	·	+	·
Other species								
Arthraxon hispidus	H	3·3	3·3	2·2	+	+	+	1·1
Leibnitzia anandria	H	+	+	+	+	+	·	+
Vitex negundo	S,H	1·2	1·1	·	1·1	·	+·2	1·1
Lespedeza bicolor	S,H	+	+·2	·	+·2	+	+	·
Carpesium cernuum	H	+	+	·	·	+	+	·
Ailanthus altissima	T,S,H	+	·	·	+	·	+	+
Cynodon dactylon	H	+	·	·	+·2	·	+	·
Indigofera kirilowii	S,H	+	·	·	·	·	·	+
Toxicodendron succedaneum	T,S	+	·	·	+	·	·	·

(continued)

Table 11.4 (continued)

Carex lanceolata	H	+	.	.	+	.	.	.
Artemisia annua	H	.	+	.	.	+	.	.

Location: running numbers 1–3 are at Lingyan Temple in Jinan city and running numbers 4–7 at Qingtan Temple in Zaozhuang city, Shandong province
Additional species occurring once in relevé reference no.1: *Boehmeria nivea* H-2·2, *Isodon glaucocalyx* H-1·1, *Tilia amurensis* T1-+·2, *Acer buergerianum* S-+,H-+; no.2: *Solanum lyratum* H-+, *Ipomoea purpurea* H-+, *Ixeris denticulata* H-+, *Humulus scandens* H-+, *Artemisia gmelinii* H-+, *Artemisia argyi* H-+, *Geum aleppicum* H-+, *Euphorbia helioscopia* H-+, *Celtis koraiensis* S-+; no.3: *Acalypha supera* H-2·2, *Morus australis* T2-1·1, *Celastrus orbiculatus* H-+·2, *Duchesnea indica* H-+, *Diospyros rhombifolia* S-+, *Angelica dahurica* H-+, *Amphicarpaea trisperma* H-+, *Adenophora wawreana* H-+, *Adenophora trachelioides* H-+, *Cornus macrophylla* H-+; no.4: *Sedum bulbiferum* H-+; no.5: *Chenopodium album* H-+·2, *Oxalis corniculata* H-+·2, *Viola collina* H-+, *Tilia miqueliana* S-+, *Lycium chinense* H-+, *Cayratis japonica* H-+, *Artemisia apiacea* H-+, *Rostellularia procumbens* H-+; no.6: *Prunus davidiana* T2-+, *Liriope spicata* H-+, *Viola selkirkii* H-+, *Lepidium virginicum* H-+; no.7: *Cotinus coggygia* T2-1·1, S-1·2, *Periploca sepium* H-+, *Lespedeza pilosa* H-+, *Ulmus macrocarpa* T2-+

Dioscorea nipponica and *Vitis amurensis*. Stratification of this subunit usually involves four layers. The tree layer reaches 10–11 m with an average cover 68%, and the dominant species are *Pteroceltis tatarinowii* and *Platycladus orientalis*. The shrub layer is 2–4 m in height and attains 10–40% in cover. The predominant species of the shrub layer include *Pteroceltis tatarinowii*, *Grewia biloba* var. *parviflora*, *Koelreuteria bipinnata* and *Vitex negundo*. The herb layer is often affected by human activities and has a cover ranging from 60% to 70%, with an average height of 0.7 m; the main species are *Arthraxon hispidus*, *Clematis pogonandra*, *Leibnitzia anandria*, *Pteroceltis tatarinowii*, *Pinellia ternata*, *Koelreuteria bipinnata*, *Arisaema erubescens*, *Clematis brevicaudata*, *Vitex negundo*, etc. The average stand richness reaches 32 species.

(2) *Ziziphus jujuba* subunit

The samples of this subunit were taken from very steep (60°–78°) south-facing slopes, at altitudes from 135 m to 177 m, with exposed-rock ratios of 98%, at Qingtan Temple in Zaozhuang city. This subunit grows on poor limy soil, with average soil pH of 7.89.

The differential species of this subunit are *Trachelospermum jasminoides*, *Zizyphus jujuba*, *Broussonetia papyrifera*, *Cocculus orbiculatus*, *Vitex negundo* var. *cannabifolia*, *Cudrania tricuspidata*, *Paederia scandens*, *Clematis kirilowii*, *Ixeris sonchifolia*, *Deyeuxia arundinacea*, *Pistacia chinensis*, *Punica granatum*, *Crataegus cuneata*, *Rubus parviflorus*, *Patrinia scabiosaefolia*, *Lespedeza cuneata* and *Rubia cordifolia*. The structure of this subunit generally consists of three layers. The tree layer is 6–10 m and has cover ranging from 75% to 90%. The tree layer is dominated by *Pteroceltis tatarinowii*, mixed with *Pistacia chinensis*, *Koelreuteria bipinnata*, *Zizyphus jujuba*, *Platycladus orientalis*, *Robinia pseudoacacia*, etc. The shrub layer attains 1.5–2 m and has cover ranging from 15% to 50%. The shrub layer is relatively rich, with a total of 22 species appearing; the dominant species are *Pteroceltis tatarinowii*, *Grewia biloba* var. *parviflora* and *Ziziphus jujuba*, and other species include *Platycladus orientalis*, *Robinia pseudoacacia*, *Cocculus orbiculatus*, *Punica granatum*, *Vitex negundo* var. *cannabifolia*, *Cudrania tricuspidata*, *Broussonetia papyrifera*, *Vitex negundo*, *Ailanthus altissim* and *Cotinus coggygria*. The herb layer reaches 0.4 m with a cover of 8–25%, and the common species are *Trachelospermum jasminoides*, *Ixeris sonchifolia*, *Arthraxon hispidus*, *Broussonetia papyrifera*, *Deyeuxia arundinacea*, *Cocculus orbiculatus*, *Rubus parviflorus*, *Patrinia scabiosaefolia*, *Lespedeza cuneata*, *Rubia cordifolia*, *Chrysanthemum indicum*, *Paederia scandens*, *Cynodon dactylon*, *Leibnitzia anandria* and *Carpesium cernuum*; seedlings of *Pteroceltis tatarinowii* were not found. The average number of species per sample is 30.

3) Pteroceltido tatarinowii-Quercetum variabilis

Diagnostic species: *Acer mono*, *Aristolochia mollissima*, *Ligustrum quihoui*, *Firmiana simplex*, *Cornus walteri* and *Dalbergia hupeana*.

Type relevé: running number 5 in Table 11.5

The Pteroceltido tatarinowii-Quercetum variabilis had been described by Tang et al. (2014) as an association in the class Quercetae variabilis. The original data were from forests dominated by *Pteroceltis tatarinowii* and by *Quercus variabilis*. Here we add new samples from only the *Pteroceltis tatarinowii* forest, for comparison.

This association was located on southeast and northeast-facing slopes (15°–30°), at low altitudes from 150 m to 201 m, in the Huangcangyu nature reserve of Xiaoxian county of Anhui. This association often occurs in valleys and hon illsides, on poor soils with an average pH of 7.1 and exposed rocks covering 5–85% of the area.

The structure of Pteroceltido tatarinowii-Quercetum variabilis can be divided into three or four layers. The tree layer has an average height of 14.7 m, with an average cover of 77%. The tree layer is usually dominated by *Pteroceltis tatarinowii* but is often mixed with other trees, such as *Quercus variablilis*, *Acer mono*, *Pistacia chinensis*, *Morus australis*, *Koelreuteria paniculata*, *Diospyros kaki* or *Dalbergia hupeana*. The height of the shrub layer attains 2–5 m, and its cover ranges from 15% to 35%. The dominant species of the shrub layer include *Ligustrum quihoui*, *Cornus walteri*, *Vitex negundo* var. *cannabifolia*, *Firmiana simplex*, *Rhamnus arguta*, *Lindera glauca*, *Grewia biloba* var. *parviflora*, *Lonicera fragrantissima*, *Tilia miqueliana*, and some saplings of taller tree species. The herb layer is 0.5–0.6 m, with cover ranging from 15% to 50%; the main species are *Aristolochia mollissima*, *Arthraxon hispidus*, *Trachelospermum jasminoides*, *Liriope spicata*, *Chrysanthemum indicum*, *Cynodon dactylon*, *Viola selkirkii*, *Ophiopogon japonicus*, *Smilax china*, *Scilla scilloides*, *Acer mono*, *Ligustrum quihoui*, *Dalbergia hupeana*, *Pteroceltis tatarinowii*, *Quercus variabilis* and *Grewia biloba* var. *parviflora*. This association occurs in the north of the transition between the deciduous and evergreen broad-leaved forest zones, and a few evergreen or semi-evergreen species can be found in the association (e.g. *Ligustrum quihoui*, *Lonicera fragrantissima*). The average number of species is about 39.

Table 11.5 Pteroceltido tatarinowii-Quercetum variabilis

Running number		1	2	3	4	5
Original relevé number		AH1	AH2	AH3	AH4	AH5
Relevé data		2011.5.14	2011.5.14	2011.7.13	2010.10.31	2010.10.31
Relevé size (m^2)		400	400	400	400	400
Altitude (m)		187	180	178	201	150
Aspect		SE	NE	SE	SE	SE
Slope (°)		21	15	30	18	24
Percentage of exposed rock (%)		30	0	65	40	0
Height(m)/cover(%) of tree layer 1(T1)		15/75	–	–	16.5/80	25/90
Height(m)/cover(%) of tree layer 2(T2)		9.0/25	8.0/65	9.0/75	8.0/15	12.0/10
Height(m)/cover(%) of shrub layer (S)		2.5/18	4.0/35	3.0/35	2.0/15	5.0/15
Height(m)/cover(%) of herb layer (H)		0.5/45	0.5/50	0.6/20	0.5/35	0.5/15
Number of species		40	50	33	39	31
Diagnostic species of Pteroceltido tatarinowii-Quercetum variabilis Tang, Fujiwara et You 2014						
Acer mono	T, S, H	2·2	2·2	+	+	+
Ligustrum quihoui	S, H	1·1	+·2	+·2	1·1	1·2
Aristolochia mollissima	H	1·1	+	.	+	+
Firmiana simplex	T, S, H	+	1·1	.	+	+
Cornus walteri	S	.	+	.	+	+
Dalbergia hupeana	T, S, H	+	1·1	+	.	.
Diagnostic species of Moro australis-Pteroceltidion tatarinowii all. nov.						
Pteroceltis tatarinowii	T, S, H	2·3	2·2	3·3	2·2	4·4
Arthraxon hispidus	H	+·2	1·2	+·2	+·2	2·2
Liriope spicata	H	+	+	.	+	+·2
Chrysanthemum indicum	H	+	+	+	+	.
Cynodon dactylon	H	+·2	.	+	+	.
Morus australis	T, S, H	+·2	.	.	1·1	.
Vitex negundo var. *cannabifolia*	S, H	+·2	.	2·2	.	.

(continued)

Table 11.5 (continued)

Viola selkirkii	H	+	.	+	.	.
Aristolochia debilis	H	+·2	.	.	+	.
Diagnostic species of higher unit of Quercetea variabilis Tang, Fujiwara et You in Box et Fujiwara 2014						
Quercus variabilis	T, S, H	3·3	1·1	1·1	3·3	+
Trachelospermum jasminoides	S, H	3·3	+·2	+	3·3	+
Grewia biloba var. parviflora	S	1·2	1·1	1·1	+	.
Pistacia chinensis	T, S, H	+·2	2·3	1·2	1·1	.
Koelreuteria paniculata	T, S, H	+·2	.	+	+·2	.
Other species						
Celtis koraiensis	T, S, H	+	+	+	+	1·1
Indigofera kirilowii	H	+	.	+	.	.
Rhamnus arguta	S, H	+	+	+	1·1	.
Ophiopogon japonicus	H	1·1	+·2	+·2	.	.
Lindera glauca	S, H	+·2	2·3	.	+·2	.
Euonymus alatus	S	+	+	.	+	.
Scilla scilloides	H	+	+	.	+	.
Diospyros kaki	T2	.	+	.	+·2	.
Lonicera fragrantissima	S, H	+·2	+	.	+	+
Smilax china	H	.	+	.	+	+
Ixeris denticulata	H	+	+	+	.	.
Polygonatum odoratum	H	+·2	1·1	.	+	.
Rubia cordifolia	H	+	+	.	+	.
Deyeuxia arundinacea	H	+	.	+·2	+	.
Liriope platyphylla	H	+	.	.	+	.
Geranium carolinianum	H	+	.	.	.	+
Bupleurum scorzonerifolium	H	+	.	+	.	.

(continued)

Table 11.5 (continued)

Allium macrostemon	H	·	+	·	+·2	·
Morus alba	S, H	·	+	+	·	·
Carex lanceolata	H	·	1·1	+	·	·
Roegneria ciliaris	H	·	+·2	·	·	2·2
Ajuga decumbens	H	·	+	·	·	+
Tilia miqueliana	S, H	·	+	·	·	+
Viola betonicifolia	H	·	+	·	·	+
Clematis kirilowii	S	1·2	+	+·2	+	·
Celtis bungeana	T, S, H	·	·	+	+·2	·

Location: Huangcangyu nature reserve of Xiaoxian County in Anhui province
Additional species occurring once in relevé reference no.1: *Pyrus betulaefolia* H-+, *Albizia kalkora* H-+, *Cudrania tricuspidata* S-+, *Carpesium cernuum* H-+, *Lysimachia clethroides* H-+, *Spodiopogon sibiricus* H-+; no.2: *Carex humilis* var. *nana* H-3·3, *Aster ageratoides* H-1·2, *Cleistogenes hancei* H-1·2, *Diospyros lotus* T-1·1, *Leptopus chinensis* S-+, *Adenophora trachelioides* H-+, *Lespedeza buergeri* H-+, *Viburnum macrocephalum* S-+, *Microstegium nudum* H-+, *Artemisia princeps* H-+, Orchidacea sp. H-+, *Paederia scandens* H-+, *Patrinia villosa* H-+, *Prunus tomentosa* S-+, *Picrasma quassioides* T2-+, S-+, *Viola variegata* H-+; no.3: *Ailanthus altissima* H-+, *Cynanchum versicolor* H-+, *Euphorbia helioscopia* H-+, *Polygonatum sibiricum* H-+, *Iris tectorum* H-+, *Lespedeza cuneata* H-+, *Lespedeza pilosa* H-+, *Ixeris sonchifolia* H-+, *Rhamnus utilis* S-+, *Setaria viridis* H-+, *Speranskia tuberculata* H-+, Moss sp. H-+; no.4: *Euonymus bungeanus* S-+·2, *Broussonetia papyrifera* S-+, H-+, *Toxicodendron succedaneum* S-+, *Adenophora wawreana* H-+, *Acer buergerianum* H-+, *Viola collina* H-+, *Oxalis corniculata* H-+; no.5: *Achyranthes bidentata* H-1·2, *Alangium chinense* T1-1·2, Granimea sp. H-+2, *Alangium platanifolium* H-+, *Crataegus cuneata* S-+, *Cynanchum auriculatum* H-+, *Akebia quinata* H-+, *Chenopodium glaucum* H-+, *Clematis ganpiniana* H-+, *Isodon inflexus* H-+, *Phryma leptostachya* var. *asiatica* H-+, *Rubus parviflorus* H-+, *Ulmus parvifolia* S-+, *Vitis ficifolia* H-+

2. Other units of deciduous *Quercus* forests in the deciduous-forest zone of eastern China

In order to compare the *Pteroceltis tatarinowii* forest with deciduous *Quercus* forests, 19 relevés of *Quercus* forest, including forests dominated by *Quercus variabilis* or *Q. acutissima*, were surveyed within Shandong and Henan. The data from the two *Quercus* forest types were summarized into five communities.

1) *Stephanandra incisa-Quercus variabilis* community

The stands of this community can be found on northwest and northeast-facing slopes (from 5° to 57°), at altitudes from 100 m to 250 m, on granite areas in the Lao-Shan Mountains of Qingdao city and the Kunyushan Mountains of Yantai city in Shandong. The soil pH ranges from 5.5 to 6.5 (Table 11.6). Differential species

Table 11.6 *Stephanandra incisa-Quercus variabilis* community

Running number		1	2	3
Original relevé number (in field)		CQ1	CQ2	CQ3
Relevé data		2000.5.7	2000.5.7	2000.9.14
Relevé size (m^2)		200	625	30 × 15
Altitude (m)		100	150	250
Aspect		NW	NW	NE
Slope (°)		20	5	57
Height(m)/cover(%) of Tree layer 1(T1)		–	17/70	–
Height(m)/cover(%) of Tree layer 2(T2)		10/60	10/10	8/60
Height(m)/cover(%) of Shrub layer (S)		4/30	3/20	3/8
Height(m)/cover(%) of Herb layer (H)		0.5/40	0.5/30	0.4/35
Number of Species		40	41	43
Diagnostic species of *Stephanandra incisa-Quercus variabilis* community				
Prunus japonica	S,H	+	+	+
Quercus serrata	T,S	4·4	1·1	1·1
Sorbus alnifolia	S,H	+	+	+
Carex ciliata-warginata	H	+·2	+·2	·
Cirsium japonicum	H	+	+	·
Lilium chindaoensis	H	+·2	·	+
Lysimachia barystachys	H	1·2	·	+
Meliosma oldhamii	T2	+	+	·
Pinus thunbegii	H	+	+	·
Prunus serralata	S	+	+	·
Sapium japonicum	S	+	+	·
Stephanandra incisa	S	+·2	+·2	·
Weigera florida	S	+	+	·
Symplocos paniculata	T,S	1·1	1·1	·
Lysimachia clethroides	H	·	1·2	+·2
Diagnostic species of higher unit of Quercetea variabilis Tang, Fujiwara et You in Box et Fujiwara 2014				
Quercus variabilis	T,S,H	4·4	4·4	3·3
Pistacia chinensis	T,S,H	+	+	·
Trachelospermum jasminum	H	1·2	1·2	·
Other species				
Carex lanceolata	H	3·3	3·3	3·3
Spodiopogon sibiricus	H	1·1	1·1	+·2
Sanguisorba officinalis	H	+·2	+·2	+
Euonymus alatus	S,H	+	+	+
Smilax china	H	1·2	1·2	·
Peucedanum wawrii	H	+	+	·
Aster ageratoides	H	1·2	1·2	·
Adenophora tetraphylla	H	+	+	·
Rubia cordifolia	H	1·1	1·1	·
Synurus deltoides	H	+	+	·

(continued)

Table 11.6 (continued)

Cynanchum chinense	H	+	+	·
Carex siderosticata	H	+	+	·
Adenophora trachelioides	H	+	+	·
Pinus densiflora	H	·	+	+
Lindera glauca	S	+	+	·
Quercus acutissima	T,S,H	2·2	1·2	·
Rhamnus utilis	S	+	+	·

Location: Running numbers 1–2 are from the Lao-Shan scenic area in Qingdao city, and running number 3 is from Kunyushan mountain of Yantai city in Shandong province

Additional species occurring once in relevé reference no.1: *Athyrium niponicum* H-1·2, *Styrax confuse* S-1·2, *Deyeuxia angustifolia* H-1·2, *Carex* sp.1 H-+·2, *Kalopanax pictus* S-+, H-+, *Dioscorea nipponica* H-+, *Isodon inflexus* H-+, *Artemisia japonica* H-+, *Hemerocallis citrine* H-+; no.2: *Styrax japonica* S-1·2, *Deyeuxia sylvatica* H-1·2, *Pteridium aquilinum* H-1·1, *Pterocarya stenoptera* T2-+, *Plectranthus glaucocalyx* H-+, *Artemisia integrifolia* H-+, *Artemisia eriopoda* H-+; no.3: *Deyeuxia pyramidalis* H-2·2, *Lespedeza bicolor* S-1·1, *Albizzia kalkora* S-1·1, *Artemisia eriopoda* H-1·1, *Lactuca indica* H-1·1, *Commelina communis* H-1·1, *Melampyrum roseum* H-1·1, *Deutzia glabrata* S-+·2, *Adenophora remotiflora* H-+·2, *Lespedeza inschanica* H-+·2, *Artemisia gmelinii* H-+, *Dianthus chinensis* H-+, *Cocculus trilobus* H-+, *Desmodium oldhami* H-+, *Broussonetia papyrifera* S-+, *Arundinella hirta* H-+, *Ixeris sonchifolia* H-+, *Indigofera kirilowii* H-+, *Fraxinus chinensis* var. *rhynchophylla* S-+, *Isodon excisus* H-+, *Rubus crataegifolius* H-+, *Celastrus orbiculatus* H-+, *Melandrium firmum* H-+, *Patrinia scabiosaefolia* H-+, *Gypsophila odlnamiana* H-+, *Vitis puinpuangularis* H-+, *Sophora flavescens* H-+, *Cynanchum atratum* H-+, *Aster tataricus* H-+, *Sedum tatarinowii* H-+, *Iris dichotoma* H-+, *Sedum bulbiferum* H-+

of the *Stephanandra incisa-Quercus variabilis* community are *Prunus japonica, Quercus serrata, Sorbus alnifolia, Carex ciliata-warginata, Cirsium japonicum, Lilium chindaoensis, Lysimachia barystachys, Meliosma oldhamii, Pinus thunbegii, Sapium japonicum, Stephanandra incisa, Weigera florida* and *Symplocos paniculata*. The structure of this community is usually not complicated and can be divided into three or four layers. The height of the tree layer is 8–17 m and the cover attains 60–70%. The main species of the tree layer are *Quercus variabilis, Pistacia chinensis, Quercus serrata, Meliosma oldhamii* and *Symplocos paniculata*. The shrub layer is 3–4 m with cover from 8% to 30%, and the dominant species include *Prunus japonica, Quercus serrata, Prunus serralata, Sapium japonicum, Stephanandra incisa, Symplocos paniculata, Weigera florida, Quercus variabilis, Pistacia chinensis, Euonymus alatus, Lindera glauca* and *Rhamnus utilis*. The herb layer is less than 0.5 m high and has an average cover of 35%. The predominant species of the herb layer are *Carex lanceolata, Spodiopogon sibiricus, Sanguisorba officinalis, Aster ageratoides, Peucedanum wawrii, Adenophora tetraphylla, Rubia cordifolia, Quercus acutissima, Lysimachia clethroides, Trachelospermum jasminum, Pinus thunbegii* and *Carex siderosticata*. The average number of species is about 42.

2) *Vitex negundo-Quercus variabilis* community

Differential species: *Aconitum henryi*, *Artemisia shangnanensis*, *Parthenocissus himalayana*, *Thalictrum przewalskii*, *Isodon inflexa*, *Akebia trifoliata*, *Amorpha fruticosa*, *Celtis biondii*, *Liriope minor*, *Stephania tetrandra* and *Juglans regia*.

This community was described by Tang et al. (2014) in a synoptic table. Stands of this community are situated on southwest-facing steep slopes (30°–32°) on Yuntaishan Mountain in Jiaozuo city of northern Henan. This community can be found at lower altitudes (from 451 m to 488 m), on poor dry soil, with exposed rocks covering 35–60% of the ground surface. The average number of species in this community is about 46.

3) Quercetum alienae-variabilis association.

Diagnostic species: *Allium anisopodium*, *Atractylodes macrocephala*, *Quercus aliena*, *Viburnum mongolicum*, *Cotinus coggygria* var. *pubescens*, *Phyllanthus glaucus*, *Carex subpediformis*, *Clematis obscura*, *Fraxinus bungeana*, *Lilium concolor*, *Lonicera ferdinandii*, *Sporobolus indicus*, *Zelkova sinica* and *Rhododendron molle*.

This association was described by Tang et al. (2014), with a synoptic table. Stands of this community occur on southwest-facing or southeast-facing gentle slopes (18°–25°), at altitudes from 982 m to 1140 m, in the Yuntaishan Mountains of northern Henan. Exposed rocks cover 40–60% of the surface, and the soil is neutral, with an average pH value of about 7. The average number of species is 48.

4) *Elaeagnus umbellata-Quercus variabilis* community

The *Elaeagnus umbellata-Quercus variabilis* community was found at altitudes from 756 m to 777 m, on gentle slopes (15°–20°), in the Tai-Shan Mountains of Shandong province (see Table 11.7).

This community can be distinguished by *Quercus variabilis*, *Koelreuteria paniculata*, *Elaeagnus umbellata*, *Rhamnus utilis*, *Humulus scandens*, *Celtis koraiensis* and *Agrostis clavata*. The structure of this community comprises four layers, and the average number of species is 37. The canopy layer attains 15 m, with cover of 80–85%, and is dominated mainly by *Quercus variabilis*, mixed with *Celtis koraiensis* and *Pistacia chinensis*. The second tree layer reaches 6–10 m, with cover of 15–20%. Common species of the second tree layer are *Quercus variabilis*, *Koelreuteria paniculata*, *Celtis koraiensis* and *Robinia pseudoacacia*, sometimes mixed with a few individuals of *Diospyros lotus*, *Quercus mongolica* and climber *Ampelopsis brevipedunculata*. The shrub layer is 3 m high, with an average cover of 50%; the main species are *Celtis koraiensis*, *Quercus variabilis*, *Grewia biloba*, *Koelreuteria paniculata*, *Elaeagnus umbellata* and *Rhamnus utilis*. The herb layer has cover of 10–25%, with an average height of 0.5 m; common species are *Quercus variabilis*, *Grewia biloba*, *Commelina communis*, *Cynanchum atratum*, *Koelreuteria paniculata*, *Rhamnus utilis*, *Humulus scandens*, *Agrostis clavata*, *Artemisia annua*, *Viola collina*, *Oplismenus undulatifolius*, *Youngia japonica*, *Clematis kirilowii* and *Solanum lyratum*.

5) *Vitex negundo* var. *heterophylla-Quercus acutissima* community

Table 11.7 *Elaeagnus umbellata-Quercus variabilis* community and *Vitex negundo* var. *heterophylla-Quercus acutissima* community

Running number		1	2	3	4	5	6	7	8
Original relevé number (in field)		ST1	ST2	ST3	ST4	ST5	ST6	ST7	ST8
Relevé data		2008	2008	2000	2008	2008	2008	2008	2008
		7	7	9	7	7	7	7	7
		4	4	16	3	3	3	3	4
Relevé size (m^2)		225	225	600	300	300	150	300	150
Altitude (m)		777	756	320	327	334	340	580	585
Aspect		SW	SE	S	SE	SW	SE	SE	NE
Slope (°)		15	20	20	20	14	5	32	27
Height(m)/cover(%) of Tree layer 1(T1)		15/85	15/80	12/70	11/65	–	–	12/75	–
Height(m)/cover(%) of Tree layer 2(T2)		10/20	6/15	9/5	–	10/75	10/75	6/15	7/75
Height(m)/cover(%) of Shrub layer (S)		2.5/40	3/60	1.5/35	1.5/20	1.5/20	2/10	1.5/25	2/50
Height(m)/cover(%) of Herb layer (H)		0.5/25	0.5/10	0.3/20	0.7/25	0.9/40	0.8/35	0.6/10	0.6/40
Number of Species		42	32	33	37	32	34	37	34
Differential species of *Elaeagnus umbellata-Quercus variabilis* community									
Quercus variabilis	T,S,H	5·4	5·5	.	.	.	.	.	.
Koelreuteria paniculata	T,S,H	1·1	+·2	.	.	.	.	.	.
Elaeagnus umbellata	S	+	1·2	.	.	.	.	.	.
Rhamnus utilis	S,H	+	+	.	.	.	.	.	.
Humulus scandens	S,H	+·2	+	.	.	.	.	.	.
Celtis koraiensis	T,S,H	2·2	4·3	.	+	.	.	.	.
Agrostis clavata	H	+·2	+	.	.	.	.	.	.
Differential species of *Vitex negundo* var. *heterophylla-Quercus acutissima* community									
Quercus acutissima	T,S,H	.	.	4·4	4·4	4·5	5·4	4·4	4·3
Vitex negundo var. *heterophylla*	S,H	+·2	.	3·3	2·2	1·2	2·2	1·1	.
Bidens parviflora	H	.	.	.	+·2	+	+	1·1	+
Spodiopogon sibiricus	H	.	.	.	+	2·2	+·2	1·1	2·2

(continued)

Table 11.7 (continued)

Species	Life form									
Ixeris denticulata	H	.	.	+	.	1·2	+	.	.	+
Dendranthema indicum	H	.	.	.	.	1·1	+	+2	+	.
Melandrium apricum	H	.	.	.	.	+2	+2	+2	+	.
Albizia julibrissin	S,H	.	.	.	.	+	+	+	+	.
Artemisia gmelinii	H	.	.	.	.	1·2	+	+	.	.
Dianthus chinensis	H	.	.	+	.	+2	+	.	.	.
Cleistogenes hancei	H	.	.	.	.	+2	+	1·2	.	.
Themeda triandra var. *japonica*	H	.	.	.	.	+	+2	+2	.	.
Diagnostic species of higher unit of Quercetea variabilis Tang, Fujiwara et You in Box et Fujiwara 2014										
Pistacia chinensis	T,S,H	.	1·1	+	.	+2	+	+	.	.
Diospyros lotus	T,S,H	+	+	.	.	.	.	.	+2	+2
Trachelospermum jasminoides	S,H	.	.	.	.	1·2	1·2	+	.	.
Forsythia suspensa	S	.	.	.	.	.	.	.	+2	3·3
Other species										
Grewia biloba	S,H	1·2	+	.	.	1·2	1·1	+	.	+
Ampelopsis brevipedunculata	T,S,H	.	+	.	.	1·2	.	+2	+	+
Commelina communis	H	+2	+2	+2	.	.	+2	+	+	.
Cynanchum atratum	H	+	+	.	.	.	+2	+	+	.
Allium tuberosum	H	.	+	.	.	.	+	+	+	.
Ampelopsis humulifolia	S,H	.	+	.	.	+2	+2	+2	.	.
Carex lanceolata	T,S,H	.	+	.	+	+2	.	.	+2	+2
Rubia cordifolia	H	+	.	.	.	+	+	+	.	+
Artemisia annua	H	+	+	.	.	.	.	.	.	+
Rhamnus parvifolius	S,H	.	.	.	+	.	.	+	.	+
Aster ageratoides	H	+	.	.	.	.	.	.	+	.
Lespedeza bicolor	S,H	.	.	2·2	.	.	.	.	.	+

Species	Layer								
Euphorbia esula	H	.	.	.	.	.	.	.	+
Vitis amurensis	H	+	.	+	.	.	.	.	.
Adenophora polyantha	H	.	.	.	.	.	.	+	+
Plectranthus inflexus	H	+·2	.	.	.	.	.	.	+·2
Pinellia ternata	H	+	.	.	+	.	.	.	.
Leibnitzia anandria	H	.	.	.	.	.	.	1·1	.
Viola collina	H	+	+	+	.	.	.	.	+
Rubus parvifolius	H	.	.	.	+·2	+	.	.	.
Cocculus trilobus	S,H	+	.	.	+	.	.	+	.
Oplismenus undulatifolius	H	1·2	+	.	.	+	.	.	.
Youngia japonica	H	+	+	+	.	.	.	.	.
Dioscorea opposita	H	1·2	.	.	+·2	+	.	.	.
Pueraria lobata	S,H	.	+	+	.	.	.	.	.
Oxalis corniculata	H	.	.	.	.	+	+	.	.
Thalictrum minus var. hypoleucum	S,H	+	.	.	.	1·1	.	.	1·1
Lespedeza pilosa	S,H	.	.	.	.	.	.	.	.
Robinia pseudoacacia	T,S,H	+·2	+	+	.	+	.	.	.
Bidens bipinnata	H	.	.	.	.	.	+	.	.
Deyeuxia pyramidalis	H	.	.	.	.	.	.	1·1	+
Paraixeris denticulata	H	.	.	.	.	.	+	+	.
Toxicodendron verniciflum	S,H	.	+	.	.	+	.	+·2	.
Clematis kirilowii	H	+	+	.	.	.	+	.	.
Solanum lyratum	H	+	+	.	.	.	.	+	.
Chenopodium ambrosioides	H	+	.	.	+	+	+	.	.
Viola betonicifolia	H	.	.	.	+	.	+	.	.
Ailanthus altissima	S,H	.	.	.	.	+	2·2	.	.
Lespedeza cyrtobotrya	H	.	.	.	.	.	.	1·1	.

(continued)

Table 11.7 (continued)

Dryopteris goeringiana	H	+	.	.	.	.	.	+	.
Setaria viridis	H	.	.	.	+	.	.	+	.

Location: Tai-Shan Mountains and Laiwu County, of Shandong province

Additional species occurring once in relevé reference no.1: *Quercus mongolica* T2-+, S-+, *Geum aleppicum* H-+, *Aster lavandulaefolius* H-+, *Microstegium nudum* H-+, *Capsella bursa-pastoris* H-+, *Prunus davidiana* S-+, *Lonicera japonica* H-+, *Phlomis umbrosa* H-+, *Phytolacca acinosa* H-+, *Armeniaca sibirica* S-+, *Crataegus pinnatifida* S-+, *Vicia unijuga* H-+, *Artemisia sylvatica* H-+; no.2: *Convallaria majalis* H-+·2, *Celastrus angulatus* S-+, *Polygonatum odoratum* H-+, *Achyranthes bidentata* S-+, *Pteroceltis tatarinowii* H-+, *Amaranthus viridis* H-+; no.3: *Cleistogenes suffruticosa* H-1·2, *Athraxon hispidus* H-1·2, *Cynanchum chinense* H-1·1, *Chenopodium album* H-+, *Achnatherum pekinense* H-+, *Rubus crataegifolius* H-+, *Galium aparine* var. *tenerum* H-+, *Chrysanthemum zawadzkii* H-+, *Arundinella hirta* H-+, *Viola mandsurica* H-+, *Ixeris sonchifolia* H-+, *Euphorbia pekinensis* H-+, *Allium senescens* H-+, *Zanthoxylum xanthoxyoides* H-+, *Artemisia anomela* H-+, *Diospyros kaki* var. *sylvestris* S-+, *Ulmus pumila* H-+, *Kalimeris incise* H-+, *Scorzonela glabra* H-+, *Setaria glauca* H-+; no. 4: *Andrachne chinensis* H-+·2, *Gypsophila oldhamiana* H-+, *Desmodium racemosum* H-+; no.5: *Impatiens nolitangere* H-1·2, *Aristolochia mollissima* H-1·2, *Cleistogenes* sp. H-+·2, *Broussonetia papyrifera* S-+, *Ulmus parvifolia* H-+, *Kalimeris indica* H-+, *Taraxacum mongolicum* H-+; no.6: *Chimaphila ambrosioides* H-+, *Artemisia argyi* H-+, *Aneurolepidium dasystachys* H-+, *Arthraxon hispidus* H-+, *Campylotropis macrocarpa* H-+; no.7: *Sophora japonica* T1-+, T2-1·1, S-+, *Viola philippica* ssp. *munda* H-1·1, *Carpesium cernuum* H-+, *Lespedeza floribunda* S-+, *Fraxinus rhynchophylla* H-+, *Plectranthus glaucocalyx* H-+, *Viola prionantha* H-+, *Ixeris polycephala* H-+, *Lespedeza tomentosa* H-+; no.8: *Deyeuxia* sp. H-2·2, *Carex humilis* var. *nana* H-1·2, *Quercus dentata* T1-1·1, *Viola coreana* H-+·2, *Spiraea fritschiana* H-+·2, *Deutzia grandiflora* S-+, *Dioscorea nipponica* H-+, *Peucedanum wawrii* H-+, *Aster* sp. H-+, *Carex siderosticata* H-+, *Artemisia eriopoda* H-+, *Sanguisorba officinalis* H-+, *Euonymus alatus* H-+, *Potentilla fragarioides* H-+, *Pinus densiflora* T1-+, *Miscanthus sinensis* H-+, *Eupatorium fortunei* H-+

The stands of *Vitex negundo* var. *heterophylla-Quercus acutissima* community were at altitudes from 320 m to 585 m, on slopes from 5° to 32°, in the Tai-Shan Mountains and Laiwu county of Shandong province (Table 11.7).

The differential species of this community are *Quercus acutissima, Vitex negundo* var. *heterophylla, Bidens parviflora, Spodiopogon sibiricus, Ixeris denticulata, Dendranthema indicum, Melandrium apricum, Albizia julibrissin, Artemisia gmelinii, Dianthus chinensis, Cleistogenes hancei* and *Themeda triandra* var. *japonica*. The community is found in plantations, with a structure that consists mostly of three layers. The height of the canopy is 7–12 m, with an average cover of about 72.5%. The tree layer is dominated mainly by *Quercus acutissima*, sometimes mixed with *Sophora japonica, Quercus dentata* and *Pinus densiflora*. The shrub layer is 1.5–2 m high and has cover of 10–50%. The main species of the shrub layer are *Vitex negundo* var. *heterophylla, Ampelopsis brevipedunculata, Grewia biloba* and *Quercus acutissima*. The average height of the herb layer is about 0.7 m, and its cover ranges from 10% to 40%. The average number of species is 35.

11.3.2 Geographical Differentiation

In the DCCA, the species-environment correlation coefficients for first four axes are 0.989, 0.957, 0.906 and 0.870 (Fig. 11.3). The first two axes explained 38.6% of the total variance, and the correlation coefficient between the two axes is low ($r = 0.010$). Therefore, the general pattern of environmental variation among communities of *Pteroceltis tatarinowii* forest is expressed most clearly by the combination of the first two axes. Axis 1 is most closely correlated with altitude ($r = 0.831$), soil pH ($r = -0.644$), latitude ($r = -0.553$), longitude ($r = -0.390$), slope ($r = -0.317$) and percentage of exposed rock ($r = -0.259$), reflecting geographical conditions. Axis 2 is most closely connected with sunshine duration ($r = 0.507$), mean annual temperature ($r = -0.333$), aspect ($r = 0.323$) and average annual precipitation ($r = 0.286$), indicating climate conditions.

In addition, the geographical distribution of the *Pteroceltis tatarinowii* forests is different, due mainly to environmental temperature and humidity, as represented by the variables mean annual temperature, average annual precipitation and slope aspect. The samples of Isodo rubescens-Pteroceltido tatarinowii fall mainly in the lower left, indicating that the association occurs usually in habitats with relatively low temperature and humidity. The samples of the *Platycladus orientalis-Pteroceltis tatarinowii* community and the Pteroceltido tatarinowii-Quercetum variabilis are mostly concentrated in the lower right, showing that the two communities generally grow in relatively warm, humid environments. The distribution of *Pteroceltis tatarinowii* forests is also influenced by soil conditions. Generally, Isodo rubescens-Pteroceltido tatarinowii and the *Platycladus orientalis-Pteroceltis tatarinowii* community grow on poor alkaline soils (average soil pH 7.6–8.1), but Pteroceltido tatarinowii-Quercetum variabilis appears on fertile, nearly neutral soils (average pH 7.1).

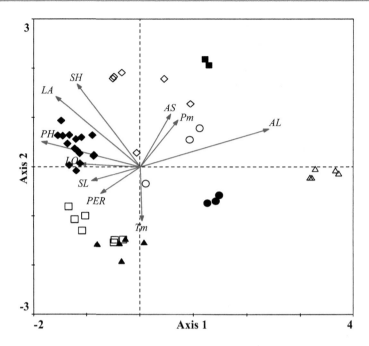

Fig. 11.3 DCCA ordination diagram. The symbols represent the individual samples: black diamonds = Isodo rubescens-Pteroceltido tatarinowii, squares = the *Platycladus orientalis-Pteroceltis tatarinowii* community, black triangles = Pteroceltido tatarinowii-Quercetum variabilis, diamonds = the *Viex negundo* var. *heterophylla-Quercus acutissima* communigy, black squares = the *Elaeagnus umbellata-Quercus acutissima* communigy, circles = the *Stephanandra incisa-Quercus variabilis* community, black dots = the *Vitex negundo-Quercus variabilis* community, and triangles = Quercetum alienae-variabilis association

11.4 Discussion

According to the Chinese zonation system, the zonal vegetation of the warm-temperate zone is deciduous broad-leaved forest, in which deciduous oak forest is thought to be the representative, typical forest type (Wu and Committee 1980; Tang et al. 2009). The deciduous oak forest is mainly dominated by *Quercus* species, including *Quercus mongolica*, *Q. wutaishanensis*, *Q. dentata*, *Q. aliena* var. *acutiserrata*, *Q. aliena*, *Q. variabilis* and *Q. acutissima* (Gao et al. 2001). However, these *Quercus* forests show different vertical distribution ranges and are mostly secondary forests. *Quercus mongolica* is a cold-resistant species, and its southern distributional boundary was recorded in the Yuntaishan Mountains in Lianyungang of Jiangsu (Wu and Committee 1980); *Quercus mongolica* forest is mostly found in the northern part of the warm-temperate zone. Both *Quercus wutaishanensis* forest and *Quercus aliena* forest occur widely at high altitude; *Quercus aliena* var. *acutiserrata* forest grows more often at altitudes of 1100–1800 m in the southern part of the area. *Quercus variabilis* forest, *Quercus dentata* forest and *Quercus acutissima* forest usually grow at low altitudes, and these oak forests were destroyed by human activities. Even *Quercus dentata* forest almost disappeared in the area (Xie and Chen 1994; Gao and Chen 1998; Gao et al. 2001).

In our field investigation, *Pteroceltis tatarinowii* in the deciduous-forest zone of eastern China is always found in *Quercus variabilis* forest. Tang et al. (2014) had determined the vegetation unit of *Quercus variabilis* forest (including *Quercus acutissima* forest) in this region to be a class of deciduous forests. Through comparing *Pteroceltis tatarinowii forest* with *Quercus variabilis* forest, we found that most diagnostic species of higher units of the Quercetae variabilis also appeared in *Pteroceltis tatarinowii* forest, e.g. *Pistacia chinensis*, *Koelreuteria paniculata*, *Quercus variabilis*, *Grewia biloba* var. *parviflora*, *Broussonetia papyrifera*, *Diospyros lotus*, *Trachelospermum jasminoides*, *Forsythia suspense*, *Toxicodendron succedaneum*, *Speranskia tuberculata*, *Vitex negundo* var. *heterophylla* and *Carpinus turczaninowii*. Therefore, *Pteroceltis tatarinowii* forest can be recognized as a basic unit of the Quercetae variabilis. The *Pteroceltis tatarinowii* forest is different from other units of the Quercetae variabilis and can be distinguished by the species *Pteroceltis tatarinowii*, *Morus australis*, *Cynodon dactylon*, *Liriope spicata*, *Aristolochia debilis*, *Chrysanthemum indicum*, *Viola selkirkii*, *Paederia scandens*, *Pinellia ternata*, *Arthraxon hispidus*, *Vitex negundo* var. *cannabifolia* and *Cudrania tricuspidata*. In addition, the *Pteroceltis tatarinowii* forest of this region contains at least two associations and one community. These species are characteristic in *Pteroceltis tatarinowii* forest, so the Pteroceltion tatarinowii was established in this paper.

Pteroceltis tatarinowii forest in the study area occurred on limestone mountains, and Chen et al. (2011) showed that the soil of *Pteroceltis tatarinowii* forest in the area is alkaline, with a calcium content that can reach 10,000 mg/kg. The *Pteroceltis tatarinowii* communities show significant differences in constituent species. On the one hand, in this study, 200 species were recorded in 28 samples of *Pteroceltis tatarinowii* forest, but 147 species (73.5%) appeared in fewer than 20% of the samples, and 56 species (28%) appeared in only one sample. This indicates that most constituent species of *Pteroceltis tatarinowii* forest are unable to adapt to all properties of limestone soils (e.g. barren, dry) and cannot occur widely in the forests. On the other hand, some drought-resistant species, which are found commonly on slopes with much exposed rock, sunny hillsides and ridges, limey soils and so on, can be found in *Pteroceltis tatarinowii* forest. For example, the species *Gleditsia heterophylla*, *Dalbergia hupeana* and *Isodon rubescens* are often found on rocky hillsides; the species *Sageretia paucicostata*, *Ziziphus jujuba*, *Ligustrum quihoui*, *Cornus walteri* and *Cudrania tricuspidata* grow generally on sunny slopes or arid shrublands; and the species *Morus australis*, *Pteroceltis tatarinowii* and *Toxicodendron succedaneum* occur commonly on limestone cliffs or slopes or lowlands. Of course, in shady valleys, lowlands or on shady slopes, the species *Arisaema erubescens*, *Clematis heracleifolia*, *Lonicera pekinensis*, *Arthraxon hispidus*, *Carpinus turczaninowii* and *Myripnois dioica* can often be found in *Pteroceltis tatarinowii* forest. Therefore, *Pteroceltis tatarinowii* can adapt very well to the habitat of high calcium and droughty, barren limey soils, and can grow and become an important species of forests on limestone mountains. Most constituent species of *Pteroceltis tatarinowii* forests, however, can only adapt to some of these habitats on limestone mountains.

The structure of *Pteroceltis tatarinowii* forest is not very complicated and can usually be divided into three or four layers; most of the forests are relatively short and reach only about 10 m in height. *Pteroceltis tatarinowii* individuals occur usually in all layers and play an important role in each layer, showing that the size of this population in the forest is large and its age structure is relatively reasonable. Especially, *Pteroceltis tatarinowii* seedlings occur in most samples (except the *Vitex negundo-Quercus variabilis* community in Qingtan temple in Shandong), indicating that the natural regeneration of *Pteroceltis tatarinowii* forest is good. The number of seedlings within the forest is not large, though, and there are many sprouts of clonal ramets growing thickly in the forests. Studies show that the natural regeneration ability of *Pteroceltis tatarinowii* is weak, but the ability of its clonal ramets is strong. This, if the top of *Pteroceltis tatarinowii* individuals is damaged (due to natural or anthropogenic disturbance) or growth is inhibited, its adventitious buds can be activated immediately (Zhang et al. 2008). This also explained why the *Pteroceltis tatarinowii* forests in this study are mostly short and only reach about 10 m.

11.5 Conclusion

The main result is that *Pteroceltis tatarinowii* forest in eastern China can be recognized as a new alliance, the Moro australis-Pteroceltidion tatarinowii, in the class Quercetea variabilis. Two associations, Isodo rubescens-Pteroceltido tatarinowii ass. nov. and Pteroceltido tatarinowii-Quercetum variabilis Tang, Fujiwara et You in Box et Fujiwara 2014, and a *Platycladus orientalis-Pteroceltis tatarinowii* community, are summarized in this new alliance.

The different geographical distributions of *Pteroceltis tatarinowii* forest and *Quercus* forests are mainly the result of altitude, slope, soil pH and the rockiness (exposed rock) of the environment. In our investigation, the distribution range of *Pteroceltis tatarinowii* forest is narrower than that of *Quercus* forests (including *Quercus variabilis* forest and *Quercus acutissima* forest). *Pteroceltis tatarinowii* forest is often found on limestone mountains and grows in relatively harsh habitats, at low altitudes (below 500 m), on steep cliffs or hillsides (average slope about 40°), on dry slopes with much exposed rock (average area 63%), and on calcium-rich limy soils (pH ranging from 7 to 8.3). *Quercus* forests, on the other hand, are more likely to grow in at higher altitudes (320–1150 m), on gentler slopes (average about 22.5°), and on relatively fertile soil with less exposed rock (average 39% in this study). In addition, the different geographical distributions among *Pteroceltis tatarinowii* forests are influenced by environmental temperature and humidity, as expressed by mean annual temperature, average annual precipitation and slope aspect. As temperature increases and humidity decreases (from north to south), the community types of *Pteroceltis tatarinowii* forest in the study area changed from Isodo rubescens-Pteroceltido tatarinowii to *Platycladus orientalis-Pteroceltis tatarinowii* community, and to Pteroceltido tatarinowii-Quercetum variabilis.

The constituent species of different *Pteroceltis tatarinowii* communities differ significantly, due mainly to different adaptation abilities of the species. Usually, *Pteroceltis tatarinowii* adapts well and can grow in dry, rocky, poor soils on limestone mountains. The companion species of *Pteroceltis tatarinowii* communities, however, are mostly unable to tolerate these unfavorable conditions and can only grow in some of the habitats, resulting in different communities. The number of species occurring in each sample of *Pteroceltis tatarinowii* forest is not large, averaging about 34; 28% of the species appeared in only one sample, and 73.5% appeared in fewer than 20% of the samples.

Most *Pteroceltis tatarinowii* forests are not tall (about 10 m), and the community structure is simple and can usually be divided into three or four layers. The natural regeneration of *Pteroceltis tatarinowii* is good, and there are many *Pteroceltis tatarinowii* individuals, including seedlings and clonal ramets growing in the forest understoreys.

Acknowledgements We are grateful to Prof. E. O. Box (University of Georgia) for editing this manuscript.

References

Braun-Blanquet J (1964) Pflanzensoziologie, Grundzüge der Vegetationskunde, 3rd edn. Springer, Berlin, p 631

Chen Y-K, Cheng Y-R, Li J-H, Zhang H, Lu C, Zhang X-P (2011) Analysis of the community types of the endangered plant *Pteroceltis tatarinowii* Maxim. J Biol 28(6):1–5. (in Chinese with English abstract)

Ellenberg H (1956) Grundlagen der Vegetationsgliederung. Part 1: Aufgaben und Methoden der Vegetationskunde. In: Walter H (ed) Einführung in die Phytosoziologie, vol 4. Eugen Ulmer, Stuttart, p 136

Fu S-L, Li H-K (1997) Community characteristics of natural *Pteroceltis tatarinowi* forests with reference of prospects for development. Econ For Res 15(1):13–15. (in Chinese with English abstract)

Fu L-K, Committee (eds) (2001) Higher plants of China, vol 4. Qingdao Publishing House, Qingdao, p 13. (in Chinese with Latin-Chinese species lists)

Fujiwara K (1987) Aims and methods of phytosociology or "vegetation science". In: Plant ecology and taxonomy to the memory of Dr. Satoshi Nakanishi. Kobe Geobotanical Society, Kobe, pp 607–628

Gao X-M, Chen L-Z (1998) Studies on the species diversity of *Quercus liaotungensis* communities in Beijing Mountains. Acta Phytoecologica Sinica 22(1):23–32. (in Chinese with English abstract)

Gao X-M, Ma K-P, Chen L-Z (2001) Species diversity of some deciduous broad-leaved forests in the warm-temperate zone and its relations to community stability. Acta Phytoecol Sin 25 (3):283–290. (in Chinese with English abstract)

Ge J-W, Wu J-Q, Zhu Z-Q, Yang J-Y, Lei Y (1998) The present status and in-situ conservation of the rare and endangered plants in Hubei province. Chin Biodivers 6(3):220–228. (in Chinese with English abstract)

Qian H, White PS, Klinka K, Chourmouzis C (1999) Phytogeographical and community similarities of alpine tundras of Changbaishan Summit, China, and Indian Peaks, USA. J Veg Sci 10:869–882

Tan W-G, Wei G-F, Tan W-N (2004) Study on the community characteristics and its diversity of *Pteroceltis tatarinowii* forest in Mulun natural reserve of Guangxi. Guangxi Forestry Sci 33 (3):126–129. (in Chinese)

Tang Q, Fujiwara K, You H-M (2009) Phytosociological study of deciduous *Quercus* forest in the warm-temperate zone of China: primary study of different kinds of *Quercus* communities. Hikobia 15(3):311–322

Tang Q, Fujiwara K, You H-M (2014) Phytosociological study of *Quercus variabilis* forest in warm-temperte China. In: Box EO, Fujiwara K (eds) Warm-temperate deciduous forests around the northern hemisphere, Geobotanical Studies. Springer, Cham

Wang W-J, He Y-Q (2001) Studies on the structural feature and species diversity of *Pteroceltis tatarinowii* forest in Baotianman national nature reserve. J Henan Agric Univ 35(4):364–367. (in Chinese with English abstract)

Wang D-P, Li J-H, Tian C-Y, Liu R-Y (2011) Structure and species diversity of wild *Pteroceltis tatarinowii* community in Daguisi national forest park of Hubei province. J Wuhan Inst Techol 33(6):50–55. (in Chinese with English abstract)

Wu Z-Y, Committee (1980) Zhongguo Zhibei [Vegetation of China]. Science Press, Beijing, p 1375. (in Chinese)

Xie J-Y, Chen L-Z (1994) Species diversity characteristics of deciduous forests in the warm-temperate zone of North China. Acta Ecol Sin 14(4):337–344. (in Chinese with English abstract)

Yang C-H, An H-P, Fang X-P (1995) The rare tree species *Pteroceltis tatarinowii*. Guizhou Forestry Sci Technol 23(2):8–11. (in Chinese)

You H-M, Yu F-Z, Yan C-H (2012) Community characteristics of *Pteroceltis tatarinowii* forests in Huangcangyu natural reserve of Anhui province. J Cent South Univ Technol 32(5):86–91. (in Chinese with English abstract)

Zhang C-H, Zheng Y-Q, Zong Y-C, Wu C, Zheng J, Jiao M, Xue X-H (2008) Morphological variation among different sources of *Pteroceltis tatarinowii* in warm-temperate zone. For Res 21(5):737–741. (in Chinese with English abstract)

Zhi Y-B, Yang C, Wang Z-S, An S-Q, Wang Z-L, Li H-L, Su Z-A, Wang Q (2008) The endangered chaacteristics and mechanism of the endemic relict shrub *Tetraena mongolica* Maxim. Acta Ecol Sin 28(2):767–776. (in Chinese with English abstract)

Zhou L, Gu J-Z, Song Y-F, Kuang R-P, Li X-J, Liu K-M (2012) Study on the Community of *Pteroceltis tatarinowii* in the Limestone Mountainous of Jiangyong County of Hunan Province. Life Sci Res 16(5):382–388. (in Chinese with English abstract)

Zhu K-Z (ed) (1984) Physical geography of China. Science Press, Beijing, p 161. (in Chinese)

12. Vegetation Ecology of *Sphagnum* Wetlands in Subtropical Subalpine Regions: A Case Study in Qi Zimei Mountains

Ting-Ting Li, Zheng-Xiang Wang, and Yun Lei

Abstract
Vegetation of *Sphagnum* moss wetlands in the Qi Zimei Mountains was surveyed by phytosociological methods (Braun-Blanquet, Pflanzensoziologie, Grundzuge der Vegetationskunde, vol 3. Springer, p 631, 1964) in May and July 2012. Results show that there are 67 families, 128 genera and 196 species (including varieties) of higher plants. Among them, there are 2 families, 2 genera and 2 species of bryophytes, 6 families, 7 genera and 8 species of pteridophytes, 59 families, 119 genera and 186 species of spermatophytes. Rosaceae exhibits the highest richness, with 25 plant species in total. Vegetation of *Sphagnum* wetlands in Qi Zimei Mountains was classified into eight associations: (1) Assn. *Juncus setchuensis* + *Carex taliensis–Sphagnum palustre* ssp. *palustre* (2) Assn. *Juncus setchuensis* + *Lycopus coreanus* var. *cavaleriei–Sphagnum palustre* ssp. *palustre* (3) Assn. *Carex taliensis–Sphagnum palustre* ssp. *palustre* (4) Assn. *Polygonum thunbergii–Sphagnum palustre* ssp. *palustre* (5) Assn. *Malus hupehensis–Carex filicina* var. *meiogyna–Sphagnum palustre* ssp. *palustre* (6) Assn. *Rhododendron auriculatum–Pteridium aquilinum* var. *latiusculum–Sphagnum palustre* ssp. *Palustre* (7) Assn. *Acorus calamus–Sphagnum palustre* ssp. *palustre* (8) Assn. *Enkianthus chinesis–Sinarundinaria nitida–Carex taliensis–Sphagnum palustre* ssp. *palustre* . The appearance of the vegetation types vary across seasons. Further research on biodiversity and ecological restoration of *Sphagnum* mire in western Hubei Province is ongoing.

T.-T. Li • Z.-X. Wang (✉)
School of Resources and Environmental Science, Hubei University, Wuhan, China
e-mail: wangzx66@hubu.edu.cn

Y. Lei
School of Life Sciences, Central China Normal University, Wuhan, China

© Springer International Publishing AG, part of Springer Nature 2018
A.M. Greller et al. (eds.), *Geographical Changes in Vegetation and Plant Functional Types*, Geobotany Studies,
https://doi.org/10.1007/978-3-319-68738-4_12

Keywords
Vegetation • *Sphagnum* wetlands • Qi Zimei Mountains • Biodiversity

12.1 Introduction

Sphagnum-dominated wetland develops in wet and cold weather conditions (Bai et al. 1999, 2004). The harsh weather and geological conditions limit its geographical distribution, and in the subtropical zone, large area of *Sphagnum* mire is uncommon (Braun-Blanquet 1964; Fang et al. 2007). In 2005, our team found a large area of *Sphagnum* mire when conducting vegetation investigations in the Qi Zimei Mountains of Hubei China. It is well preserved and has unique natural scenery as well as important ecological value (Gorham 1991). Since then, further observations and studies on this *Sphagnum* mire were carried out annually by our research team and the results published in a series of articles (Lally et al. 2012; Lin et al. 2002; Ma et al. 2008; Mao et al. 2009).

Qi Zimei Mountains National Nature Reserve is located on the east of Xuan'en County in En'shi City, Hubei Province. The geographical location ranges from 109°37′ to 109°51′ E longitude, and 29°39′ to 30°5′N latitude, covering a total area of 551.8 km^2. The climate of Qi Zimei Mountains National Nature Reserve is mid-Humid, Subtropical Monsoon; and there is a gradient of elevation. Average annual temperature of the low mountain belt, with altitude less than 800 m, is 15.8 °C; average annual precipitation is 1491.3 mm; 294 days are freeze-free; and there are 1136.2 h of annual sunshine. In "alpestrine" belt, where altitude ranges from 800 to 1200 m, the average annual temperature and precipitation are 13.7 °C and 1635.3 mm, respectively; 263 days are freeze-free; and there are 1212.4 h of annual sunshine. In the high mountain belt where the altitude reaches more than 1200 m, the average annual temperature and precipitation are respectively 8.9 °C, 1876 mm; with 203 days freeze-free; and there are 1519.9 h of annual sunshine. Natural vegetation at Qi Zimei Mountains is dominated by mixed evergreen-deciduous, broad-leaved forests that possess the typical mid-subtropical montane vegetation forms. It is a tertiary relict forest ecosystem, sheltering epibiotic plants.

Qi Zimei Mountains *Sphagnum*-dominated wetland is located at Shai'ping of Chun Muying in Qi Zimei Mountains National Nature Reserve. Altitudes range from 1650 to 1950 m and wetlands are distributed in patches. Combining remote sensing image with field investigation, we have found that there are 60 patches of *Sphagnum* wetland with a total area of 30.64 hectares. The biggest of these patches, is 2.07 hectares. Surrounding there is a transition around every patch from shrubland to the larger area of *Larix kaempferi* forests which were manually planted in early 1970s. Average dbh of *Larix kaempferi* is 20 cm and average height is 18 m, with a canopy density of about 70%.

12.2 Investigation Methods

12.2.1 Plant Species Identification Methods

An integrated survey of *Sphagnum* wetlands in the Qi Zimei Mountains was carried out in May and July of 2012. Specimens were collected along the main patchy belts and identified in the field. In addition, species and communities were located and photos taken. Combining data from 2006 to date, plant species in Qi Zimei Mountains *Sphagnum* wetlands could be summarized.

12.2.2 Vegetation Investigation Methods

To assess the specificity of vegetation in the *Sphagnum* wetlands of the Qi Zimei Mountains, vegetation was surveyed by phytosociological methods (Mcneil and Waddington 2003). We selected a couple of homogeneous quadrats of pre-determined area (area of tree quadrat >200 m^2, area of shrub quadrat >25 m^2, area of herb quadrat >4 m^2) according to different topographies and altitudes in different sites. Vegetation was divided into several layers, namely tree layer (T1), tree sublayer (T2), shrub layer (S), herb layer (H) and moss layer (M), respectively. Plant species in each layer were recorded along with their cover, abundance, locations (longitude and latitude), topographies, aspects, gradients, altitudes, edaphic and geological conditions, wind strength, disturbance, etc. Community composition tables were completed based on data from field vegetation investigation. Moreover, coenotype was determined, as well as character species and dominant species.

12.2.3 Association Type Nomenclature

Dominant species nomenclature was adopted. Dominant species of each association were used as the name of an association: a series of dominant species from upper to lower according to the vegetation stratum. Dominant species in the same layer were connected by "+", while different layers were connected by "–".

12.3 Results and Analysis

12.3.1 Plant Species Composition in Qi Zimei Mountains *Sphagnum* Wetlands

Field investigation shows that there are 67 families, 128 genera and 196 species (including varieties) of higher plants. Among them, there are 2 families, 2 genera and 2 species of bryophytes, 6 families, 7 genera and 8 species of pteridophytes, 59 families, 119 genera and 186 species of spermatophytes. Among 67 families,

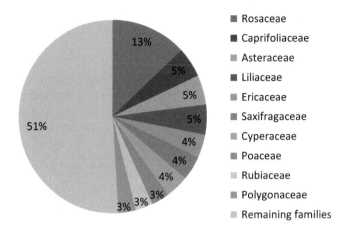

Fig. 12.1 Percentage of species of the top ten families in total plant species in Qi Zimei Mountains *Sphagnum* wetlands

Rosaceae has the most taxa;, there are 25 species in total. Some other families have 5 or more species, namely Caprifoliaceae (10 species), Asteraceae (9 plant species), Liliaceae (9 plant species), Ericaceae (8 plant species), Saxifragaceae (7 species), Cyperaceae (7 species), Poaceae (6 species), Rubiaceae (5 species), Polygonaceae (5 species). The above-mentioned ten families almost constituted 70% of the total species (Fig. 12.1).

Among 193 species, some have higher degree of dominance (cover) and develop into synusia. In the tree layer, the common species are: *Enkianthus chinensis, Rhododendron auriculatum*, and *Salix wallichiana*. In the shrub layer, *Sinarundinaria nitida* and *Malus sieboldii* have highest dominance, The herb layer has more species of high cover, including *Juncus setchuensis, Lycopus coreanus, Viola verecumda, Carex taliensis, C. fargesii, Pteridium aquilinum* var. *latiusculum, Acorus calamus, Scirpus lushanensis, Oenanthe dielsii* var. *stenophylla, Polygonum thunbergii, Polygonum sieboldii, Luzula effusa, Carex filicina* var. *meiogyna, Scolochloa festucacea*. In the moss layer, *Sphagnum palustre, Lycopodium japonicum*, and *Ceratodon purpureus* are dominant species.

12.3.2 Vegetation Classification of Qi Zimei Mountains *Sphagnum* Wetlands

Based on the methods of Braun-Blanquet, species in each quadrat, dominance and abundance of each species, and average height were recorded. Vegetation can be divided into eight associations in terms of species composition.

Assn. *Juncus setchuensis* + *Carex taliensis-Sphagnum palustre* ssp. *palustre*
This association is located at Luo Er'ping with an altitude of 1807 m. Species such as *Rhododendron auriculatum, Enkianthus chinensis, Cotoneaster bullafus*, and

Litsea cubeba are distributed in the quadrat sporadically, but do not develop into synusia. Herb layer is about 60 cm in height, and has 90% cover. *Juncas setchuensis* possesses the highest dominance of all herbs, with 65% cover. Other dominant species are *Carex taliensis, Luzula effusa, Viola verecunda, Pteridium aquilinum,* and *Oenanthe dielsii. Sphagnum* has a depth of 15 cm, and approximately 100% cover. Besides, a small pool was found in this surveyed area, and small areas of *Brasenia schreberi* was found here.

Assn. *Lycopus coreanus* var. *cavaleriei* + *Juncas setchuensis–Sphagnum palustre* ssp. *palustre*

This association is located at Luo Er'ping, at an altitude of 1805 m. Quadrat observations show that *Malus hupehensis, Cotoneaster bullafus, Rosa helenae, Viburnum erubescens* var. *gracilipes, Litsea pungens, Populus lasiocarpa, Lyonia ovalifolia* var. *elliptica* are distributed in the quadrat sporadically, but do not develop into synusia. Herb layer is about 50 cm in height, and has more than 95% cover. *Lycopus coreanus* var.*cavaleriei* and *Juncus setchuensis* are co-dominant species of the community with a total cover of over 80%, in which *Lycopus coreanus* var.*cavaleriei* has a slightly higher dominance. Average height of these two species are about 25 cm and 50 cm, respectively. Main accompanying species are *Scirpus lushanensis, Oenanthe dielsii* var. *stenophylla, Fragaria orientalis, Lysimachia stenosepala, Hosta ventricosa, Lycopodium obscurum, Ligularia intermedia, Lonicera acuminata, Viola verecunda,* etc.. *Sphagnum* has a depth of 15 cm, with approximately 100% cover

Assn.*Carex taliensis–Sphagnum palustre* ssp. *palustre*

This association is located at Shi'pai with altitude of 1796 m. Shrub layer is of 2 m in height, dominated by *Sinarundinaria nitida* and *Rhododendron auriculatum*. The height and cover of the herb layer are about 40 cm and 75%, respectively. *Carex taliensis* is dominant and has a cover of about 60%. Main companion species are *Pteridium aquilinum* var. *latiusculum* and *Lycopodium japonicum*. The depth of *Sphagnum* layer is 15 cm and cover approaches 100%.

Assn.*Polygonum thunbergii–Sphagnum palustre* ssp. *palustre*

This association is located at Diao Baogou (N30°01'28.74", E109°45'40.13") in Chun Muying, at an altitude of 1824 m. Some trees and shrubs occur in the quadrat occasionally; *Agapetes lacei* has a high dominance. Herb layer has a greater number of species, and *Polygonum thunbergii* is dominant. Main companion species are *Oenanthe dielsii* var. *stenophylla, Carex cinerascens, Hydrocotyle wilsonii, Oxalis corniculata, Luzula effusa, Gentianopsis scabromarginata,* and *Aletris spicata*. Cover of the *Sphagnum* layer reaches 75%.

Assn. *Malus hupehensis–Carex filicina* var. *meiogyna–Sphagnum palustre* ssp. *palustre*

This association is located at Da Ping (N30°01'21.46", E109° 45'23.94") in Chun Muying at an altitude of 1863 m. Trees and shrubs have developed into synusia. Tree layer is dominated by *Malus hupehensis. Corylus ferox* var. *thibetica, Corylus heterophylla* var. *sutchuenensis, Lyonia ovalifolia* var. *elliptica* and

Hydrangea strigosa are common species. Herb layer is dominated by *Carex filicina* var. *meiogyna*. Main companion species are *Polygonum thunbergii*, *Carex taliensis*, *Luzula effusa*, *Pteridium aquilinum* var. *latiusculum*, *Galium aparine* var. *tenerum*, *Hosta ventricosa*, *Pyrola calliantha*, *Carex doniana*, *Lycopodium obscurum*, *Juncus bufonius* and so on. Cover of *Sphagnum* layer has reached over 95%.

Assn. *Rhododendron auriculatum–Pteridium aquilinum* var. *latiusculum–Sphagnum palustre* ssp. *palustre*

This association is located at Da Ping (N30°01′13.20″, E109° 45′ 14.30″) in Huo Shaobao with altitude of 1863 m. Tree layer has formed a noticeable synusia with dominating *Rhododendron auriculatum*, of which the cover is 60%. *Malus hupehensis* and *Enkianthus chinensis* are the main companion species. Shrub layer is sparse. *Pteridium aquilinum* var. *latiusculum* and *Lycopodium obscurum* are dominant species in herb layer. In addition, there are some other common species including *Juncus setchuensis*, *Polygonum thunbergii*, *Deyeuxia hakonensis*, *Galium aparine* var. *tenerum*, *Osmunda japonica*, *Lysimachina punctatilimba*, *Cyclosorus acuminatus*, *Oenanthe dielsii* var. *stenophylla*, *Lycopus coreanus*, *Carex taliensis*, *Parathelypteris japonica*, *Polygonum alatum* etc. The *Sphagnum* layer is well-developed with 85%.cover.

Assn. *Acorus calamus–Sphagnum palustre* ssp. *palustre*

This association is located at Niu Siyanwu (N 29°58′54.81″, E109°45′55.77″) in Chun Muying at an altitude of 1752 m. Trees and shrubs don't form synusia. Common tree species are *Salix wallichiana*, *Malus sieboldii* and *Rhododendron auriculatum*. Herb layer has a greater number of species; *Acorus calamus* and *Scolochloa festucacea* are dominant. Main companion species in this layer are *Lycopus coreanus*, *Parathelypteris japonica*, *Gentianopsis paludosa* var. *ovatodeltoidea*, *Oenanthe dielsii* var. *stenophylla*, *Polygonum sieboldii*, *Ceratodon purpureus*, *Carex taliensis*, *Astilbe chinensis*, *Polygonum alatum*, *Pteridium aquilinum* var. *latiusculum* etc. The *Sphagnum* layer is well-developed, with 85% cover.

Assn. *Enkianthus chinensis–Sinarundinaria nitida–Carex taliensis–Sphagnum palustre* ssp. *palustre*

This association is located at Tian Huaban (N 30°00′57.84″, E109°44′39.24″) in Chun Muying at an altitude of 1963 m. Tree layer is dominated by *Enkianthus chinensis*, *Rhododendron sutchuenense*, *Rhododendron stamineum*, *Lyonia ovalifolia* var. *elliptica* and *Corylus ferox* var. *thibetica*. In shrub layer, *Sinarundinaria nitida* possesses a cover of 65. Other species, however, are sparse. *Carex taliensis* is the dominating species in herb layer with 20% cover. Companion species are rare. *Sphagnum* layer develops well, with 85% cover.

12.4 Conclusion and Outlook

12.4.1 Conclusion

(1) Qi Zimei Mountains *Sphagnum* wetlands possess a high biodiversity. Our survey shows that there are 196 species (including varieties) of higher plants from 128 genera and 67 families. Rosaceae, Caprifoliaceae, Asteraceae, Liliaceae, Ericaceae, Saxifragaceae, Cyperaceae, Poaceae, Rubiaceae and Polygonaceae have the highest species richness.
(2) As for the vegetation in Qi Zimei mountains *Sphagnum* wetlands, three tree species (*Enkianthus chinesis*, *Rhododendron auriculatum*, *Malus hupehensis*) have developed into synusia in small areas. *Sinarundinaria nitida* forms a shrub layer in some high altitude regions. Meanwhile, herbs like *Juncus setchuensis, Lycopus coreanus, Carex taliensis, Pteridium aquilinum* var. *latiusculum, Oenanthe dielsii* var. *stenophylla, Polygonum thunbergii,* and *Carex filicina* var. *meiogyna* tend to develop into an herb layer, which was conducive to *Sphagnum* development in the moss layer. In flooded areas, *Acorus calamus* and *Scirpus triangulates* make up the herb layer. The *Sphagnum* layer here does not develop well.
(3) Qi Zimei mountains *Sphagnum* wetlands are rich in coenotypes, which can be recognized as eight associations. In some regions, *Sphagnum* mires are in an oligotrophic state and have developed into apparent *Sphagnum* hummock. While in other regions, *Sphagnum* mires are at medium nutrient level with an herb layer dominant.In addition, on some upper slopes, trees and shrubs occur sparsely, grow poorly, and are typically short even though they can survive. Poor growth may be related to the limitation of acidic wet soil conditions in *Sphagnum* mires. On the whole, *Sphagnum* mires in Qi Zimei Mountains are in an oligotrophic or medium nutrient state. The vegetation structure and species composition are influenced by hydrological status of wetlands (Quinty and Rochefort 2003).

12.4.2 Outlook

Sphagnum mire is a type of wetland with unique landscapes. It has important ecological, economic and societal value (Quinty and Rochefort 2003; Rochefort et al. 2003; Wang et al. 2005, 2013). The Qi Zimei mountains *Sphagnum* wetlands are a new-found, large-scale subtropical montane *Sphagnum* community. They were first discovered as extensive subtropical montane *Sphagnum* wetlands in Da Jiuhu in Shennongjia (Hubei) by Lang Huiqing et al, In the 1980s. Our survey found that there is a widespread tendency to usurp natural wetlands to grow alpine vegetables or to harvest *Sphagnum* illegally for economic interest. Qi Zimei Mountains *Sphagnum* wetlands are subjected to intense interference due to the economic benefits, which would lead to varying degrees of damages to these precious resources. Without strong restrictions, the *Sphagnum* wetlands will experience significant changes, and the regional ecosystem will degenerate and possibly disappear. Therefore, further researches are being conducted to explore the mechanism of sustainable development for *Sphagnum* wetlands.

References

Bai G-R, Wang S, Leng X-T et al (1999) Bio-environmental mechanism of herbaceous peat forming. Acta Geograph Sin 54(3):247–254

Bai G-R, Wang S-Z, Gao J et al (2004) Liquid-heat conditions and microbic decomposition on the forming of Turf Deposits. J Shanghai Norm Univ (Nat Sci) 33(3):92–97

Braun-Blanquet J (1964) Pflanzensoziologie, Grundzuge der Vegetationskunde, vol 3 Aufl. Springer, Wien, p 631

Fang Y-P, Liu S-X, Wang Z-X et al (2007) Quantitative assessment of priority for conservation of the national protected plants in Qizimeishan Mountain Nature Preserve. Acta Bot Boreal-Occident Sin 27(2):0348–0355

Gorham E (1991) Northern Peatlands: role in the carbon cycle and probable responses to climatic warming. Ecol Appl 1:182–195

Lally H, Gormally M, Higgins T, Colleran E (2012) Evaluating different wetland creation approaches for Irish cutaway peatlands using water chemical analysis. Wetlands 32:129–136

Lin P, Liu Q, Wu Y et al (2002) Water holding capacity of moss and litter layers of subalpine coniferous plantations in Western Sichuan, China. Chin J Appl Environ Biol 8(3):234–238

Ma G-L, Lei Y, Wang Z-X et al (2008) Plant diversity of Sphagnum Mire at Qizimei Mountains in Western Hubei Province. J Wuhan Bot Res 26(5):482–488

Mao R, Wang Z-X, Lei Y et al (2009) Profile characteristics and element vertical distribution in Sphagnum Wetland in Qizimei Mountains Nature Reserve, Hubei. Acta Pedol Sin 46 (1):160–163

Mcneil P, Waddington JM (2003) Moisture controls on sphagnum growth and CO_2 exchange on a cuto-ver bog. J Appl Ecol 40(2):354–367

Quinty F, Rochefort L (2003) Peatland restoration guide, 2nd edn. Canadian Sphagnum Peat Moss Association, and New Brunswick Department of Natural Resources and Energy, St. Albert, pp 4–5

Rochefort L, Quinty F, Campeau S, Johnson K, Malterer T (2003) North American approach to the restoration of *Sphagnum* dominated peatlands. Wetl Ecol Manag 11:3–20

Wang Z-X, Lei Y, Liu S-X et al (2005) One subalpine sphagnum wetland being discovered Qizimei Mountains Nature Reserve, Hubei. J Cent China Norm Univ (Nat Sci) 39(3):387–388

Wang Z-X, Lei Y, Xiong K-C (2013) Comprehensive scientific investigation and research on Sphagnum wetlands in Qizimei Mountains Nature Reserve, Hubei[M]. Chinese Forestry Publishing House, Bei Jing, pp 5–56

Part IV
Vegetation and Plant Ecology

High-Resolution Aerial Imagery for Assessing Changes in Canopy Status in Hawai'i's 'Ōhi'a (*Metrosideros polymorpha*) Rainforest

Linda Mertelmeyer, James D. Jacobi, Hans Juergen Boehmer, and Dieter Mueller-Dombois

Abstract

'Ōhi'a Lehua (*Metrosideros polymorpha*) is the most abundant tree species in the native wet and mesic forests throughout the main Hawaiian Islands. In the late 1960s and early 1970s large areas on the wet, eastern side of Hawai'i island appeared to have extensive defoliation and death of the 'ōhi'a trees. The dieback on Hawai'i island extended to approximately 49,000 ha of which 24,000 ha was considered to be in heavy to severe dieback (>50% of the canopy trees dead or defoliated), and 25,000 ha characterized as having slight to moderate dieback (25–50% of the canopy trees dead or defoliated). Research was initiated in 1976 by a team led by Professor Dieter Mueller-Dombois to assess both extent and ecological characteristics of the forest impacted by canopy dieback relative to areas that did not experience dieback in this same forest zone. To assess the spread or recovery of the 'ōhi'a dieback forest over time, twenty-six permanent plots were established across the study area. The results from the monitoring of the 26 permanent plots indicate that many of the original dieback sites are now showing strong recovery of the 'ōhi'a tree canopy through recruitment of new seedlings that have now grown into saplings and even taller trees (Boehmer et al.

L. Mertelmeyer (✉)
Technical University of Munich, München, Germany
e-mail: l.mertelmeyer@onlinehome.de

J.D. Jacobi
U.S. Geological Survey, Pacific Island Ecosystems Research Center, Hawai'i National Park, HI, USA

H.J. Boehmer
School of Geography, Earth Science & Environment (SGESE), University of the South Pacific (USP), Suva, Fiji

D. Mueller-Dombois
Department of Botany, University of Hawai'i at Manoa, Honolulu, HI, USA

J Veg Sci 24(4):639–650, 2013). However, it was not clear if these results truly represented the conditions across the entire original dieback area. Therefore, we conducted a much larger survey of response of the 'ōhi'a forest to that dieback event, across the entire wet forest region on the eastern side of the island of Hawai'i. We did this by analyzing very high-resolution aerial imagery (<10 cm pixels) taken by Pictometry International (POL), to assess both canopy and understory change throughout this region. The POL imagery proved to be an effective and efficient tool to use for assessing the status of 'ōhi'a forest across the eastern Hawai'i Island study area. The results of this large area survey, using the POL imagery, agree closely with the conclusions presented by Boehmer et al. (J Veg Sci 24(4):639–650, 2013), that most of the 'ōhi'a forests on the eastern side of the island of Hawai'i that were affected by canopy dieback in the 1960s and 1970s have started to recover their tree canopy, as a new cohort of young trees are growing back in these sites.

Keywords

Dieback · Forest decline · Hawai'i · Imagery · Mapping · *Metrosideros polymorpha* · Montane rainforest · Vegetation dynamics

13.1 Introduction

'Ōhi'a Lehua (*Metrosideros polymorpha*) is the most abundant tree species in the native wet and mesic forests throughout the main Hawaiian Islands (Wagner et al. 1999, 2005–present). Although 'ōhi'a has many characteristics typical of a pioneer species (Burton and Mueller-Dombois 1984), it dominates plant communities that range from sea level to tree-line, from very dry habitats on young lava flows to the summit of the island of Kaua'i, one of wettest areas in the world (Mueller-Dombois et al. 2013).

In the late 1960s and early 1970s a great deal of concern was expressed about the future of the 'ōhi'a forest on the island of Hawai'i, as large areas on the wet eastern side of this island appeared to have extensive defoliation and death of the 'ōhi'a trees (Burgan and Nelson 1972; Petteys et al. 1975; Mueller-Dombois 1985; Jacobi 1993). A similar canopy "dieback" had been documented in the wet forest on the northeast slope of Haleakala volcano on Maui Island around 1900 (Lewton-Brain 1909; Lyon 1909; Holt 1983), but there was no clear conclusion as to the cause of the widespread death of trees at that time and no objective assessment of the status of this forest over time (Holt 1983). The dieback on Hawai'i island extended to approximately 49,000 ha, of which 24,000 ha was considered to be in heavy to severe dieback (>50% of the canopy trees dead or defoliated), and 25,000 ha was characterized as having moderate to no dieback (<50% of the canopy trees dead or defoliated) (Jacobi 1990, 1993).

Research was initiated in 1976 by a team led by Professor Dieter Mueller-Dombois to assess both extent and ecological characteristics of the forest impacted by canopy dieback relative to areas that did not experience dieback in this same forest zone (Mueller-Dombois 1980, Mueller-Dombois et al. 1980). As part of that project 62 relevé plots were established in both dieback and non-dieback sites to serve as the basis for a detailed study of the trees and their associated plant communities (Mueller-Dombois et al. 1980). Ever since they were first established, 26 of these plots have been resampled at approximately 5 year intervals to track changes in the trees and associated vegetation over time (Jacobi et al. 1983, 1988), with the most recent analysis of the data from these plots completed in 2013 (Boehmer et al. 2013).

The results from the reassessment of the 26 permanent plots indicate that many of the original dieback sites are now showing strong recovery of the 'ōhi'a tree canopy through recruitment of new seedlings that have now grown into saplings and even taller trees (Boehmer et al. 2013). However, it was not clear if these results truly represented the conditions across the entire original dieback area. Therefore, we felt it was necessary to conduct a much larger survey of the response of 'ōhi'a forest to that dieback event across the entire wet forest region on the eastern side of the island of Hawai'i. To do this, we analyzed very high-resolution aerial imagery (<10 cm pixels) to characterize both canopy and understory change throughout this region.

13.2 Data and Methods

13.2.1 Original Data Sets

The original vegetation map that identified the full extent of the distribution of 'ōhi'a dieback on the eastern side of the island of Hawai'i (Fig. 13.1) was produced by Jacobi (1990) using a visual analysis of black and white aerial photographs taken of this area by the U.S. Geological Survey in 1977 (USGS VEED series images). Since a stereoscope was used to examine the photos it was relatively easy to identify areas that had experienced canopy dieback because they were viewed in a three-dimensional optical display. Boundaries for the plant communities and dieback areas were identified on the images and drawn directly on the photographs. The mapped boundaries were then compiled onto scale-stable base maps using an optical planimeter (Jacobi 1990).

The 26 permanent plots that were established in 1976–1980 were distributed across the dieback region on the island of Hawai'i to allow for a reassessment of the vegetation over time (Mueller-Dombois et al. 1980) (Fig. 13.1). The detailed locations of these plots and the descriptions of the vegetation have been documented elsewhere (Jacobi et al. 1983, 1988; Boehmer et al. 2013). These relevé plots were 20 × 20 m in size and all of the trees >5 m tall were permanently marked, DBH measured, and assessed for their crown foliage vigor using the five tree vigor classes described by Mueller-Dombois et al. (1980). Additionally, data were recorded on the abundance of all woody species <5 m tall in subplots within the relevé, as well as cover for all plant species within the plot in 1 m height classes.

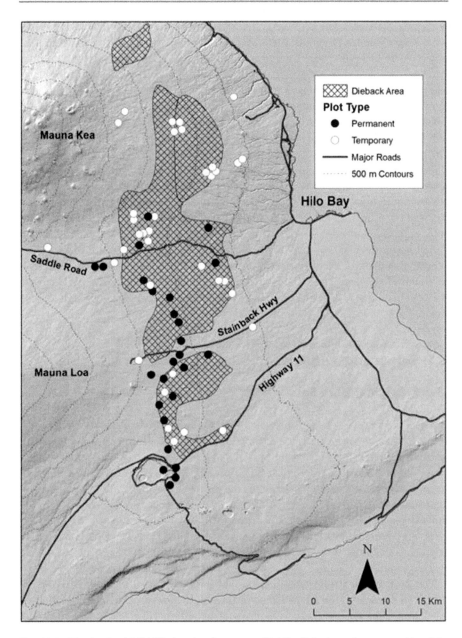

Fig. 13.1 Map by Jacobi (1990) showing the extent of 'ōhi'a dieback on the eastern side of the island of Hawai'i based on aerial photographs taken in 1977. The circle symbols indicate the approximate location of temporary (white) and permanent (black) relevé plots established across this study area to monitor the vegetation in these forests

Tree vigor classes included: (1) fully foliated crowns; (2) some defoliated branches, 10–50% of the crown dead; (3) most (>50%) of the upper crown branches defoliated, but with some foliated adventitious branches along the trunk; (4) recently dead trees with small branches and most of the bark remaining; and (5) dead trees, with only major branches remaining and most of the trunk without bark. A dieback index (DI) for each plot was calculated by dividing the sum of the number of trees in vigor classes 3, 4, and 5 by the total number of trees. Plots with a DI >50% were considered to be in dieback condition; plots with DI <50% were considered to be in non-dieback condition.

13.2.2 New Data Collection

To assess the current status of the 'ōhi'a forest canopy across the eastern side of the island of Hawai'i we used very high-resolution imagery collected by Pictometry International (POL) (EagleView Technologies, Inc.), between the years 2008 and 2012. The POL imagery consists of 3-band (RGB) digital photographs taken from a fixed-wing aircraft. Five images were taken at each POL photo point, one straight down (vertical), and four oblique images that were aimed left, right, forward, and to the rear of the path of the aircraft along its flightline. The POL imagery was then georectified and a mosaic was made of the images before distribution for use. The POL images can be viewed in either a vertical or oblique orientation using the POL Connect viewer or in an interactive window through a POL add-in program for the ESRI ArcMap 10 GIS (ESRI 2011). The POL Connect viewer includes tools for establishing circular boundaries (plots) of a given diameter around a specific location on the imagery. The circular plot boundary may appear to be distorted when viewed in oblique mode due to corrections made for site terrain. POL Connect additionally has a tool to measure the height of an object when viewing the image obliquely.

We used the random points tool in ArcMap 10 to generate a set of 1331 virtual sampling locations across the study area; 560 plots were located in areas that were mapped by Jacobi (1990) as in heavy to severe dieback on the 1977 imagery; and 771 plots located in similar and adjacent wet forest habitat in this same area, but not mapped as in dieback condition from the 1977 imagery. The coordinates for all of these points were imported as a KMZ file into the POL Connect viewer for analysis. A 100 m radius perimeter was established around each of the point locations using the POL Connect circle tool. These circular plots were used to visually reassess the current status of the 'ōhi'a forest canopy and understory at each location. For each plot we estimated the percent of trees >5 m tall that were dead or heavily defoliated (i.e., vigor class 3–5). If the plot had >50% of its trees in this condition it was considered to be in dieback condition; otherwise, it was classified as non-dieback condition. All five of the views (vertical and four oblique) were used to make the assessment for each plot.

To calibrate our assessment process we also constructed virtual plots around the locations for each of the 25 permanent plots that had been established in 1976–1980 (one plot could not be relocated for reassessment after 1985), and compared our new assessment of canopy dieback for these locations with the most recent ground-based resurvey of these plots that was conducted in 2003 (Boehmer et al. 2013).

13.3 Results and Discussion

The POL imagery proved to be an effective and efficient tool to use for assessing the status of 'ōhi'a forest across the eastern Hawai'i Island study area. There was adequate spatial and spectral resolution in the imagery to recognize the crown condition in each plot, particularly when viewed in the oblique image mode using the POL Connect viewer. Figures 13.2 and 13.3 show examples of how the vegetation appeared from different vantage points (oblique versus vertical) using the POL Connect viewer. Having the opportunity to select the four different oblique views was extremely helpful for the canopy assessment, particularly since each angle presented a different sun and shade perspective.

For our calibration set of 26 permanent plots that were sampled on the ground in 2003 by Boehmer et al. (2013) and with the POL assessment, 76% (19) of the plots showed no difference in DI for the two methods, 12% (3) of the plots showed a

Fig. 13.2 Example of the Pictometry imagery taken in 2010 that was used for our analysis of the current status of 'ōhi'a forests on the eastern side of the island of Hawai'i. This image is shown in the Pictometry Connect online viewer

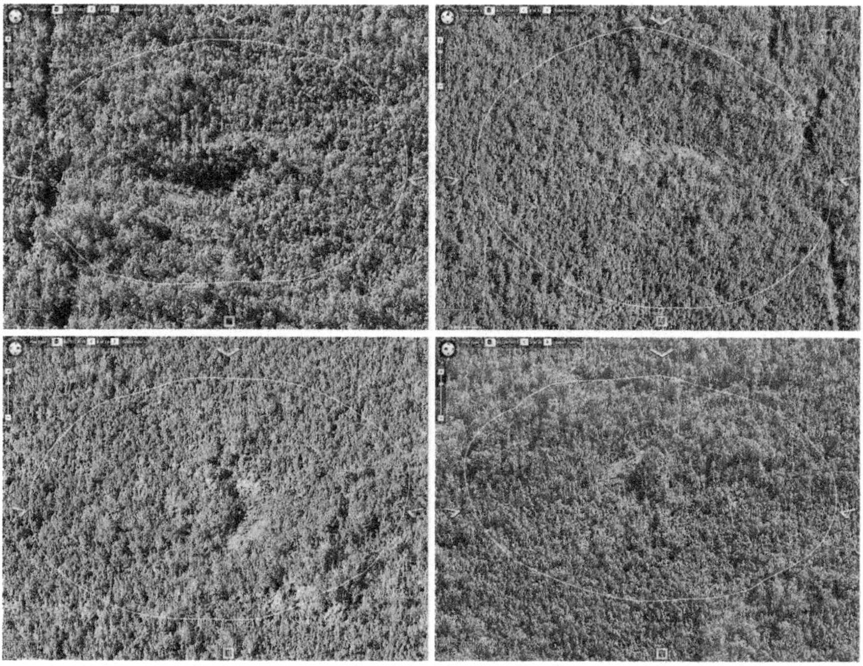

Fig. 13.3 Oblique views of the same forest location that is shown in Fig. 13.2. Orientation of the images is (clockwise from the upper left image): viewed from the north, south, west, and east

reduction in the DI, and another 12% (3) showed an increase in the DI. The difference in time between the two sampling times may explain why we saw these slight differences. However, most of the plots remained in the same dieback class for both times.

All of the 'ōhi'a forest mapped as dieback by Jacobi in 1990 had a DI of ≥ 3 (heavy to severe). Therefore, all of the random plots we allocated to this mapping stratum would have been in dieback condition. Likewise, all of the areas and their associated random plots not mapped as dieback in this area at that time would have been in non-dieback condition with a DI <3. In our 2013 assessment of this area using the POL imagery, only 20% of the plots originally found in the dieback areas were still in dieback condition. This means that the 'ōhi'a canopy in 80% of the plots had recovered to the point that they could no longer be considered in dieback. For the part of the study area that was previously mapped in non-dieback condition, only 8% of the POL plots were found to now be in dieback, indicating a slight increase of canopy dieback into these areas. Examples of POL plots from each of these types of forest are given in Figs. 13.4 and 13.5. A much more detailed analysis of these results and their relationships to various environmental factors will be published elsewhere (L. Mertelmeyer et al. in prep.).

The results of this large area survey of the distribution and intensity of 'ōhi'a canopy dieback on the eastern side of the island of Hawai'i agree closely with the conclusions presented by Boehmer et al. (2013). It appears that the large-scale

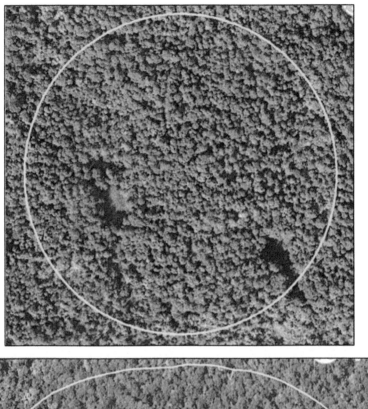

Fig. 13.4 Pictometry image taken in 2010 of an 'ōhi'a forest on the eastern side of Hawai'i Island that was mapped as being in heavy canopy dieback on the 1977 aerial photographs (dieback index >50%), but now shows very strong regeneration of a new cohort of 'ōhi'a trees. The dieback index for this site is now <5%. The upper image is a vertical view of the forest and the lower image shows an oblique view from the south

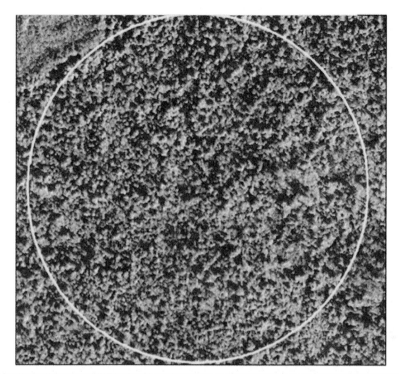

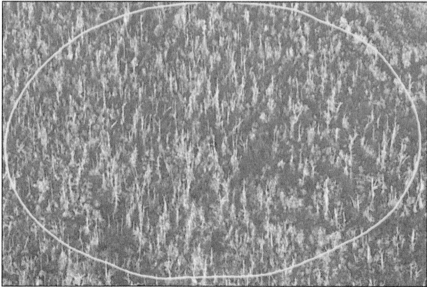

Fig. 13.5 Pictometry image taken in 2010 of an 'ōhi'a forest on the eastern side of Hawai'i Island that was mapped as being in heavy canopy dieback on the 1977 aerial photographs (dieback index >50%), and still shows many standing dead 'ōhi'a trees. The dieback index for this site is still >50%. The upper image is a vertical view of the forest and the lower image shows an oblique view from the south

dieback event that was first noticed in this area in the mid-1960s has slowed down its spread considerably and most of the former dieback areas are now showing a clear recovery of their canopy as a result of the vigorous growth of a new cohort of young 'ōhi'a trees once the canopy was opened by the dieback event. However, a small percentage (20%) of the former dieback forest has not recovered its canopy and an even smaller percentage (8%) of the forest that was healthy in the 1970s is now showing signs of dieback.

References

Boehmer HJ, Wagner HH, Jacobi JD, Gerrish GC, Mueller-Dombois D (2013) Rebuilding after collapse: evidence for long-term cohort dynamics in the native Hawaiian rain forest. J Veg Sci 24(4):639–650

Burgan RE, Nelson RE (1972) Decline of ohia lehua forests in Hawaii. General Technical Report PSW-3, U.S. Department of Agriculture, Forest Service, Pacific Southwest Forest and Range Experiment Station

Burton PJ, Mueller-Dombois D (1984) Response of *Metrosideros polymorpha* seedlings to experimental canopy opening. Ecology 65(3):779–791

ESRI (2011) ArcGIS desktop: release 10.x. Environmental Systems Research Institute, Redlands, CA

Holt RA (1983) The Maui forest trouble: a literature review nd proposal for research. Hawaii Botanical Science Paper 42, University of Hawaii, Department of Botany, Honolulu

Jacobi JD (1990) Distribution maps, ecological relationships, and status of native plant communities on the island of Hawai'i. Ph.D. dissertation, University of Hawaii at Manoa, Honolulu, HI, 290 p

Jacobi JD (1993) Distribution and dynamics of *Metrosideros* dieback on the island of Hawai'i: implications for management programs. In: Huettl RF, Mueller-Dombois D (eds) Forest decline in the Atlantic and Pacific region. Springer, Berlin, pp 236–242

Jacobi JD, Gerrish G, Mueller-Dombois D (1983) 'Ohi' dieback in Hawaii: vegetation changes in permanent plots. Pac Sci 37(4):327–337

Jacobi JD, Gerrish G, Mueller-Dombois D, Whiteaker LD (1988) Stand-level dieback and *Metrosideros* regeneration in the montane rainforest of Hawai'i. GeoJournal 17(2):193–200

Lewton-Brain L (1909) The Maui forest troubles. Hawaii Plant Rec 1:92–95

Lyon HL (1909) The forest disease on Maui. Hawaii Plant Rec 1(4):151–159

Mueller-Dombois D (1980) The ohia dieback phenomenon in the Hawaiian rain forest. In: Cairns J Jr (ed) The recovery process in damaged ecosystems. Ann Arbor Science Publishers, Ann Arbor, pp 153–161

Mueller-Dombois D (1985) Ohia dieback in Hawaii: 1984 systhesis and evaluation. Pac Sci 39 (2):150–170

Mueller-Dombois D, Jacobi JD, Cooray R, Balakrishnan N (1980) 'Ohi' rain forest study: ecological investigations of the 'ohi' dieback problem in Hawaii. Technical Report Miscellaneous Publication 183, University of Hawaii, Hawaii Agricultural Experiment Station, Honolulu, Hawaii

Mueller-Dombois D, Jacobi JD, Boehmer HJ, Price J (2013) 'Ohi'a Lehua rainforest: born among Hawaiian volcanoes, evolved in isolation. Friends of the Joseph Rock Herbarium, Honolulu

Petteys EQP, Burgan RE, Nelson RE (1975) Ohia forest decline: its spread and severity in Hawaii. Technical Report Research Paper PSW-105, U.S. Department of Agriculture, Forest Service, Pacific Southwest Forest and Range Experiment Station, Berkeley, California

Wagner WL, Herbst DR, Sohmer SH (1999) Maunal of the flowering plants of Hawaii, 2nd edn. University of Hawaii Press, Bishop Museum Press, Honolulu

Wagner WL, Herbst DR, Lorence DH (2005–present) Flora of the Hawaiian islands website. Smithsonian Institution, Department of Botany, Washington, DC http://botany.si.edu/pacificislandbiodiversity/hawaiianflora/index.htm

Plant Assemblages of Abandoned Ore Mining Heaps: A Case Study from Roşia Montană Mining Area, Romania

Anamaria Roman, Dan Gafta, Tudor-Mihai Ursu, and Vasile Cristea

Abstract

Plant assemblages and successional pathways were studied on large, abandoned ore mining heaps located around the open-cast pits from Roşia Montană, Romania. Four differently aged mining spoils with relatively homogenous substrate and one control plot were investigated using the chronosequence approach.

The effects of the waste dump age, slope steepness, position on slope, terrain curvature and potential solar radiation on the plant species composition of different assemblages from spontaneously revegetated primary sites were evaluated. Relevés were grouped into floristically similar vegetation types using non-hierarchical cluster analysis (Fuzzy c-Means). The ecological interpretation of the plant assemblages was performed through indicator species values (IndVal) and non-metric multidimensional scaling (NMDS).

This study reinforces several important concepts about the deterministic patterns of primary succession. First, the multiple comparisons between plant assemblages reveal that the highest differences are caused by the age of the spoil heap and substrate acidity. Further differences, lower but significant, are determined by terrain curvature and potential solar radiation. Also, environmental factors acting prior to the establishment of the observed plant assemblages (age, substrate pH, slope steepness and position on slope), have subsequently induced a structural differentiation in terms of species richness, vegetation cover and

---

A. Roman • T.-M. Ursu
Institute of Biological Research, National Institute of Research and Development for Biological Sciences, Cluj-Napoca, Romania
e-mail: anamaria.roman@icbcluj.ro; tudor.ursu@icbcluj.ro

D. Gafta (✉) • V. Cristea
Department of Taxonomy and Ecology, Babeş-Bolyai University, Cluj-Napoca, Romania
e-mail: dan.gafta@ubbcluj.ro; vasile.cristea@ubbcluj.ro

© Springer International Publishing AG, part of Springer Nature 2018
A.M. Greller et al. (eds.), *Geographical Changes in Vegetation and Plant Functional Types*, Geobotany Studies,
https://doi.org/10.1007/978-3-319-68738-4_14

relative cover of the N-fixing species. There are two main possible successional pathways, determined, most likely, by the long term changes in substrate pH under the influence of both abiotic and biotic factors. Two successional series were clearly distinguished: the weakly acidophilous series, comprising communities of *Poo compresae-Tussilaginetum farfarae, Festuco rubrae-Agrostietum capillaris* and *Carpino-Fagetum*, and respectively the acidophilous series, comprising plant assemblages of *Deschampsietum flexuosae, Pinetum sylvestris* sensu lato, *Festuco rubrae-Genistetum sagittalis* and *Vaccinio-Callunetum vulgaris*. The spontaneous succession progresses towards woodland and appears to be an ecologically suitable way of restoring the studied disturbed sites, because species typical of natural and semi-natural vegetation have become dominant over time. Within our study area, spontaneous vegetation succession resulted in plant assemblages that resemble the original semi-natural vegetation.

Keywords
Ecological ordination • Habitat chronosequence • Indicator species analysis • Plant community classification • Primary succession • Successional pathways • Topographic variables • Vegetation series

14.1 Introduction

Ore extraction via open cast mining produces a huge amount of waste rock that is deposited in the surroundings and thereby irreversibly transforms the landscape. Worldwide, only some of these mining areas have been restored, while most opencast pits and waste dumps have been abandoned after the ore extraction activities have ceased. This is also the case of Roşia Montană (Alba County, Romania), where gold mining activities undertaken since the Roman Period in Dacia (106–275 AD) have scarred and fragmented the landscape, affecting the natural habitats, the land-use potential, and the attractiveness of the landscape. It is therefore an important task to understand the natural processes related to vegetation succession for a cost-effective, successful restoration of such areas (Cristea et al. 1990; Sänger and Jetschke 2004; Cadenasso et al. 2009; Prach and Walker 2011; Prach et al. 2014). Ecologists consider restoration to be the practical field testing method for the drivers and concepts of the ecological succession theory (Bradshaw 1987; Cadenasso et al. 2009). Nevertheless, Walker and del Moral (2009) noted that although many studies have provided useful examples for restoration, the technicians have not fully taken advantage of them.

The abandoned mining spoil heaps are an ideal research area for the study of primary vegetation succession due to the complete absence of soil and propagule sources, exact dating and relatively homogenous anthropogenic substrate. Thus, the spontaneous primary succession can be followed and the changes in the specific structure and composition of different plant assemblages, as a result of extreme anthropogenic disturbance, can be analyzed in order to test hypotheses within the theory of ecological succession (Walker et al. 2010). These hypotheses state that three main interconnected causes influence the dynamics of plant communities: differentiation in habitat suitability, differential species availability and differential species performance (Pickett et al. 2009, 2011; Meiners and Pickett 2011). Although plant succession was initially explained by progressive and gradual change based on the facilitation of late successional species by the early successional species, these hypotheses have not been confirmed by studies carried out at fine scales or in highly disturbed areas (Pickett and McDonnell 1989). There are many papers on the vegetation of metal contaminated areas (Punz and Mucina 1997; Banásová et al. 2006; Skubała 2011; Rola and Osyczka 2014; Rola et al. 2015), but detailed studies concerning successional pathways linking various seral stages are rather scarce (Martínez-Ruiz et al. 2007; Prach et al. 2013).

The main aim of this study is to reveal some of the environmental factors that drive the successional trajectories on the mining waste heaps from Roşia Montană by investigating the plant assemblages that have spontaneously colonized the primary sites that form a chronosequence of 60 years. Inferring the successional pathways and the processes that drive vegetation changes in such habitats has important practical applications, such as the sustainable planning of ecological restoration for highly disturbed areas (Bradshaw 1983; Pickett and Cadenasso 2002; Walker et al. 2007; Pickett et al. 2009; Prach et al. 2014). Within this context, the following objectives were set: (1) distinguishing plant assemblages with relatively homogeneous floristic composition by classifying the vegetation relevés performed on abandoned mining spoil heaps of different ages; (2) detecting the main topographic factors generating differences in species composition among plant assemblages from the primary sites; (3) inferring the pathways followed by plant assemblages along primary succession on the mining spoil heaps.

14.2 Materials and Methods

14.2.1 Study Area

Roşia Montană (46°18′0″N, 23°08′0″E) is situated in the south-eastern Carpathians, within the Apuseni Mountains (Fig. 14.1a), Transylvania (Romania), in an area known as the Golden Quadrilateral. The climate is temperate, moderate continental, with the mean minimum and maximum temperatures of the coldest and respectively, warmest months reaching −6.3 °C (in January) and 20.2 °C (in July), respectively. The average precipitation ranges between 800 and 1000 mm per year.

The gold ore deposit from Roşia Montană, considered the largest in Europe, was already being exploited in Roman-Dacian times, the mining settlement being called *Alburnus Major* (Duma 2008; Sântimbrean 2012). The ore deposits are of epithermal and mesothermal type, associated with Neogene volcanic and sub-volcanic andesite-dacite bodies intruded in a varied lithological assemblage (Tămaş 2002). The waste materials (dacite or breccia) have a high content of heavy

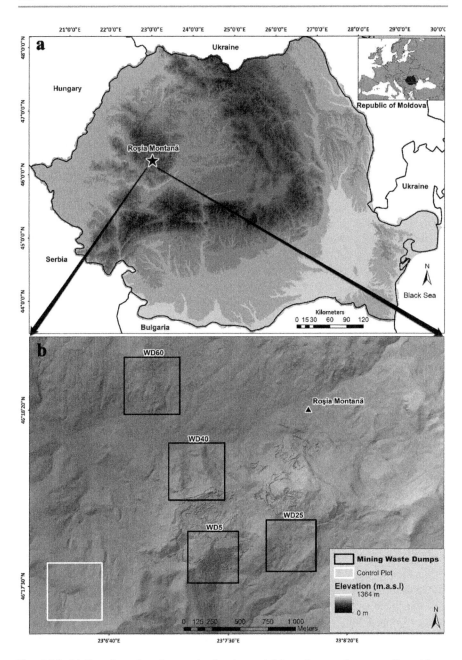

Fig. 14.1 (a) Location of study area in the Apuseni Mountains, south-eastern Carpathians; (b) Spatial distribution of the study plots, a chronosequence encompassing almost 60 years (abbreviations in text)

metals such as Cu, Zn and Fe (Roman 2013) and consequently the mining waste dumps are a toxic environment for colonizing plants, which are thus restricted to a narrow group of species (Roman et al. 2009; Roman and Cristea 2012). The potential natural vegetation in the study area is beech dominated forest—*Symphyto cordati-Fagetum* (Pop 1976), but the woodland patches from the oldest waste dump are composed mainly of pioneer species like *Betula pendula, Populus tremula, Salix caprea* and *Pinus sylvestris*. Other patches consist of heathlands (*Calluna vulgaris, Vaccinium myrtillus* and *V. vitis-idaea*) or semi-natural grasslands dominated by *Agrostis capillaris, Festuca rubra, Poa pratensis* and *Deschampsia flexuosa* (Ghişa et al. 1960, 1970). The semi-natural vegetation surrounding the dumps comprises mostly meadows, but also woodlands, and plays an important role for plant community assembly on mining spoil heaps (Roman and Gafta 2013).

In order to identify plant species assemblages (community types) in different successional stages, plots were established on spoil heaps of different ages, encompassing a chronosequence. We restricted our investigation to dry spoil heaps, where the waste rock, with too low concentrations of silver and gold to be extracted at profit, was deposited. Dry spoil heaps are very different in regard to physical and chemical properties from the fine sand tailings, which derive from the chemical processing of the ore.

Field investigations were carried out between 2007 and 2010 on four study sites, each covering an area of 25 ha of abandoned gold mining wastes placed around the open-cast pits (Cetate and Cârnic). The study plots shared homogenous substrates and similar management histories, but were abandoned at different times: WD5, ~5 years old; WD25, 20–25 years old; WD40, 38–45 years old; and WD60, ~60 years old (Fig. 14.1b). In addition, a control plot of 25 ha, undisturbed by mining activities (CP), was taken into account as a reference. Its role was to help assess the floristic similarities between the plant assemblages that colonized the heaps and the surrounding, undisturbed semi-natural vegetation, but also to explore the possible directions of vegetation dynamics.

14.2.2 Data Collection and Analysis

Vegetation sampling was performed within all study plots that compose the chronosequence (WD60, WD40, WD25, WD5 and CP).

Detailed floristic inventories (relevés) were recorded in square plots of 25 m^2, in meadows, scrubs and other open habitats, and 400 m^2 in woodlands. The number of relevés within each study plot was established *a priori* in order to keep the sampled area proportional to the size of various vegetation patches, as follows: 36 (CP), 15 (WD5), 25 (WD25), 37 (WD40) and 37 (WD60). A total of 150 vegetation relevés (comprising 259 vascular species) were performed in the study plots.

Within each relevé, the abundance-dominance (AD) was recorded for each vascular plant species. The AD ranks were converted to central percentage values according to the Braun-Blanquet scale, supplemented by Tüxen and Ellenberg and retrieved from Cristea et al. (2004). Plant taxa nomenclature follows the Flora

Europaea (RBGE 2011). Distinction of syntaxa based on their diagnostic species was done according to Coldea (1991), Sanda (2002) and Coldea et al. (2012).

The following environmental variables were measured in the center of every vegetation relevé, using a GPS terminal (Garmin 60S): longitude and latitude (metric units, Stereo 1970 Projection System), altitude (meters) as an indicator of position on slope and aspect. Subsequently, for every vegetation relevé, on the basis of a DEM (Digital Elevation Model) with 2 m pixel ground resolution, the following parameters were computed: average slope (degrees), curvature (negative values for concave, positive values for convex and zero for flat) and potential solar radiation (kilowatt-hour/m^2). In order to extract the geomorphological parameters from the DEM, the following functions in ArcGIS 9.3.2 software (ESRI 2009) were employed: Slope, Curvature and Points Solar Radiation.

The differentiation of plant assemblages with relatively homogenous floristic composition was performed through a non-hierarchical cluster analysis, called Fuzzy c-Means (FCM—Bezdek 1981), which was proved to be effective in vegetation classification (Marsili-Libelli 1989; Mucina 1997; De Cáceres et al. 2010). The optimal number of groups (clusters) and their subsequent validation was achieved by using both internal and external assessment criteria, according to specific literature recommendations (Cristea et al. 2004; De Cáceres et al. 2009; Tichý et al. 2011). A fuzziness coefficient of 1.2 was used as input parameter in FCM, as this value was suggested for the analysis of ecological data that do not display a normal distribution (Cristea et al. 2004; De Cáceres et al. 2010). The floristic dissimilarities between relevés were expressed by means of Steinhaus distance. Since the number of clusters needed to be specified *a priori,* we ran the FCM requesting successively all group numbers from two up to ten. Following this step, the optimum relevé partitioning was determined based on the local maxima or minima of several coefficients. The indices used for the validation of clusters were: the normalized Dunn coefficient, the partition entropy and the non-parametric silhouette. The optimum number of groups corresponds to the maximal and minimal values of the normalized Dunn coefficient and partitioning entropy, respectively. The values of silhouette are comprised between -1 and 1, but the closer they are to the positive upper limit, the crisper is the classification. In addition, the silhouette value for each relevé was extracted, which, if negative, indicates an inaccurate group membership (De Cáceres et al. 2007).

The ecological interpretation of the clusters was performed using the indicator values of species (IndVal) and non-metric multidimensional scaling (NMDS). The analysis of diagnostic species is useful since the higher their number and discriminating power, the higher the distinctiveness of relevé groups (clusters) is expected to be. IndVal was calculated as product of the relative frequency of the species and its relative cover in that respective cluster (Dufrêne and Legendre 1997). The statistical significance of the IndVal calculated for every species was estimated on the basis of 9999 permutations (Monte Carlo test). NMDS was run from 2 dimensions (axes) to 6, using as input data the floristic distance matrix between relevés. The three-dimensional ordering solution was chosen on the basis of the badness of fit (stress) and Shepard diagram (Legendre and Legendre 1998). The stability of this solution was evaluated through 25 runs with different starting configurations and a very low value (0.0001) of the convergence criterion. In order to associate ecological gradients to the NMDS axes, the Spearman correlation coefficients between the latter and the environmental variables were calculated.

The multiple comparisons between relevé clusters in regard to the environmental factors in hand were accomplished through the non-parametric Savage test, which offers a high detection power for exponential type distributions. The statistical significance was assessed through a Monte-Carlo simulation on the basis of 10^6 permutations.

The most probable successional pathways followed by the observed plant species assemblages were deduced by generating the Minimum Spanning Tree (MST) across the clusters distinguished. The distance between the nodes of the graph (the length of the graph segments) is proportional to the Euclidean floristic distance between the synthetic relevés that characterize the plant assemblages (Podani 2001). The synthetic relevé of a cluster (community type) was obtained by calculating, for each species, the mean relative cover weighted by its frequency in the target cluster.

FCM analysis and the calculation of cluster separation indices, as well as the IndVal analysis, were performed using the GINGKO 1.7.0 software (De Cáceres et al. 2007). The NMDS analysis and the Savage tests were run in SAS/STAT 9.2 (SAS Institute 2008). The MST was produced using the SYN-TAX 2000 software (Podani 2001). All tests were considered statistically significant at less than 5% threshold of alpha probability.

14.3 Results and Discussions

14.3.1 Classification of Plant Assemblages

The variation of the three coefficients of cluster validation indicates that the optimum solution corresponds to eight clusters (groups), for which there is a maximum of the Dunn coefficient and average Silhouette as well as a minimum of partitioning entropy (Fig. 14.2).

Relevés with negative or close-to-zero, positive values of the silhouette ($\leq$0.05) were considered outliers and were eliminated from all subsequent analyses. Of the initial 150 relevés, 132 (comprising 244 plant species) were retained and distributed within the eight clusters as follows: 25 in C1, 14 in C2, 16 in C3, 14 in C4, 15 in C5, 17 in C6, 13 in C7 and 18 in C8.

The IndVals of at least one species were statistically significant in seven of the eight identified clusters, C6 lacking diagnostic species (Table 14.1). The species presenting the highest IndVals were employed for the subsequent ecological and

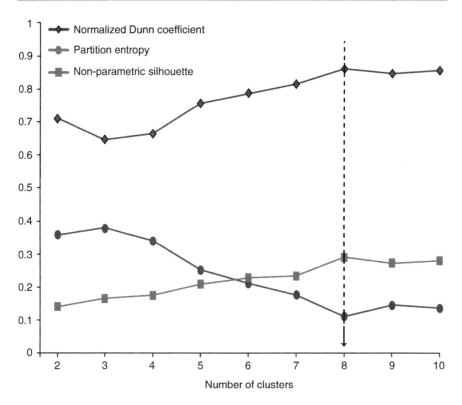

Fig. 14.2 Distribution of three indices for validation of the optimal number of clusters

syntaxonomical interpretation of the clusters. Since this study has mainly focused on spontaneous vegetation in its early and mid-seral stages, such as small isolated vegetation patches lacking a well-structured core species assemblage, the syntaxonomic assignment of relevés was very difficult. Actually, the plant assemblages distinguished (excepting those in the control area) represent different transitional dynamic stages that converge towards the syntaxa that were already described in the literature (Coldea 1991; Sanda 2002; Coldea et al. 2012). As an exception, for three clusters (C2, C3 and C6), no correspondent syntaxa were found, these communities being, most probably, transitional stages towards *Populo tremulae-Betuletum pendulae*. According to the syntaxonomical nomenclature, the clusters were labelled as follows: C1: *Festuco rubrae-Agrostietum capillaris* (FA); C2: *Deschampsietum flexuosae* sensu lato (FD); C3: *Pinetum sylvestris* sensu lato (PI); C4: *Festuco rubrae-Genistetum sagittalis* (FG); C5: *Poo compressae-Tussilaginetum farfarae* (PT); C6: Exposed Mining Spoils (ES); C7: *Vaccinio-Callunetum vulgaris* (VC); C8: *Carpino-Fagetum* (CF).

Table 14.1 Synoptic table of reléve clusters distinguished through numerical classification

Cluster	FA	FD	PI	FG	PT	ES	VC	CF	IndVal	p
Number of relevés	25	14	16	14	15	17	13	18		
Diagnostic species										
FA: *Festuco rubrae-Agrostietum capillaris*										
Agrostis capillaris	V$^{2-5}$	III^{+-1}	II^{+-1}	IV^{+-2}	II$^+$	II$^+$	II$^+$	.	0.93	0.001
Festuca rubra	V^{+-2}	I$^+$	.	IV$^+$	.	I$^+$	I$^+$	.	0.86	0.001
Centaurea phrygya	IV$^+$	.	.	I$^+$	.	.	.	.	0.76	0.001
Achilea millefolium	V$^+$	.	.	IV$^+$	.	.	.	.	0.74	0.001
Antoxanthum odoratum	V^{+-2}	I$^+$	.	III$^+$	I$^+$	I$^+$	I$^+$	.	0.74	0.001
Succisa pratensis	III^{+-1}	.	.	I$^+$	.	.	.	.	0.71	0.001
Briza media	IV$^+$	.	.	III$^+$	.	.	.	.	0.61	0.001
Trifolium pratense	IV^{+-1}	.	I$^+$	III$^+$	.	.	.	.	0.61	0.001
Leucanthemum vulgare	IV$^+$	.	I$^+$	III$^+$	.	.	.	.	0.60	0.001
Daucus carota	III$^+$	.	.	I$^+$	.	.	.	.	0.60	0.001
Peucedanum oreoselinum	III$^+$	.	.	.	.	.	I$^+$	.	0.59	0.001
Prunella vulgaris	III$^+$	.	.	II$^+$	.	.	I$^+$	.	0.59	0.001
Stachys officinalis	III$^+$	.	.	I$^+$	.	.	.	.	0.58	0.001
Lotus corniculatus	IV^{+-1}	II$^+$	I$^+$	III$^+$	I$^+$	.	I$^+$	.	0.57	0.001
Hypericum maculatum	IV$^+$	.	.	III$^+$	.	.	II$^+$	.	0.57	0.001
Ranunculus polyanthemos	III$^+$	.	.	II$^+$	.	.	I$^+$	.	0.57	0.001
Dianthus carthusianorum	III$^+$	.	I$^+$	I$^+$	.	.	I$^+$	.	0.53	0.001
Carex pallescens	II$^+$	.	.	.	.	.	.	.	0.53	0.001
Helianthemum nummularium	II$^+$	.	.	.	.	.	.	.	0.53	0.001
Thymus pulegioides	II$^+$	.	.	.	.	.	.	.	0.53	0.001
Leontodon autumnalis	III$^+$	.	.	II$^+$	.	.	I$^+$	.	0.52	0.001
Astrantia major	II$^+$	.	.	.	.	.	I$^+$	.	0.52	0.001

(continued)

Table 14.1 (continued)

Cluster	FA	FD	PI	FG	PT	ES	VC	CF	IndVal	p
Number of relevés	25	14	16	14	15	17	13	18		
FD: *Deschampsietum flexuosae*										
Deschampsia flexuosa	III$^+$	V$^{1-2}$	IV^{+-3}	V^{+-3}	.	V$^+$	IV$^+$	I$^+$	0.53	0.01
PI: *Pinetum sylvestris*										
Pinus sylvestris	.	III^{+-1}	V^{+-3}	II^{+-2}	I$^+$	III$^+$	II$^+$	I$^+$	0.91	0.001
FG: *Festuco rubrae-Genistetum sagittalis*										
Chamaespartium sagittale	III^{+-1}	II^{+-1}	I$^+$	V$^{1-2}$	.	II$^+$	I$^+$	.	0.87	0.001
Poa pratensis	II^{+-1}	I$^+$	.	IV^{+-3}	.	.	.	.	0.83	0.001
Plantago lanceolata	V$^+$	I$^+$	.	V^{+-1}	I$^+$	I$^+$	.	.	0.64	0.001
Arrhenatherum elatius	II$^+$	I$^+$	.	III^{+-1}	.	.	.	.	0.63	0.001
Thymus x porcii	I$^+$	I$^+$	I$^+$	III$^+$	.	.	I$^+$	.	0.62	0.001
Taraxacum officinale	II$^+$	.	I$^+$	III$^+$	II$^+$	.	.	.	0.57	0.001
Galium verum	III$^+$	.	.	IV$^+$	.	.	I$^+$	.	0.56	0.001
Luzula campestris	III^{+-1}	I$^+$	.	IV^{+-1}	.	.	.	.	0.55	0.001
Cruciata glabra	III$^+$	I$^+$	I$^+$	IV$^+$	.	.	.	II$^+$	0.54	0.001
PT: *Poo compresae-Tussilaginetum farfarae*										
Tussilago farfara	I$^+$	I$^+$	I$^+$	I$^+$	V^{+-2}	.	.	I$^+$	0.89	0.001
Poa compressa	I$^+$	I$^+$	I$^+$	.	IV^{+-1}	.	.	.	0.78	0.001
Polygonum aviculare	.	.	.	.	II$^+$	.	.	.	0.58	0.001
ES: *Exposed Mining Spoil*										
Nil	.	.	.	.	.	.	.	.	Nil	Nil
VC: *Vaccinio-Callunetum vulgaris*										
Betula pendula	I$^+$	IV^{+-1}	V^{+-1}	I$^+$	I$^+$	IV$^+$	V$^{1-3}$	II$^+$	0.85	0.001
Calluna vulgaris	.	V^{+-2}	IV^{+-1}	II^{+-1}	.	V^{+-1}	V$^{1-4}$	II$^+$	0.74	0.001
Populus tremula	.	III^{+-1}	III^{+-1}	.	.	IV^{+-1}	V^{+-2}	II^{+-1}	0.74	0.001

14 Plant Assemblages of Abandoned Ore Mining Heaps: A Case Study from...

Species	1	2	3	4	5	6	7	8	9	IndVal	p
Vaccinium vitis-idaea	.	III^{+-1}	.	.	.	.	.	III$^+$	II^{+-2}	0.68	0.001
Deschampsia cespitosa	III^{+-1}	.	II$^+$	II^{+-2}	I$^+$	.	IV^{+-2}	II^{+-4}	II$^+$	0.57	0.001
CF: *Carpino-Fagetum*											
Fagus sylvatica	.	.	I$^+$	.	.	.	I$^+$	.	**V^{+-4}**	0.97	0.001
Carpinus betulus	.	.	II$^+$	.	.	.	IV^{+-2}	.	**V$^{1-4}$**	0.96	0.001
Anemone nemorosa	.	.	.	.	.	.	.	.	**V^{+-2}**	0.92	0.001
Aposeris foetida	.	.	I$^+$	.	.	.	.	.	**IV^{+-2}**	0.82	0.001
Acer pseudoplatanus	.	.	I$^+$	.	.	.	.	.	**III$^+$**	0.73	0.001
Cardamine bulbifera	.	.	.	.	.	.	.	.	**III$^+$**	0.73	0.001
Polygonatum odoratum	.	.	I$^+$	.	.	.	.	.	**III$^+$**	0.69	0.001
Corylus avellana	.	.	I$^+$	.	.	.	I$^+$	.	**III$^+$**	0.62	0.001
Carex brizoides	.	.	.	.	.	.	I$^+$	.	**II$^+$**	0.60	0.001
Maianthemum bifolium	.	.	I$^+$	.	.	.	.	.	**II$^+$**	0.59	0.001
Allium ursinum	.	.	.	.	.	.	.	.	**II^{+-1}**	0.56	0.001
Calamagrostis arundinacea	.	.	.	.	.	.	.	.	**II$^+$**	0.56	0.001
Glechoma hederacea	.	.	.	.	.	.	.	.	**II$^+$**	0.53	0.001
Carex sylvatica	.	.	.	.	.	.	.	.	**II$^+$**	0.51	0.001

The cluster discriminated by each species, based on the strength and significance of its IndVal, is marked by boldface. The constancy class of species is indicated with Roman numerals, whereas the abundance–dominance range is written as exponent. Only species with IndVal > 0.5 are presented

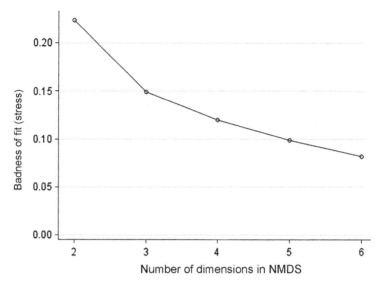

Fig. 14.3 Decrease in badness of fit (stress) with increasing number of dimensions in NMDS

14.3.2 Ecological Gradients Inferred from the Ordination Space

The solution in three dimensions (axes) was retained for the ordination of clusters, based on the decrease in stress criterion (Fig. 14.3).

Most clusters can be well distinguished in the species' space determined by the three NMDS axes (Figs. 14.4 and 14.5). In the bidimensional space determined by the axes 1 and 2 (Fig. 14.4), it is evident that the best separated cluster is FA, followed by FG, both situated towards the positive end of axis 1, whereas the clusters CF, PI and VC are grouped towards the negative end of the same axis. This configuration suggests a clear floristic gradient from dominantly herbaceous vegetation towards woody vegetation. The ES and FD clusters partially overlap in the positive part of axis 2 (Fig. 14.4), reflecting certain floristic similarities between them. The PT cluster is well discernible at the positive end of the third NMDS axis (Fig. 14.5).

The main ecological gradients are emphasized by correlating the relative species cover, the environmental factors and the age of the mining spoil heaps with the axes of the NMDS ordination (Table 14.2). Axis 1 may be associated to a floristic gradient spanning from woody plant communities to herbaceous ones, since many trees/shrubs and forest herbs are negatively correlated with that axis, whereas many grassland species are positively correlated with the same axis (Table 14.2). Axis 1 may also overlap on a gradient of soil fertility (given the good correlation with nitrogen-fixing species), which in turn promotes species richness (Table 14.2). Such a relationship was also reported in other studies (van der Heijden et al. 1998; Klironomos et al. 2000; van der Heijden and Horton 2009; Gómez-Aparicio 2009).

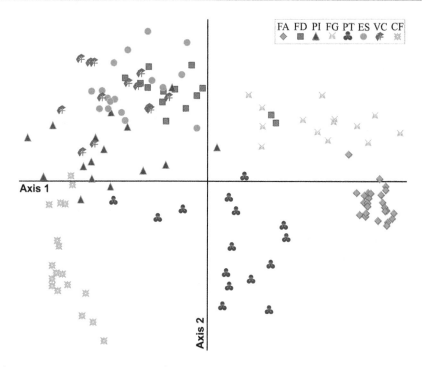

Fig. 14.4 NMDS ordination of relevés in the space determined by axis 1 *vs.* axis 2. Abbreviations as in Table 14.1

The significant negative correlation between the NMDS axis 2 and the relative cover of species with preferences for weakly acidic soils, such as *Acer pseudoplatanus, Aposeris foetida, Cardamine bulbifera, Carpinus betulus* and *Corylus avellana*, on one side, and the positive correlation of the same axis with strongly acidophilous species (*Deschampsia flexuosa* and *Calluna vulgaris*), on the other side, suggests a soil acidity gradient. Among topographic variables, the position on slope shows the strongest (negative) correlation with axis 2, suggesting that acidophilous species colonize particularly the waste dump ridges and upper slopes, which are prone to base cation and nutrient leaching. On the contrary, weakly acidophilous species prefer the base of slopes, where nutrients washed by surface water flows accumulate.

NMDS axis 3 is negatively correlated with the relative cover of mid-late successional species (*Carpinus betulus, Vaccinium myrtillus, V. vitis-idaea*), and positively correlated with the relative cover of pioneer species (*Poa compressa, Polygonum aviculare, Populus tremula, Solidago virgaurea* and *Tussilago farfara*). The gradients associated with this axis are slope, age of the waste dump and vegetation cover (Table 14.2). While the positive correlation between waste dump age and vegetation cover is straightforward, the negative correlation between slope and waste dump age might be a consequence of pluvial erosion, which

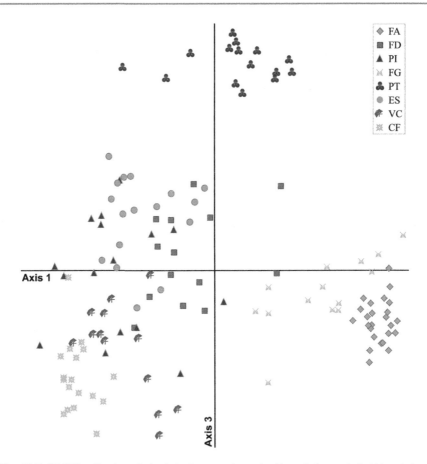

Fig. 14.5 NMDS ordination of relevés in the space determined by axis 1 *vs.* axis 3. Abbreviations as in Table 14.1

determines, in time, spoil accumulation at the dump base and the decrease of slope steepness. All these considerations support the association of the axis 3 with a successional maturity gradient.

14.3.3 Ecological Differentiation of the Plant Assemblage Types

There were statistically significant differences between at least two of the eight clusters in regard to every single variable considered (mining waste dump age, plant species richness, vegetation cover, N-fixing species cover, position on slope, slope steepness, terrain curvature and potential solar radiation) (see Chi-square and p values in Figs. 14.6–14.13).

Significant differences in terms of waste dump age were revealed between all plant assemblages types, excepting FA vs. CF (Fig. 14.6). The PT communities

Table 14.2 Correlation of environmental variables and relative species cover with the three NMDS axes (rho—Spearman coefficient; p—transgression probability)

Variable or Species	rho/p	Axis 1	Axis 2	Axis 3
Vegetation cover	rho	0.183	−0.280	**−0.828**
	p	0.035	0.001	**<0.0001**
Waste Dump Age	rho	0.136	−0.243	**−0.638**
	p	0.121	0.005	**<0.0001**
Species richness	rho	**0.490**	−0.237	−0.572
	p	**<0.0001**	0.006	<0.0001
N-fixing species cover	rho	**0.665**	0.190	0.013
	p	**<0.0001**	0.029	0.882
Position on slope	rho	0.054	**−0.476**	−0.057
	p	0.537	**<0.0001**	0.516
Slope steepness	rho	−0.180	0.257	**0.407**
	p	0.039	0.003	**<0.0001**
Terrain curvature	rho	−0.013	0.025	0.053
	p	0.884	0.778	0.548
Potential solar radiation	rho	−0.149	−0.038	−0.167
	p	0.088	0.665	0.056
Agrostis capillaris	rho	**0.712**	−0.021	−0.249
	p	**<0.0001**	0.808	0.004
Anemone nemorosa	rho	−0.344	**−0.512**	−0.193
	p	**<0.0001**	**<0.0001**	0.026
Anthoxanthum odoratum	rho	**0.631**	−0.047	−0.216
	p	**<0.0001**	0.590	0.013
Aposeris foetida	rho	−0.298	**−0.484**	−0.101
	p	0.001	**<0.0001**	0.251
Arrhenatherum elatior	rho	**0.421**	0.068	0.014
	p	**<0.0001**	0.442	0.874
Betula pendula	rho	**−0.594**	0.144	−0.230
	p	**<0.0001**	0.099	0.008
Briza media	rho	**0.583**	−0.101	−0.213
	p	**<0.0001**	0.250	0.014
Calluna vulgaris	rho	**−0.500**	**0.565**	−0.221
	p	**<0.0001**	**<0.0001**	0.011
Campanula patula	rho	**0.582**	−0.037	−0.216
	p	**<0.0001**	0.678	0.013
Carpinus betulus	rho	**−0.481**	**−0.540**	−0.399
	p	**<0.0001**	**<0.0001**	<0.0001
Chamaespartium sagittale	rho	**0.547**	0.295	−0.115
	p	**<0.0001**	0.001	0.188
Deschampsia flexuosa	rho	0.013	**0.752**	−0.027
	p	0.885	**<0.0001**	0.762
Fagus sylvatica	rho	−0.333	**−0.561**	−0.253
	p	**<0.0001**	**<0.0001**	0.003

(continued)

Table 14.2 (continued)

Variable or Species	rho/p	Axis 1	Axis 2	Axis 3
Festuca rubra	rho	**0.714**	−0.064	−0.282
	p	**<0.0001**	0.467	0.001
Frangula alnus	rho	**−0.359**	−0.104	−0.256
	p	**<0.0001**	0.236	0.003
Galium verum	rho	**0.483**	−0.008	−0.149
	p	**<0.0001**	0.924	0.089
Lotus corniculatus	rho	**0.535**	−0.062	−0.107
	p	**<0.0001**	0.481	0.220
Pinus sylvestris	rho	**−0.402**	0.326	0.155
	p	**<0.0001**	0.000	0.076
Plantago media	rho	**0.411**	−0.035	−0.183
	p	**<0.0001**	0.688	0.036
Poa compressa	rho	0.119	−0.253	**0.473**
	p	0.174	0.003	**<0.0001**
Poa pratensis	rho	**0.444**	0.110	−0.018
	p	**<0.0001**	0.208	0.838
Populus tremula	rho	**−0.570**	0.174	−0.251
	p	**<0.0001**	0.046	0.004
Sorbus aucuparia	rho	**−0.407**	−0.138	−0.217
	p	**<0.0001**	0.113	0.012
Trifolium pratense	rho	**0.586**	−0.048	−0.134
	p	**<0.0001**	0.584	0.126
Tussilago farfara	rho	0.181	−0.201	**0.464**
	p	0.038	0.021	**<0.0001**
Vaccinium myrtillus	rho	**−0.575**	0.071	**−0.394**
	p	**<0.0001**	0.415	**<0.0001**
Vaccinium vitis-idaea	rho	**−0.423**	0.281	**−0.309**
	p	**<0.0001**	0.001	**0.000**

The statistically significant correlations are marked by the boldface marking of cells (only species with rho > 0.4 are presented)

(but not ES) appeared to be the most pioneer among all. Species richness was significantly different between groups, excepting the pairs in which similar successional stages were contrasted e.g., mid-successional (FD vs. PI) or early successional (ES vs. PT) (Fig. 14.7). The comparison of species richness between the CF cluster and the other plant assemblages needs to be interpreted with caution since the sizes of sampling plots (relevés) were very different (400 m^2 vs. 25 m^2). Nevertheless, the difference in species numbers between the clusters FA and CF is not statistically significant, indicating that the early, open stage of beech and hornbeam woodlands still retains numerous species that are typical to the nearby grasslands.

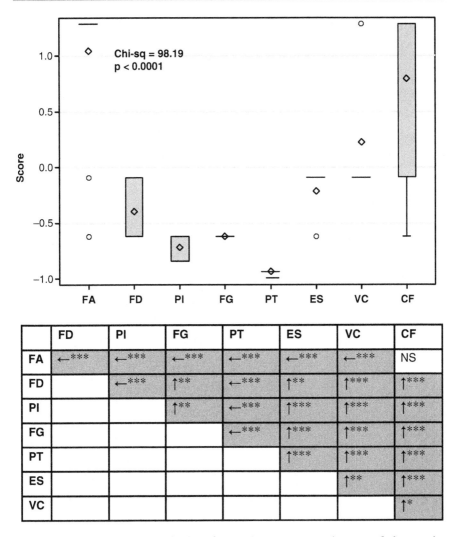

Fig. 14.6 Box and whiskers distribution of waste dump age across the types of plant species assemblages. The segment within boxes represents the median, while the position of the rhombus indicates the mean. The lower table presents the results of the multiple comparisons through Savage test, the grayed cells marking significant differences: (*): $5\% \geq p > 1\%$; (**): $1\% \geq p > 0.1\%$; (***): $0.1\% \geq p > 0.01\%$. The arrow indicates the plant assemblage type with higher values of the tested variable. NS (non significant). Other abbreviations as in Table 14.1

Vegetation cover is significantly different between the eight clusters (Fig. 14.8), excepting the cases where the early stages were contrasted (PT vs. ES), or some mid-late seral stages were compared to each other (FA vs. CF, FG vs. CF, PI vs. FG and PI vs. CF).

The relative cover of N-fixing species is significantly higher in FG cluster compared to all other groups (Fig. 14.9). The FG communities also have a higher

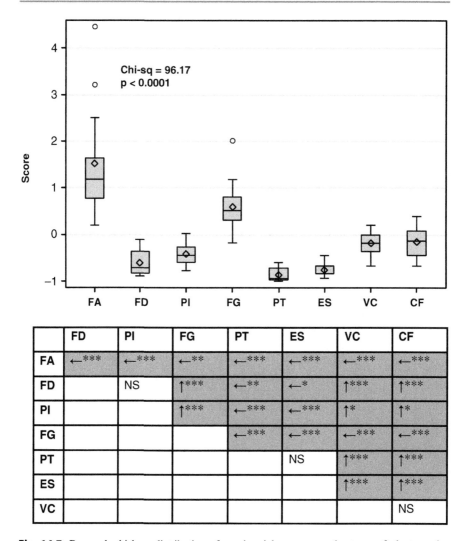

Fig. 14.7 Box and whiskers distribution of species richness across the types of plant species assemblages. Abbreviations and symbols as in Fig. 14.6

species richness compared to other groups (apart from FA; see Fig. 14.7) and are exposed to the lowest potential solar radiation (Fig. 14.13). While the association between the low insolation and high species richness was expected on coarse-grained spoils that dry out easily via evaporation, the possible negative relationship between insolation and the N-fixing species cover needs further clarifications. This is because higher cover of the N-fixing species was observed in open habitats as compared to healthlands or woodlands (Fig. 14.9).

Regarding position on slope, many non-significant differences were observed between the types of plant assemblages distinguished (Fig. 14.10). The prevalence

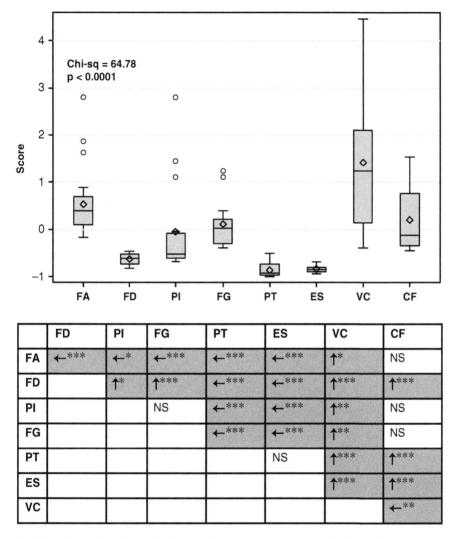

Fig. 14.8 Box and whiskers distribution of vegetation cover across the types of plant species assemblages. Abbreviations and symbols as in Fig. 14.6

of PI and PT assemblages on the stressful upper slopes was expected given the pioneer character of the dominant species. The FA communities also occur towards the upper slope but only on small terraces.

The slope distribution in Fig. 14.11 suggests that slightly acidophilous grasslands (FA), heather heathlands (VC) and hornbeam-beech woodlands (CF) are situated in microhabitats with lower slope than those used by the other communities. The other differences are not statistically significant. Mild slopes allow organic matter accumulation and decrease water runoff, promoting faster

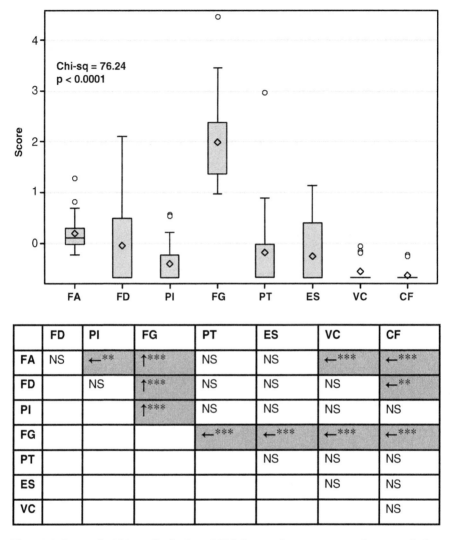

Fig. 14.9 Box and whiskers distribution of N-fixing species cover across the types of plant species assemblages. Abbreviations and symbols as in Fig. 14.6

advancement of successions towards woody communities, such as heathlands and woodlands.

Slight but statistically significant differences were observed in terms of terrain curvature (Fig. 14.12). It's worth noting that FD and CF, which are the most mesic among the plant assemblages distinguished, usually occupy concave microhabitats.

The multiple comparisons between the eight plant assemblage types revealed that the largest differences were caused by the age of the mining spoil heap, terrain declivity, position on slope and substrate acidity. Terrain curvature and potential

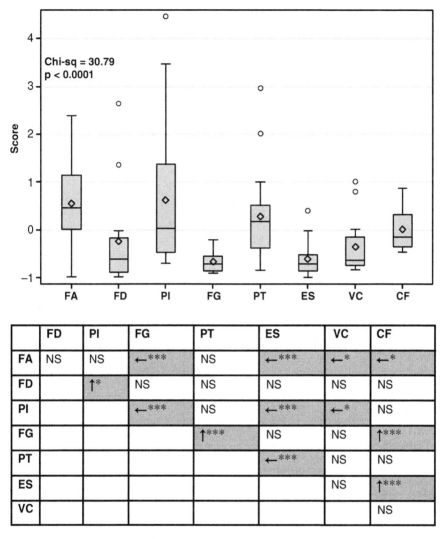

Fig. 14.10 Box and whiskers distribution of position on slope across the types of plant species assemblages. Abbreviations and symbols as in Fig. 14.6

solar radiation induce, in turn, smaller but still significant differences. Moreover, the environmental differences between habitats, which existed prior to the establishment of plants (dump age, slope steepness, position on slope, substrate pH) have subsequently promoted a structural differentiation of plant assemblages in terms of species richness, vegetation cover and relative cover of N-fixing species.

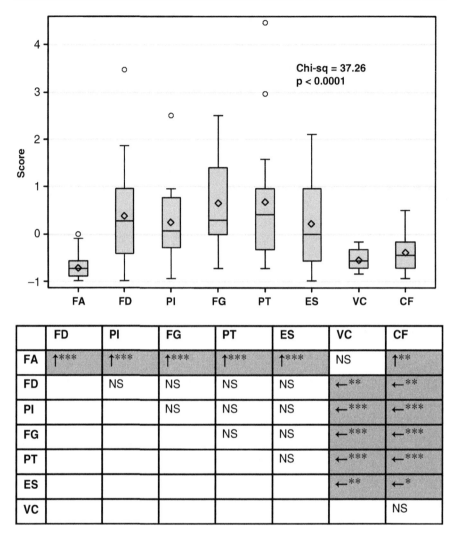

Fig. 14.11 Box and whiskers distribution of slope steepness across the types of plant species assemblages. Abbreviations and symbols as in Fig. 14.6

14.3.4 Successional Pathways Linking the Plant Assemblage Types

The minimum spanning tree (MST) illustrates the probable, past successional pathways that were followed by the observed plant assemblages on the waste dumps, and the possible future trajectories that they might pursue (Fig. 14.14). Spontaneous colonization may start from ES, in which seedlings of woody species such as *Calluna vulgaris*, *Betula pendula*, and *Populus tremula* have a higher frequency. However, woody plant assemblages occur only in small patches

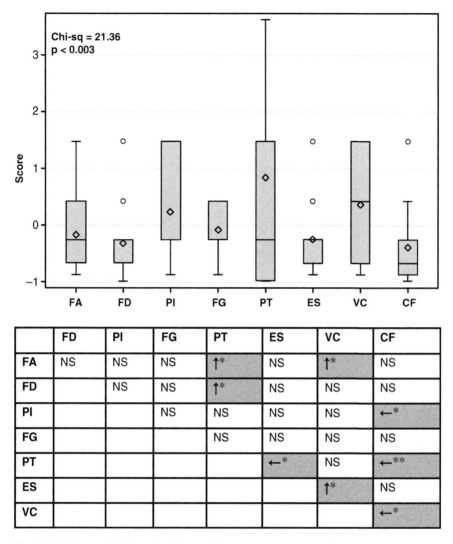

Fig. 14.12 Box and whiskers distribution of terrain curvature across the types of plant species assemblages. Abbreviations and symbols as in Fig. 14.6

(1–10,000 m²) on the 40 and 60 years old dumps. Starting from the ES stage there are two main possible successional pathways, determined, most likely, by the changes in substrate pH during time, under the influence of abiotic and biotic factors.

The two resulting dynamic series are: the weakly acidophilous series, comprising the communities PT, FA and CF, and respectively the acidophilous series, comprising the plant assemblages FD, PI, FG and VC.

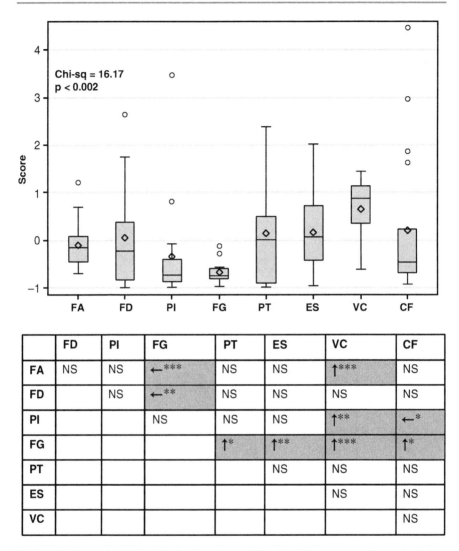

Fig. 14.13 Box and whiskers distribution of potential solar radiation across the types of plant species assemblages. Abbreviations and symbols as in Fig. 14.6

Within the weakly acidophilous series, the succession can advance from ES towards PT, but only on particular topographic conditions. PT established on the flat waste dump tops and upper third of the slopes and can be replaced by either mesic grasslands (FA) or hornbeam-beech woodlands (CF), depending on seed availability from the nearby areas. Within the acidophilous series, the vegetation develops from ES towards VC or FD, depending on slope and terrain curvature. Thus, on gentle slopes with convex terrain, the succession advances towards VC. On the contrary, on steeper slopes with concave curvature, the succession

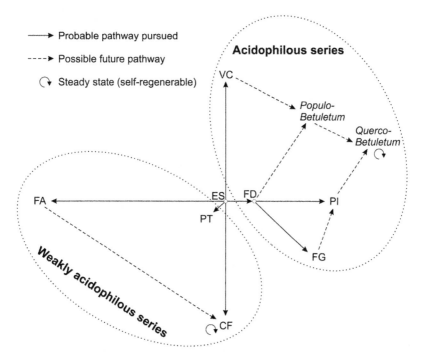

Fig. 14.14 Minimum spanning tree linking the plant assemblage types from the mining spoil heaps of Roşia Montană. The solid arrows indicate the probable successional trajectories followed until present, while the dashed arrows represent the possible future pathways. Abbreviations as in Table 14.1

progresses towards the FD stage, dominated by *Deschampsia flexuosa*, but with a high frequency of *Betula pendula*, *Populus tremula* and *Pinus sylvestris* seedlings. From FD, the vegetation might develop towards either of the two transitional stages: FG or PI, most probably depending on seed availability. The successional dynamics of the acidophilous series towards *Populo-Betuletum pendulae* is supported by the floristic composition of the FD, PI, and VC assemblages. At the scale of hundreds of years, it is possible that *Populo-Betuletum pendulae* communities converge dynamically towards beech forests (e.g., *Hieracio rotundati-Fagetum*) or sessile oak forests (e.g., *Querco petraeae-Betuletum*), depending on site micro-topography, aspect, pedological conditions and seed availability. The *Querco-Betuletum* stage was also observed on acidic, waste disposal sites that resulted from uranium mining in Germany (Sänger 1995; Sänger and Jetschke 2004).

The comparisons performed between the relevés grouped in the two series (acidophilous and weakly acidophilous) indicate that they are different in terms of waste dump age, position on slope and slope declivity (Table 14.3). Therefore, the plant assemblages belonging to the weakly acidophilous series are more likely to colonize the slightly inclined slopes located in the upper third of the older waste dumps.

Table 14.3 Mean rank scores of response variables used in comparisons between successional series through multiple Savage tests

Environmental variable	Weakly acidophilous series	Acidophilous series	p
Mining spoil heap age	0.457	−0.358	<0.0001
N-fixing species cover	−0.158	0.124	0.09
Position on slope	0.310	−0.243	<0.0001
Slope steepness	−0.256	0.200	0.006
Terrain curvature	0.024	−0.019	0.7
Potential solar radiation	0.050	−0.039	0.6

The spontaneous colonization of the recently abandoned heaps (WD5) started in their upper third, namely on the top of plateaus where ruderal communities belonging to *Poo compressae-Tussilagetum farfarae* are established. The waste heap base and the lower third of the slopes are for the moment still barren, probably because of the intense colluvium accumulation. Due to its ruderal life history traits, the first species to colonize the spoils is *Tussilago farfara*, which dominates the barren waste dumps.

The 25-year old heap (WD25) is currently dominated by *Pinus sylvestris* open woodlands, which may represent a transitional stage towards those composed of *Betula pendula* and *Populus tremula*. In this case, the facilitation process, corresponding to the homonymous model of succession, seems to prevail.

The 40-year old heap (WD40) is dominated by communities of *Festuco rubrae-Genistetum sagittalis*, which established mostly on shady slopes. On the contrary, the sunny slopes are colonized by plant assemblages dominated by *Deschampsia flexuosa*, or remain in the stage of exposed mining spoil. These may represent transitional stages towards communities of *Populo-Betuletum*. On patches with moister and less acidic substrate, plant assemblages that tend towards *Festuco rubrae-Agrostietum capillaris* are established. The prevailing process driving the succession is, again, the facilitation.

On the 60-year old heap, the forest communities of *Carpino-Fagetum* cover the upper third part of the slopes and the upper terraces, while in the lower third, on convex surfaces, heaths of *Vaccinio-Callunetum vulgaris* are present. Although this spoil heap is the oldest in the studied area, it still exhibits patches of exposed mining spoil (between 25 and 10,000 m^2), located on steep slopes and convex terrain and scattered as gaps within heathlands. In these areas, prior to heather dominance, the succession was probably directed by the facilitation of late successional species by early successional species (e.g., N-fixing species that enrich the substrate in organic matter). After the establishment of a compact *Calluna vulgaris* heathland, the colonization by other species is inhibited, except for pioneer tree species (birch and trembling poplar). It can be assumed, therefore, that in the future, the encroachment of these tree species will cause the subsequent wilting of the shrubs.

Most of the relevés of *Festuco rubrae-Agrostetum capillaris* and a few of *Carpino-Fagetum* were performed in the control plot. These communities, in addition to those of *Vaccinio-Callunetum vulgaris,* were previously reported in the proximity of Roşia Montană by other researchers, such as Ghişa et al. (1960) and Pop (1970, 1976). Their presence on the oldest waste dump, although in smaller patches, suggests that, in time and in the absence of major disturbance, vegetation succession is going to restore the original semi-natural vegetation types that existed before the start of intensive mining activities (successional convergent pathways).

The current plant assemblages are in different seral stages and may follow different trajectories in the future, depending on the mechanisms that come into action over time. The facilitation process leading to advanced successional stages are dominant and evident in the case of the 25-, 40- and 60-year mining waste dumps. On the youngest spoil heap, the successional pathway is rather difficult to foresee. However, since only 5 years after abandonment seedlings of *Pinus sylvestris, Salix capraea* and *Betula pendula* are already present, and considering the adjacent ecotone with the forest habitats of *Carpino-Fagetum*, it is possible that tolerance mechanisms would govern the succession towards woodland in a shorter timespan than was needed in the case of Orlea mining waste dump (60 years old). Finally, on the 40-year spoil heap, facilitation mechanisms seem to have been dominant as a result of early colonization with N-fixing species, such as *Chamaespartium sagittale,* which might have facilitated the establishment of other species in later stages.

Evidence of facilitation, tolerance and inhibition were all observed on the acidic waste heaps, resulted from lignite mining in the Czech Republic, in different stages of the succession or even within the same stages (Prach 1987). The modern theories consider these processes, defined by Connell and Slayter (1977), as specific mechanisms that can act and direct the successional trajectories differently depending on site conditions, seed availability and species traits (Pickett and Cadenasso 2005; Pickett et al. 2009). On the acidic mining heaps that result from coal extraction in the region of Lusatia (Lausitz, Germany), Wiegleb and Felinks (2001) revealed that environmental variables such as pH, phosphorus availability, organic carbon and substrate age do not influence the floristic composition or the vegetation cover, while the distance from propagule-source habitats has an important role in spontaneous colonization. On the contrary, Prach et al. (2007) observed a greater importance of the pH and climatic factors, compared to the age of the deposits or their nutrient content, for the spontaneous colonization of waste heaps. The order of arrival of the first colonizing species can influence the succession pathway, by determining the subsequent order of interspecific interactions. Usually, the priority effect is often noticed in habitats where nutrients and resources are scarce (Lockwood 1997).

The complex deterministic patterns observed in vegetation succession developing on ore mining spoils can provide useful insights for a better understanding of the ecological mechanisms underlying the natural restoration of disturbed areas. This is because vegetation dynamics is the outcome of complex effects from particular ecological conditions and processes across multiple temporal and spatial scales.

References

Banásová VO, Hora M, Ciamporová M, Nadubinská M, Lichtscheidl I (2006) The vegetation of metalliferous and non-metalliferous grasslands in two former mine regions in Central Slovakia. Biologia 61:433–439

Bezdek JC (1981) Pattern recognition with fuzzy objective functions. Plenum, New York

Bradshaw AD (1983) The reconstruction of ecosystem. Presidential address to the British Ecological Society. J Appl Ecol 20:1–17

Bradshaw AD (1987) Restoration: the acid test for ecology. In: Jordan WR, Gilpin ME, Aber JD (eds) Restoration ecology: a synthetic approach to ecological research. Cambridge University Press, Cambridge, pp 23–29

Cadenasso ML, Meiners SJ, Pickett STA (2009) The success of succession: a symposium commemorating the 50th anniversary of the Buell-Small Succession Study. Appl Veg Sci 12:3–8

Coldea G (1991) Prodrome des associations végétales des Carpates du sud-est (Carpates Roumaines). Doc Phytosociol 13:326–522

Coldea G, Oprea A, Sârbu I, Sîrbu C, Ştefan N (eds) (2012) Les associations végétales de Roumanie. Les associations anthropogènes, 2. Presa Universitară Clujeană, Cluj-Napoca

Connell JH, Slayter RO (1977) Mechanisms of succession in natural communities and their role in community stability and organization. Am Nat 111:1119–1144

Cristea V, Hodişan I, Pop I, Bechiş E, Groza G, Gălan P (1990) Reconstrucţia ecologică a haldelor de steril minier. I. Dezvoltarea vegetaţiei spontane. Contrib Bot 30:33–37

Cristea V, Gafta D, Pedrotti F (2004) Fitosociologie. Presa Universitară Clujeană, Cluj-Napoca

De Cáceres M, Oliva F, Font X, Vives S (2007) Ginkgo, a program for non-standard multivariate fuzzy analysis. Adv Fuzzy Sets Syst 2:41–56

De Cáceres M, Font X, Vicente P, Oliva F (2009) Numerical reproduction of traditional classifications and automated vegetation identification. J Veg Sci 20:620–628

De Cáceres M, Font X, Oliva F (2010) The management of numerical vegetation classifications with fuzzy clustering methods. J Veg Sci 21:1138–1151

Dufrêne M, Legendre P (1997) Species assemblages and indicator species: the need for a flexible asymmetrical approach. Ecol Monogr 67:345–366

Duma S (2008) Impact of mining activity upon environment in Roşia Montană. Rom Rev Reg Stud 4:87–96

ESRI (2009) ArcGIS 9.3.1. Environmental Systems Research Institute, Redlands

Ghişa E, Pop I, Hodişan I, Ciurchea M (1960) Vegetaţia Muntelui Vulcan-Abrud. Studii şi Cercetări de Biologie (Cluj) 11(2):255–264

Ghişa E, Resmeriţă I, Spârchez Z (1970) Contribuţii la studiul callunetelor din Munţii Apuseni. Contrib Bot 10:183–190

Gómez-Aparicio L (2009) The role of plant interactions in the restoration of degraded ecosystems: a meta-analysis across life-forms and ecosystems. J Ecol 97:1202–1214

Klironomos JN, McCune J, Hart M, Neville J (2000) The influence of arbuscular mycorrhizae on the relationship between plant diversity and productivity. Ecol Lett 3:137–141

Legendre P, Legendre L (1998) Numerical ecology. Developments in environmental modelling, vol 20. Elsevier, Amsterdam

Lockwood JL (1997) An alternative to succession: assembly rules offer guide to restoration efforts. Restor Manag Notes 15:45–50

Marsili-Libelli S (1989) Fuzzy clustering of ecological data. Coenoses 4:95–106

Martínez-Ruiz C, Fernández-Santos B, Putwain PD, Fernández-Gómez MJ (2007) Natural and man-induced revegetation on mining wastes: changes in the floristic composition during early succession. Ecol Eng 3:286–294

Meiners SJ, Pickett STA (2011) Succession. In: Simberloff D, Rejmanek M (eds) Encyclopedia of biological invasions. University of California Press, Berkeley, pp 651–657

Mucina L (1997) Classification of vegetation: past, present and future. J Veg Sci 8:751–760

Pickett STA, Cadenasso ML (2002) Ecosystem as a multidimensional concept: meaning, model and metaphor. Ecosystems 5:1–10
Pickett STA, Cadenasso ML (2005) Vegetation succession. In: van der Maarel E (ed) Vegetation ecology. Blackwell, Oxford, pp 172–198
Pickett STA, McDonnell MJ (1989) Changing perspectives in community dynamics: a theory of successional forces. Trends Ecol Evol 4:241–245
Pickett STA, Cadenasso ML, Meiners SJ (2009) Ever since Clements: from succession to vegetation dynamics and understanding to intervention. Appl Veg Sci 12:9–21
Pickett STA, Meiners SJ, Cadenasso ML (2011) Domain and propositions of succession theory. In: Scheiner SM, Willig MR (eds) The theory of ecology. University of Chicago Press, Berkeley, pp 185–216
Podani J (2001) SYN-TAX 2000—Computer program for data analysis in ecology systematics. User's manual. Scientia, Budapest
Pop I (1970) Făgetele şi făgeto-molidetele de pe dealul Şturţ-Abrud (Jud. Alba). Studia Universitatis Babeş-Bolyai (Biologia) 2:9–13
Pop I (1976) Contribuţii la cunoaşterea vegetaţiei munceilor din împrejurimile Abrudului (jud. Alba). Contrib Bot 16:123–132
Prach K (1987) Succession of vegetation on dumps from strip coal mining, northwest Bohemia, Czechoslovakia. Folia Geobotanica et Phytotaxonomica 22:339–354
Prach K, Walker LR (2011) Four opportunities for studies of ecological succession. Trends Ecol Evol 26:119–123
Prach K, Pyšek P, Jarošík V (2007) Climate and pH as determinants of vegetation succession in Central European man-made habitats. J Veg Sci 18:701–710
Prach K, Lencová K, Rehounková K, Dvoráková H, Jírová A, Konvalinková P, Mudrák O, Novák J, Trnková R (2013) Spontaneous vegetation succession at different central European mining sites: a comparison across seres. Environ Sci Pollut Res 20:7680–7685
Prach K, Řehounková K, Lencová K, Jírová A, Konvalinková P, Mudrák O, Student V, Vaněček Z, Tichý L, Petřík P, Šmilauer P, Pyšek P (2014) Vegetation succession in restoration of disturbed sites in Central Europe: the direction of succession and species richness across 19 seres. Appl Veg Sci 17:193–200
Punz W, Mucina L (1997) Vegetation on anthropogenic metalliferous soils in the eastern Alps. Folia Geobotanica et Phytotaxonomica 32:283–295
RBGE (2011) Flora Europaea database—PANDORA taxonomic database system of the Royal Botanic Garden Edinburgh. http://rbgweb2.rbge.org.uk/FE/fe.html
Rola K, Osyczka P (2014) Cryptogamic community structure as a bioindicator of soil condition along a pollution gradient. Environ Monit Assess 18:5897–5910
Rola K, Osyczka P, Nobis M, Drozd P (2015) How do soil factors determine vegetation structure and species richness in post-smelting dumps? Ecol Eng 75:332–342
Roman A (2013) Vegetation structure and dynamics from former silver-gold mining sites from Roşia Montană (Alba County). PhD Thesis, Babeş-Bolyai University, Cluj-Napoca
Roman A, Cristea V (2012) Spontaneous vegetation development on mining waste dumps from Roşia Montană (România). Stud Biol UBB 57(1):15–20
Roman A, Gafta D (2013) Proximity to successionally advanced vegetation patches can make all the difference to plant community assembly. Plant Ecol Divers 6(2):269–278
Roman A, Gafta D, Cristea V, Mihuţ S (2009) Small-scale structure change in plant assemblages on abandoned gold mining dumps (Roşia Montană, România). Contrib Bot 44:83–91
Sanda V (2002) Vademecum ceno-structural privind covorul vegetal din România. Editura Vergiliu, Bucureşti
Sänger H (1995) Flora and vegetation on dumps of uranium mining in the southern part of the former GDR. Acta Soc Bot Pol 64:409–418
Sänger H, Jetschke G (2004) Are assembly rules apparent in the regeneration of a former uranium mining site? In: Temperton VM, Hobbs RJ, Nuttle T, Halle S (eds) Assembly rules and restauration ecology—Bridging the gap between theory and practice. Island, Washington, pp 305–324

Sântimbrean A (2012) Roşia Montană deposit and its associated mineral substances. In: Pompei C (ed) Roşia Montană in Universal History. Cluj University Press, Cluj-Napoca, pp 18–23

SAS Institute (2008) SAS/STAT 9.2 User's guide. SAS Institute, Cary

Skubała K (2011) Vascular flora of sites contaminated with heavy metals on the example of two post-industrial spoil heaps connected with manufacturing of zinc and lead products in Upper Silesia. Arch Environ Prot 37:55–74

Tămaş C (2002) Structuri de breccia pipe asociate unor zăcăminte hidrotermale din România. PhD Thesis, Babeş-Bolyai University, Cluj-Napoca

Tichý L, Chytrý M, Šmarda P (2011) Evaluating the stability of the classification of community data. Ecography 34:807–813

van der Heijden MGA, Horton TR (2009) Socialism in soil? The importance of mycorrhizal fungal networks for facilitation in natural ecosystems. J Ecol 97:1139–1150

van der Heijden MGA, Klironomos JN, Ursic M, Moutoglis P, Streitwolf-Engel R, Boller T (1998) Mycorrhizal fungal diversity determines plant biodiversity, ecosystem variability and productivity. Nature 396:72–75

Walker LR, del Moral R (2009) Lessons from primary succession for restoration of severely damaged habitats. Appl Veg Sci 12:57–67

Walker LR, Walker J, Hobbs R (2007) Linking restoration and ecological succession. Springer, New York

Walker LR, Wardle DA, Bardgett RD, Clarkson BD (2010) The use of chronosequences in studies of ecological succession and soil development. J Ecol 98:725–736

Wiegleb G, Felinks B (2001) Predictability of early stages of primary succession in post-mining landscapes of lower Lusatia. Appl Veg Sci 4:5–18

Autoecological and Synecological Resilience of *Angelica heterocarpa* M.J. Lloyd, Observed in the Loire Estuary (France)

15

Kevin Cianfaglione and Frédéric Bioret

Abstract

This study aims to characterize the ecology, behavior and responses of *Angelica heterocarpa* up to the level of the vegetation series and of the human modification of its habitats, in the Loire estuary. Original observations and an extensive bibliography review follows.

Angelica heterocarpa is classified as a Sublittoral Eu-Atlantic (Franco-Atlantic endemic) taxon, living in clay-muddy estuary banks. The locus classicus of the species is from Loire River, in the area of Nantes. Its natural biotope corresponds to the oligohaline domain of the West French Atlantic estuaries. In many ways, *Angelica heterocarpa* is a very adaptable species. The plant develops also in epiphytic conditions and in "stand-by mode" strategy, allowing a hydrochoral dispersion. Vegetation series, Ecological, Phytosociological, Phytodynamic features and management advices are described.

While the geographic distribution area is large enough not to be of concern, and the plant shows a good adaptability, the global population trend shows a huge regression linked to the destruction of habitats. Due to these threats and its endemic status, this species is considered worthy of protection, and conservation policies and *in situ* experimentations are being carried out.

Keywords

Angelica heterocarpa • Adaptability • Ecology • Endemic • Megaforb community • Oligohaline • Plant landscape • Phytodynamics • Population regression • Vegetation series

---

K. Cianfaglione (✉) • F. Bioret
EA 2219 Géoarchitecture, UFR Sciences & Techniques, Université de Bretagne Occidentale,
6 Avenue Victor Le Gorgeu - CS 93837, 29238 BREST Cédex 3, France
e-mail: kevin.cianfaglione@univ-brest.fr

© Springer International Publishing AG, part of Springer Nature 2018
A.M. Greller et al. (eds.), *Geographical Changes in Vegetation and Plant Functional Types*, Geobotany Studies,
https://doi.org/10.1007/978-3-319-68738-4_15

15.1 Introduction

Angelica heterocarpa (Apiaceae) was discovered by Dr. Moriceau, botanist and physician in Nantes, than described by Lloyd (1860a, b). Many works followed about its characteristics (e.g., Bernard 1991; Reduron and Muckenstrum 2005; Reduron 2007). *Angelica heterocarpa* is a typical monocarpic species which blooms after 2 or 3 years of optimal ecological conditions. After seed dispersal, it completely dies, or it can leave small vegetative propagules at the stem base.

According to Dupont (1962), it is a Sublittoral Eu-Atlantic (Franco-Atlantic endemic) entity; Tutin et al. (1968) consider it as South-West French endemic; Dupont (2015) specifies the chorotype as West and South-West France. It is considered a sub-nitrophilous and sub-halophilous taxon.

The locus classicus of this species is the Loire River, in the Nantes zone. It is a hygrophyte, living in the clay-muddy estuaries banks. Its natural biotope is the oligohaline portion of the French West Atlantic estuaries. Gentle slopes comprise the most favorable habitats, especially when wooded. The substrate is soft, loamy to clay, frequently flooded by tides; the organic matter inputs are mainly provided by tide floods, which deposit organic debris accumulated by estuarine currents. The best populations are found in relatively stabilized mud (cf., Malinvaud 1903; Magnanon et al. 1998; Bensettiti et al. 2002). The plant is more salt-tolerant than salt-preferring (cf., Bernard 1991) and its presence in the estuary oligohaline portion could be just opportunistic: limited downstream from the increase in salinity, and upstream from the competition with other species, especially *A. sylvestris* (Reduron 2007). The presence downstream of its optimum conditions must be facilitated by less presence of predators (i.e.: terrestrial Gastropoda, and fostered by local fresh water influences.

About its distribution (cf., Lloyd 1868; Corillion 1961; Figureau et al. 1998; Bensettiti et al. 2002; Blanchard et al. 2005), the northern part starts in the Loire estuary (Paimbœuf-Frossay to Cellier, up to Vertou in the Sèvre Nantaise affluent). It is present in the Charente estuary (around Rochefort and Saint-Savinien); down to Gironde estuary (Dordogne River up to Fronsac, and Garonne River up to Bordeaux); and in the Adour estuary at the south, from Bayonne (Nive affluent portion incl.) to Urt. Sought unsuccessfully in Nivelle and Bidasoa, it should be actively sought in Spain, at the mouth of the Bidasoa (Lesouëf 1986; Figureau 1995). Former indication from Spain was never confirmed (Reduron 2007). It needs more study up to the Laïta River (at the north) where recently it was doubtfully observed by the authors (very rare) near Quimperlé, up to the North Finistère, between dune slacks, occasional and very rare (Fig. 15.1); right now it was not possible to find complete diagnostic sample to confirm these findings.

From a botanical point of view, *A. heterocarpa* is close to *A. sylvestris*; intermediate forms are known, showing a huge variability (cf., Metais et al. 2008). Hybrids with *Angelica sylvestris* were obtained experimentally in botanical garden (Bernard 1991). These two species differ in some morphological and ecological traits: *A. heterocarpa* flower earlier, with narrower leaflets, bigger fruits, with their distinctive shape variability. At the ecological level, *A. heterocarpa* is more

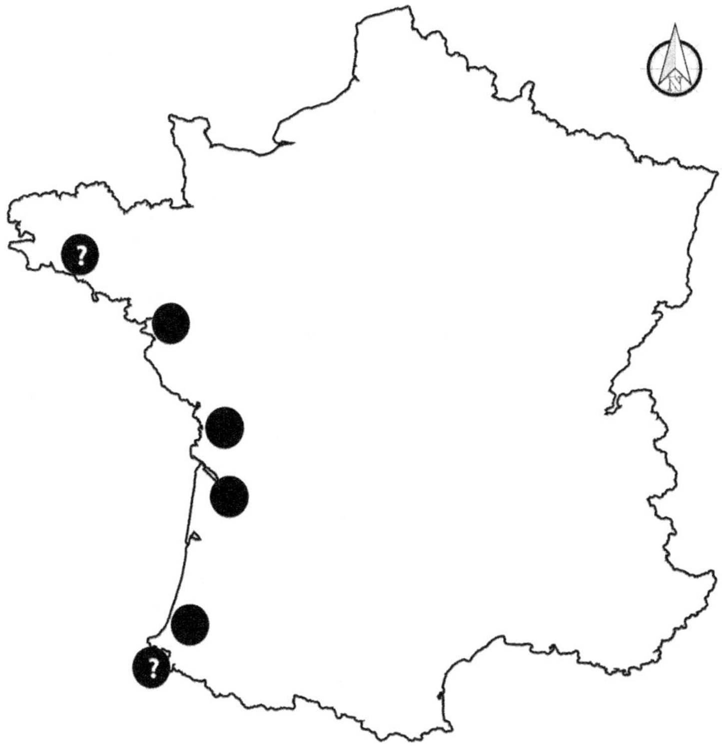

Fig. 15.1 *Angelica heterocarpa* distribution. The black dots indicate the presence in France, dots with question mark indicate the probable presence

salt-tolerant, so it can be considered as vicariant taxon. Another close species is *A. archangelica*, which is distinguished from *A. heterocarpa* by a larger size and a different phenology.

From an ecological point of view, *A. archangelica* subsp. *archangelica* is closer to *A. sylvestris*, while *A. archangelica* subsp. *litoralis* is closer to *A. heterocarpa*. In France, *Angelica archangelica* subsp. *litoralis* is characteristic of the northern estuaries, as *A. heterocarpa* is *characteristic* of the western ones.

Angelica heterocarpa is rare or worthy of protection by many authors (Dupont 1983, 1989a; Hoarher 1984; Lesouëf 1986; Clément and Touffet 1989; Magnanon 1993; Danton et al. 1995; Figureau 1995; Magnanon et al. 1998; Lahondère 1998; Buord and Lesouëf 2006). That species is mentioned in the IUCN World Red List as "Least Concern" (Juillet 2013) and, in the Pays de Loire regional Red List, as "Vulnerable" (Lacroix et al. 2008). It is listed as a priority species in Annex II of the "Habitats Directive" (Bensettiti et al. 2002) and in Appendix I of the Convention on the Conservation of European Wildlife and Natural Habitats (Bern Convention); it is also on the list of protected species in France (Annex I). The conservation actions plans are actually coordinated by the Brest and Sud-Atlantique botanical conservatories.

The present paper aims to characterize the behavior and the responses of *A. heterocarpa* to the level of human disturbance of its habitats in the Loire estuary, in particular. When not specified otherwise, taxonomy is based on "The Plant List" database (http://www.theplantlist.org).

15.2 Discussion

Angelica heterocarpa was found on muddy deposits, on more solid edges, in woodlands, in artificial banks, even in cracks and crevices of cemented places. In Nantes area, it is in the potential zone of the *Salix alba* communities and of the *Phragmites australis* communities. Figuerau and Richard (1990) noticed its epiphytic behavior. We observed this phenomenon on *Salix alba* (according to: Figureau 1995), and also on *S. fragilis* (s.l.), *S. atrocinerea* and occasionally in other woody species. It was observed up to one meter from the soil line, covering trunk bases, branches or other portions subjected to tides. In this condition *A. heterocarpa* show an epiphytic "stand-by activity"; the plant extends its cycle, prolonging the sapling stage, strongly decreasing the development, waiting for better opportunities.

In this condition, the plants can, with difficulty, reach a full development; but they may be able to live more years, permitting a hydrochoral dispersion, if driven off by floods (cf., Figureau and Richard 1990). In that case, the small plants can settle in the bank sediments, where they can start full development. Between the minimum and maximum range of tides (in the Loire, between 2.50 and 3.70 m), the Aerial Plant phenomenon was observed on rocks, poles, dead trunks, fallen branches, walls, thrown objects and on shipwrecks; indicating a strong colonization ability. In each case, the small plants are facilitated by encrustations mainly formed by algae, bryophytes or *Eleocharis bonariensis*, and trapped mud.

When the reed belt is too dense and high (especially in secondary conditions), *Angelica heterocarpa* can survive in the *Phragmites australis* coenosis, although occurring there more rarely, appearing opportunistically, or growing less vigorously.

According to Reduron (2007), despite its monocarpic status, its cycle could exceed the period of 3 years: either in stand-by mode, or when the bloom is interrupted (i.e., a special dry event, predation, or cutting). This reiteration is made possible by production of vegetative lateral sprouts in the stem, or by occasional root crown propagules. Excessive predation or cutting can easily kill the plant (according to Reduron 2007).

After disturbance to shore up the banks, the tidal zone can be re-colonized quickly if propagules dispersion can occur. In the Nantes area, in urbanized habitats, we found two opposite conditions linked to the management of *A. heterocarpa*. In one case it is favored by garden cultivation for public ornament (i.e.: on the Île de Nantes), and in the other surrounding cases the plant is threatened by cleaning of the banks and by cutting herbaceous growth.

Its resilience is also manifested in functional trait dynamics (plant architecture), following a gradient from artificial to natural environments, where we have noticed a variation both in phenology and in the rules of functional traits. In the most anthropic habitats, the phenology is precocious: leaf area is evidently reduced, the height is about 10–50 cm, the stalk basal diameter is about 0.2–1 cm, and the fruits are smaller, frequently less able to keep for a long time or to sprout. *Angelica heterocarpa* expresses the maximum vegetative growth in the most natural habitats, where it shows its largest leaf area, greatest development, which can exceptionally exceed 2 m in height and 4 cm in stalk foot diameter; and where the fruits are bigger, with late phenology (Fig. 15.2). The phenological distance between individuals from anthropic to natural areas, can reach up to 1.5 months.

In oligohaline megaforb-, and in non-excessively dense reed communities, it develops halfway, with intermediate phenology and medium (to medium/good) functional trait size. Following a gradient from saline (downstream) to freshwater (upstream), it was noted that the plant size increases at the oligohaline sector, in a continuum from disturbed to natural conditions. Analyzing the samples density and functional traits, the ecological optimum zone corresponds to the most natural portions of the oligohaline sector. Phenological trends are to delay with approaching to the optimum. The epiphytic (Aerial Plant) phenomenon was more commonly observed in the optimum zone. At the phytodynamic level, we found *Angelica heterocarpa* in various estuarine oligohaline stages, from the substitutional ones to the more natural ones (series heads).

Fig. 15.2 Two different samplings of *Angelica heterocarpa*. To the left: in a young woodland. To the right: an example of stand-by form, epiphytic on *Salix atrocinerea* (Photo: K. Cianfaglione, June, 2015)

At the geobotanical level, *A. heterocarpa* is typical of the oligohaline megaforb community, which manifests several plant groupings (Dupont 1978, 1986, 1989b; Géhu and Géhu 1978; Géhu and Géhu Franck 1982; Géhu 1995; Magnanon et al. 1998; Bioret 2002; Lazare 2006; Lazare and Bioret 2006; De Foucault 2011; Geffray and Kaupe 2012); and it characterize the *Calystegio sepium-Angelicetum heterocarpae* Géhu and Géhu-Franck 1978, phytosociological association.

In natural conditions, from the lowest water level to the upper part of the banks, the zonation is complete. We can distinguish the following transect, as is in Fig. 15.3 - from the river to the bank: ① *Apio nodiflori-Heleocharetum amphibiae* Géhu & Géhu-Franck 1972, or communities belonging to the class *Bidentetea tripartitae* Tüxen, Lohmeyer & Preising *ex* von Rochow 1951; ② *Scirpetum compacti* Van Lagendonck 1931 *corr.* Bueno and F. Prieto *in* Bueno 1997, or *Scirpetum triquetri* Géhu & Biondi 1988, or *Scirpetum tabernaemontani* Passarge 1964; ③ *Phragmites australis* communities (*Phragmiti australis-Magnocaricetea elatae* Klika *in* Klika & V. Novák 1941); ④ *Salicetum albae* Issler 1926, or *Salici albae-Populetum nigrae* (Tx. 1931) Meyer-Drees 1936, or *Salicetum albo-fragilis* Tüxen *ex* Moor 1958, or *Salicetum fragilis* Passarge 1957; ⑤ woodlands from *Querco roboris-Fagetea sylvaticae* Br.-Bl. & J. Vlieger *in* J. Vlieger 1937.

Angelica heterocarpa normally can be found in ③ and ④ levels, rarely also in ② and ⑤, occasionally in the upper ①. Level ⑤ is characterized by *Fraxinus excelsior, F. angustifolia*, with sometimes *Ulmus minor, U. laevis, Alnus glutinosa, Populus tremula, Prunus avium, Quercus robur*, and even *Acer pseudoplatnaus, A. platanoides, Castanea sativa, Fagus sylvatica*, and other *Quercus* species (more rare, in driest conditions) presences. In ④, we can also find groups of *Salix x rubens* and of *S. atrocinerea* accompanied by *Crataegus monogyna* and *Sambucus nigra*; while presence of *S. cinerea, S. purpurea, S. viminalis, S. triandra, S. pentandra, Populus alba, P. canescens, Betula pendula*, and *Betula pubescens* are scattered and rare.

In ① level is possible to find some simple and fragmented grouping (even very simplified or ruderalized) characterized by *Polygonum* and *Bidens* species or by *Eleocharis bonariensis*.

Between the upper ①, and lowest ⑤ levels, some *Phragmites australis* communities can develop as primary (in few stripes), or in secondary succession. In that case, they compete with the open Megaforbs community. Generally, the Oligohaline Megaforbs community can be found in forest (as primary) and outside. It lives especially in ③ and ④, more rarely in upper ② and in ⑤ levels: in any case favourished under woody cover. The *Phalaris arundinacea* community, in secondary succession, tends to compete mainly with the open Megaforbs community (by light availability), less with the *Scirpetum* communities (by habitat constraint) and lesser with reed communities, that can develop more height and more biomass. *Phragmites australis* also compete with the open Oligohaline Megaforbs community, by light availability occurrence.

Generally, the Oligoaline megaforb community is very species-rich, characterized by *Agrostis stolonifera, Angelica heterocarpa, A. sylvestris* (to the upper limit), *Apium nodiflorum, Arctium* sp.pl., *Bidens tripartita, Callitriche* sp.pl., *Carex* sp.pl,

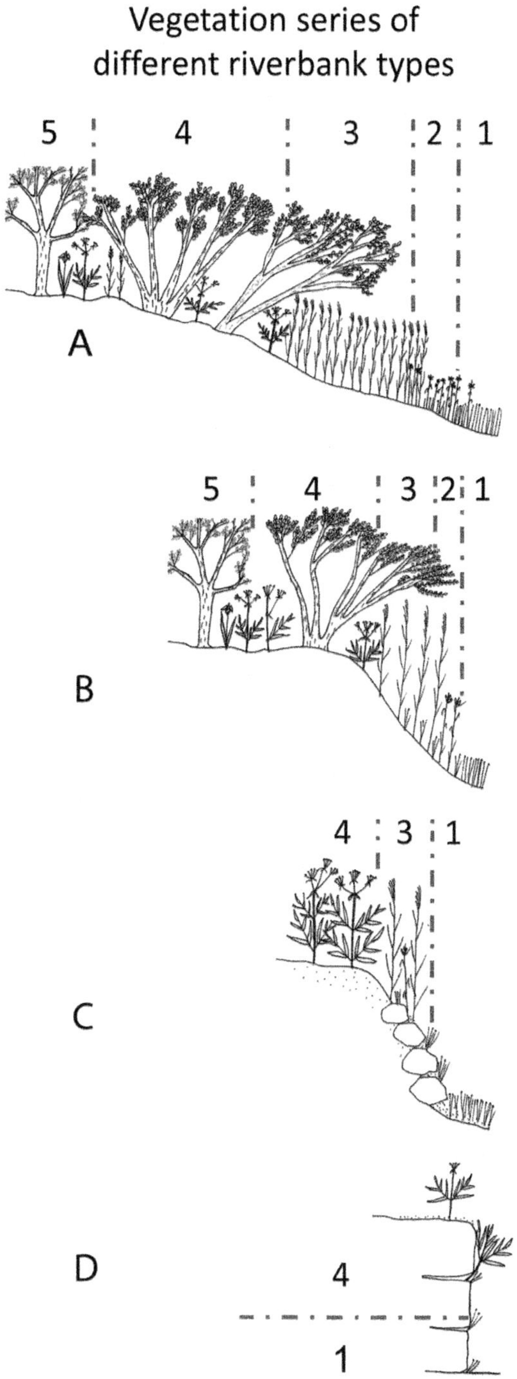

Fig. 15.3 Schema of *Angelica heterocarpa* series transect. Type of riverbank: (**A**) Natural gentle slope riverbank; (**B**) Natural steep riverbank; (**C**) Artificial steep riverbank; (**D**) Artificial riverbank (Masonry or Cement). Type of series: 1) *Apio nodiflori-Heleocharetum amphibiae*; 2) *Scirpetum* communities; 3) *Phragmites australis* communities; 4) *Salix* communities; 5) *Querco roboris-Fagetea sylvaticae* communities

Epilobium hirsutum, E. obscurum, Eupatorium cannabinum, Equisetum sp.pl., *Filipendula ulmaria, Galium elongatum, G. palustris, Glechoma hederacea, Glyceria fluitans, Hedera helix, Iris pseudacorus, Jacobaea aquatica, Juncus effusus, Lychnis flos-cuculi, Lycopus europaeus, Lysimachia vulgaris, Lythrum salicaria, Mentha aquatica, Oenanthe crocata, Persicaria amphibia, Persicaria maculosa, Petasites hybridus, Rubus* sp.pl., *Samolus valerandi, Solanum dulcamara, S. nigrum, Valeriana officinalis, Veronica anagallis-aquatica, V beccabunga*. Are typical, although underrepresented: *Arum italicum* subsp. *neglectum, Atriplex hastata, Bellis perennis, Bolboschoenus maritimus* subsp. *maritimus* (alias: *Scirpus compactus*), *Calystegia sepium, C. sylvatica, Circaea lutetiana, Cirsium arvense, C. palustre, Dipsacus fullonum, Eleocharis bonariensis, Galium aparine, Glyceria maxima, Iris foetidissima, Juncus* sp.pl., *Lamium maculatum, L. purpureum, Mentha arvensis, Myosoton aquaticum, Oenanthe lachenalii, Persicaria hydropiper, Phalaris arundinacea, Phragmites australis, Potentilla anserina, P. reptans, Ranunculus* sp.pl., *Rorippa aquatica, R. sylvestris, Rumex* sp.pl., *Schoenoplectus lacustris, S. tabernaemontani, S. triqueter, Scutellaria galericulata, Stellaria media, S. neglecta, Taraxacum officinale, Urtica dioica*, and *Veronica* sp.pl.

In more grazed/mowed areas some grassland develops (rich in reptant species like *Agrostis stolonifera*); while in more nitrophilous/ruderal places, other herbaceous communities are present, characterized mainly by *Urtica dioica, Galium aparine, Cirsium arvense, Calystegia sepium, Phragmites australis, Phalaris arundinacea, Dipsacus fullonum, Galega officinalis, Artemisia verlotiorum, Sambucus ebulus* and *Reynoutria japonica*.

In steepest shores, we observed a zonation compression, with a smaller representation of the vegetation series (smaller surface, simplification of the structure, degeneration/regression, fewer series, etc.). In the most disturbed conditions, *A. heterocarpa* can develop in the masonry/cement banks, fragments of ①, till the ⑤ level. In the last case, the communities are similar to those that grow on tree trunks, where *Angelica heterocarpa* lives as epiphyte; and frequently accompanied by scattered species from ruderal vegetation and those from secondary grasslands communities, mainly as: *Agrostis stolonifera, Epilobium* sp.pl., *Hordeum murinum, Impatiens balsamina, Lythrum salicaria, Persicaria maculosa, Polygonum* sp.pl., *Ranunculus* sp.pl., *Rumex* sp.pl., *Potentilla reptans* and *Trifolium repens*. In that conditions, the tree species are mainly represented by seedlings or saplings of *Salix atrocinerea, Populus* sp.pl. (incl. *P.* × *canadensis*) and *Alnus glutinosa. Ailanthus altissima, Pterocarya* sp.pl., and *Robinia pseudoacacia* can rarely form secondary small root sprouts coenosis in the most disturbed areas, without any real ecological concerns right now.

15.3 Threats

The global population of *Angelica heterocarpa* is still large enough, despite the drastic ecological changes linked to management operations carried out during last centuries in the French estuaries, that represent a serious threat. Sometimes, the largest concentrations of this species are located in the most urbanized part of estuaries (for geographical coincidence) making it difficult to assess rewilding and strong conservation policies (Dupont 1981; Magnanon et al. 1998; Figureau and Ferard 2001, 2002; Blanchard 2003; Figureau et al. 2004).

In many ways, *A. heterocarpa* is a very adaptable species. But despite its wide adaptive capacity, the species is seriously threatened by human activities, especially those that can alter (or suppress) the estuarine dynamics. Its riversides range was thinned and compressed towards the river banks, by the reclamation for land use purposes (urbanization, agriculture, livestock). Forest logging and tree elimination is another environmental problem. Another threat, described in the Loire estuary, is the excessive erosion of the top portion of the estuary, and the related excessive deposition in the lowest part, which is caused by navigation works (Geffray and Kaupe 2012).

In the Loire estuary, *A. heterocarpa* and *Scirpus triqueter* monitoring programs showed the shifting of their distribution to upstream, following the salinity front shift and changes in the sedimentation regime, after excavations and channeling works was carried out in the estuary for navigation purposes (cf., Mignot and Le Hir 1997; Magnanon et al. 1998; Dupont 2001; Le Bail and Lacroix 2005; Geffray and Kaupe 2012).

This demonstrates how these species are dependent on their environment and its evolution. *Angelica heterocarpa* range has decreased overall a dozen kilometers downstream and over 4 km upstream; in the valley of the Sèvre Nantaise River, it is present in a few kilometers upstream of the dam at Pont-Rousseau (cf., Dupont 1981, 1989c; Lebot 2006) (Fig. 15.4).

A similar situation has been described for the Charente estuary (Terrisse and Champion 2008), where *A. heterocarpa* followed the same trends of the Thermo-

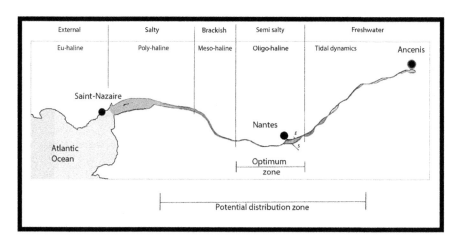

Fig. 15.4 Distribution in the Loire estuary. The lateral distribution reduction is linked to canalization, reclamation and land use. The upstream-downstream distribution reduction is linked to navigation purposes, and water flows changes. In Erdre (*E*) and Sévre Nantaise (*S*) affluents, the distribution interruption is linked to the barrages. The Potential areal was extrapolated from historical data and by geomorphologic homogeneity. The actual and the optimum areas, were both extrapolated from original findings (2015) following also the available bibliography

Atlantic sub-salty meadows (habitat 1410), because of the disappearance of a regular brackish water submersion period. The disappearance of these two typical estuarine components, resulting in the replacement by freshwater vicarious elements (*A. sylvestris* and hygrophilous eutrophic grasslands), has its probable origin in the pressure of agricultural requirements, navigation and urbanization purposes, with modification of land forms and natural dynamics (i.e.: dams; canalization; reclamation, deforestation; changes in soil salt and water contents).

Loire Forests are still very deteriorate, too young, ruderalized, degraded in structure (cf: Cornier, 2002); the best and more structured forest are often of very modest size and highly fragmented. Urbanization, dams, deforestation, and channeling of the estuaries, represent the main general threats for *A. heterocarpa*. For that reason, over time, some protection plans were carried out in order to reduce the plant extinction risk and the decline of populations. Many efforts, and project hypotesis were made to study and re-establish estuarine dynamics, such as doing experiments and *in situ* interventions, re-naturalization of cemented areas, reshaping of artificial embankments, specimen removal during bank renovation works and replanting in specific areas, and restoration of hygrophilous woodlands (cf., Leroy 1989; Barreau 1996; Mignot and Le Hir 1997; Figureau and Ferard 2001, 2002; Lebret 2002; Guitton et al. 2003; Blanchard 2003; Figureau et al. 2004; Lacroix and Guitton 2006; Lacroix et al. 2009; Lacroix and Figureau 2009; Figureau and Lacroix 2009a, b).

15.4 Conclusions

Angelica heterocarpa shows remarkable adaptation and high resilience capacity, being able to adjust its forms and rhythms to environmental changes. It tolerates highly constraining conditions (anthropic, epiphytic, distribution limits, etc.); and it can even live between the cracks in cement. When in trouble, this species showed an epigenetic plasticity, with biomass investment reduction, different plant architecture, and a consequent shorter phenology period, as reported also in others species (cf., Nilsen and Orcutt 1996; Chinnusamy and Zhu 2009).

This supports the idea of how much the human pressure exerted in estuaries during centuries was strong enough to eradicate (sometimes), or to change the distribution, of this species. To promote this species, it seems necessary to increase awareness of its existence in the general public and in administrations; and to use trees to fight bank erosion. If artificial bank stabilization is required, it is necessary to do it less as possible; preferring to be less invasive as possible, using not walled shores, and non-jointed rock, laying them on a gentle slope, to allow better the mud deposition and the plant recovery (see Fig. 15.3d).

When possible, re-wilding, re-naturalization or specific reshaping of banks can be good practices (cf., Figureau et al. 2004). In addition, it is necessary to think about reforestation, and the delineation of uncutable ("forever wild") forests, light-cutable forest, and ungrazed or low-grazing spaces, up to and behind the banks cordons, in order to reduce their landward erosion. Leaving trees can be useful also by undertaking megaforbs facilitation, in order to thin, to dominate and to control the reed communities or preventing it expansion.

References

Barreau F (1996) Etude de la végétation du lit majeur endigué de la Loire de Nantes à Ancenis, caractéristiques, dynamique et intérêt patrimonial. DESS Gestion des ressources naturelles. Cons. Rég. des rives de la Loire, Eq. pluridiscipl. PLGN, IEA Angers, France

Bensettiti F, Gaudillat V, Malengreau D, Quéré E (coord) (2002) Cahiers d'habitats Natura 2000. Connaissance et gestion des habitats et des espèces d'intérêt communautaire. Connaissance et gestion des habitats et des espèces d'intérêt communautaire. Espèces végétales. MATE/MAP/MNHN. Éd. La Documentation française. Paris, France 6:149–152

Bernard C (1991) Etude biosystématique, caryosystématique et phytochimique des taxons du genre Angelica L., se développant en France. Thèse de Doctorat en Sciences, spécialité Biologie végétale. Muséum National d'Histoire Naturelle, Paris, France

Bioret F (2002) Mégaphorbiaies oligohalines. In: Cahiers d'habitats Natura 2000. Habitats humides. La documentation française 3:294–297

Blanchard F (2003) Étude sur la restauration des habitats de l'Angélique des estuaires (Angelica heterocarpa Lloyd) dans le cadre d'un projet de reprofilage des berges érodées de la Dordogne (Libourne, Gironde). Association syndicale autorisée de la plaine de Condat and Conservatoire botanique national sud-atlantique, Audenge, France

Blanchard F, Caze G, Castagne H (2005) Premier bilan sur les populations d'angélique des estuaires en Gironde et dans le sud-ouest français. Conservatoire Botanique Sud-Atlantique, Conseil Général de la Gironde, Synthèse du séminaire angélique des estuaires du 7 octobre 2005 à Nantes. Nantes Métropole, DIREN Pays de la Loire, Conservatoire Botanique National de Brest, Jardin Botanique de Nantes, Nantes, France

Buord S, Lesouëf JY (2006) Consolidating knowledge on plant species in need for urgent attention at European level. Museum National d'Histoire Naturelle/European Topic Centre on Biological Diversity and Conservatoire Botanique National de Brest, Paris, France

Chinnusamy V, Zhu J-K (2009) Epigenetic regulation of stress responses in plants. Curr Opin Plant Biol 12(2):133–139

Clément B, Touffet J (1989) Les espèces végétales menacées ou protégées des zones humides de Bretagne. Plantes sauvages menacées de France. Bilan et protection. Actes du colloque de Brest, 8–10 octobre 1987. Conservatoire Botanique de Brest, Bureau des Ressources Génétiques, Association Française pour la Conservation des Espèces Végétales, pp 109–118

Corillion R (1961) Phytogéographie des halophytes du nord-ouest de la France (Phanérogames). Penn ar bed 25:48–80

Cornier T (2002) La végétation alluviale de la Loire entre le Charolais et l'Anjou: essai de modélisation de l'hydrosystème. Thèse de doctorat, Université de Tours. Discipline: Sciences de la Vie; Spécialité: Ecologie végétale, Tours, France

Danton P, Baffray M, Reduron J-P (1995) Inventaire des plantes protégées en France. Nathan, Paris; AFCEV, Mulhouse, France

De Foucault B (2011) Contribution au prodrome des végétations de France: les Filipendulo ulmariae-Convolvuletea sepium Géhu & Géhu-Franck 1987. J Bot Soc Bot France 53:73–137

Dupont P (1962) La flore atlantique européenne. Introduction à l'étude phytogéographique du secteur ibéro-atlantique. 19 décembre 1960. Thèse de doctorat, Université de Toulouse. Imp. Edouard Privat, Toulouse, France

Dupont P (1978) Etude générale d'environnement de l'estuaire de la Loire (contrat O.R.E.A.M.). La végétation des zones humides bordant l'estuaire de la Loire. Laboratoire d'Ecologie et de Phytogéographie, Université de Nantes, Nantes, France

Dupont P (1981) La végétation de l'estuaire de la Loire; intérêt, modifications récentes, urgence de mesures de protection. Fédération Régionale des Associations de Protection de l'Environnement du Centre, Ministère de l'Environnement, pp 123–144

Dupont P (1983) Remarques sur les espèces végétales protégées ou méritant de l'être en Loire-Atlantique et en Vendée. Bulletin de la Société des Sciences Naturelles de l'Ouest de la France, N.S 5(2):94–105

Dupont P (1986) Principaux aspects de la végétation des zones humides de l'estuaire de la Loire. Bull Soc Bot France LB, 133(1):41–60

Dupont P (1989a) La flore endémique du littoral atlantique français, du Morbihan au Pays basque. Remarques sur le micro-endémisme. Bulletin de la Société des Sciences Naturelles de l'Ouest de la France, N.S. 11(2):90–97

Dupont P (1989b) Etude de la végétation de quelques zones humides de la ville de Nantes. Laboratoire d'Ecologie et de Phytogéographie, Université de Nantes, Nantes

Dupont P (1989c) Quelques problèmes de protection des espèces végétales. Exemples en Loire-Atlantique et en Vendée. Remarques sur les responsabilités individuelles et collectives. Actes du colloque «Plantes sauvages menacées de France. Bilan et protection, Brest, 8–10 octobre 1987», Conservatoire Botanique de Brest, Bureau des Ressources Génétiques, Association Française pour la Conservation des Espèces Végétales, pp 61–77

Dupont P (2001) Atlas floristique de la Loire-Atlantique et de la Vendée. Etat et avenir d'un patrimoine. Conservatoire Botanique National de Brest, Société des Sciences Naturelles de l'Ouest de la France

Dupont P (2015) Les plantes vasculaires atlantiques, les pyrénéo-cantabriques et les éléments floristiques voisins dans la péninsule ibérique et en France. Bulletin Société Botanique du Centre-Ouest, N. S 45:1–495

Figureau C (1995) Angelica heterocarpa Lloyd. In: Olivier L, Galland JP, Maurin H, Roux J-P (coord) Livre rouge de la flore menace de la France. Espèces prioritaires. Collection «Patrimoines naturels», MNHN, CBN de Porquerolles, Ministère de l'environnement, direction de la nature et des paysages. I(20):29

Figureau C, Ferard P (2001) Projet d'arrêté de protection de biotope Couëron. Dossier scientifique: Quais de la Loire à Angelica heterocarpa, Nantes. Société des Sciences Naturelles de l'Ouest de la France

Figureau C, Ferard P (2002) Expérimentations sur la création d'un biotope boisé à Angelica heterocarpa à l'occasion d'un déplacement de plantes menacées sur un quai en voie d'aménagement. Essais de «levée au champ». Jardin Botanique de la Ville de Nantes (Service des Espaces Verts), Port Autonome Nantes Saint-Nazaire, France

Figureau C, Lacroix P (2009a) Cahier des Clauses Techniques Particulières type. Opérations de création, restauration de berges à angélique des estuaires et opérations de déplacement de pieds d'angélique des estuaires. Jardin Botanique de la Ville de Nantes, Conservatoire Botanique National de Brest, Nantes Métropole, Nantes, France

Figureau C, Lacroix P (2009b) Catalogue des savoir-faire et des pratiques favorables à l'angélique des estuaires. Jardin Botanique de la Ville de Nantes, Conservatoire Botanique National de Brest, Nantes Métropole, Nantes, France

Figureau C, Richard P (1990) Angelica heterocarpa: écologie et répartition dans le Sud armoricain. Service espaces verts et environnement. Jardin botanique, ville de Nantes, Index Seminum, pp 17–24

Figureau C, Cuche A, Arthuys P, Zepele M (1998) Carte de répartition et d'importance des populations d'Angelica heterocarpa. Jardin botanique, SEVE, Nantes, France

Figureau C, Marsac M, Garcia-Melgares J (2004) Restauration d'un habitat à Angelica heterocarpa dans le cadre de l'opération Ile de Nantes secteur 1. Ville de Nantes – SEVE/ Jardin Botanique, Nantes, France

Geffray O, Kaupe S (2012) Inventaire et cartographie des berges de Loire entre Saint-Nazaire et Montsoreau. Conservatoire régional des rives de la Loire et de ses affluents, Nantes

Géhu JM (1995) Résumé typologique des milieux littoraux de France. Schéma synoptique hiérarchisé des végétations côtières. Centre international de Phytosociologie, Bailleul, France

Géhu J-M, Géhu J (1978) Les groupements à Angelica heterocarpa des estuaires atlantiques français. «La végétation des prairies inondables (Lille, 1976)». Colloques phytosociologiques V:359–362

Géhu J-M, Géhu-Franck J (1982) Etude phytocoenotique analytique et globale de l'ensemble des vases et prés salés et saumâtres de la façade atlantique française. Bull Ecol 13(4):357–386

Guitton H, Lacroix P, Brindejonc O (2003) Etude préalable à un plan de conservation en faveur de l'angélique des estuaires (Angelica heterocarpa Lloyd). Conservatoire Botanique National de Brest, DIREN Pays de la Loire, Communauté Urbaine de Nantes, FEDER, Nantes, France

Hoarher J (1984) Ces plantes de Bretagne sont protégées. Penn ar Bed 116:26–31

Juillet N (2013) Angelica heterocarpa. The IUCN red list of threatened species 2013. On-line version: e.T162184A5554781. https://doi.org/10.2305/IUCN.UK.2011-1.RLTS.T162184A5554781.en. Downloaded on 28 Sept 2015

Lacroix P, Figureau C (2009) Bilan des connaissances sur l'angélique des estuaires (Angelica heterocarpa Lloyd). Conservatoire Botanique National de Brest, Jardin Botanique de Nantes, Nantes Métropole, Nantes, France

Lacroix P, Guitton H (2006) Vers un plan de conservation en faveur de l'angélique des estuaires (Angelica heterocarpa Lloyd) dans l'estuaire de la Loire. Actes du colloque de l'AFCEV à Troyes (13–15 novembre 2003). La biodiversité végétale. Des plantes pour l'avenir, Nancy, pp 221–233

Lacroix P, Le Bail J, Hunault G, Brindejonc O, Homassin G, Guitton H, Geslin J, Poncet L (2008) Liste rouge régionale des plantes vasculaires rares et/ou menacées en Pays de la Loire. CBN and Région Pays de la Loire, Nantes, France

Lacroix P, Figureau C, Garcia-Melgares J (2009) L'angélique des estuaires, un enjeu de la biodiversité. De la conservation d'une espèce à la préservation du milieu estuarien. Guide à l'attention des acteurs de l'aménagement et des gestionnaires. Conservatoire Botanique National de Brest, Jardin Botanique de Nantes, Nantes Métropole, Nantes, France

Lahondère C (1998) Liste rouge de la flore menacée en Poitou-Charentes. Bulletin de la Société Botanique du Centre-Ouest, NS 29:669–686

Lazare JJ (2006) Les habitats à Angelica heterocarpa Lloyd de la Nive (Pyrénées-Atlantiques). Journal de Botanique 36:63–70

Lazare JJ, Bioret F (2006) Associations végétales nouvelles du littoral du Pays basque. Journal de Botanique de la Société Botanique de France 34:71–80

Le Bail J, Lacroix P (2005) État des lieux des populations de scirpe triquètre (Scirpus triqueter L.) dans l'estuaire de la Loire. Propositions de conservation. Conservatoire Botanique de Brest, Antenne régionale des Pays-de-la-Loire, Nantes, France

Lebot K (2006) La dynamique de la vie. Présence des espèces protégées dans la couverture végétale. In: Cahier 2002 indicateurs. Gip Loire Estuaire, Nantes 1(L2–B3):1–7

Lebret S (2002) Etude de la végétation et des habitats d'intérêt communautaire de la Loire, en aval d'Ancenis, en relation avec la dynamique estuarienne. Rapport de DESS, Cellule de Mesures et de Bilans Loire Estuaire, Nantes, France

Leroy R (1989) Intérêt et vulnérabilité du patrimoine écologique de la vallée de la Loire face aux facteurs d'évolution du milieu. DRAE des Pays de la Loire, Nantes, France

Lesouëf J-Y (1986) Les plantes endémiques et subendémiques les plus menacées de France (partie non méditerranéenne). Conservatoire botanique national de Brest, Brest, France

Lloyd MJ (1860a) Sur une nouvelle espèce de angelica (Nantes, 29 septembre 1859). Bulletin de la Société botanique de France 6(9):709–710

Lloyd J (1860b) Sur une espèce nouvelle de angelica. Mémoires de la Société Académique de Maine-et-Loire 8:22–23

Lloyd J (1868) Flore de l'Ouest de la France, ou description des plantes qui croissent spontanément dans les départements de: Charente-Inférieure, Deux-Sèvres, Vendée, Loire-Inférieure, Morbihan, Finistère, Côtes-du-Nord, Ille-et-Vilaine, Nantes, deuxième édition

Magnanon S (1993) Liste rouge des espèces végétales rares et menacées du Massif armoricain. ERICA 4:1–22

Magnanon S, Bioret F, Dupont P (1998) Angelica heterocarpa dans l'estuaire de la Loire: répartition, écologie, menaces. Proposition de mesure de gestion. DIREN Pays de la Loire, conservatoire botanique national de Brest, Brest, France

Malinvaud E 1903 Notules floristiques. Bulletin de la Société botanique de France 50:471–474

Metais I, Simo P, Lambert E (2008) Caractérisation de la diversité génétique chez Angelica heterocarpa Lloyd (Pays de la Loire) à l'aide de marqueurs moléculaires (isoenzymes, RAPD). Centre d'Etudes et de Recherche sur les Ecosystèmes Aquatiques. Université Catholique de l'Ouest, Angers, France

Mignot C, Le Hir P (1997) Estuaire de la Loire. Rapports de synthèse de l'APEEL 1984–1994. Tome I, Hydrosédimentaire. Edition de l'APEEL, Nantes 1, pp 10–15

Nilsen ET, Orcutt DM (1996) Physiology of plants under stress. Abiotic factors. Wiley, Hoboken

Reduron J-P (2007) Ombellifères de France. Bulletin de la Société Botanique du Centre-Ouest, N. S 26(1):314–321

Reduron J-P, Muckenstrum B (2005) Angelica heterocarpa: une espèce d'intérêt chimique. Conservatoire Botanique de la Ville de Mulhouse. Synthèse du séminaire angélique des estuaires du 7 octobre 2005 à Nantes. Nantes Métropole, DIREN Pays de la Loire, Conservatoire Botanique National de Brest, Jardin Botanique de Nantes, Nantes, France

Terrisse J, Champion E (2008) L'Angélique des estuaires (Angelica heterocarpa): statut et répartition sur la ZSC n°FR5400-472. DIREN Poitou-Charentes, Ligue pour la Protection des Oiseaux, Rochefort, France

Tutin TG, Heywood VH, Burges NA, Valentine DH, Walters SM, Webb DA (1968) Flora europaea. University Press, Cambridge, UK, p 357

Environmental Mapping: From Analysis Maps to the Synthesis Map

16

Marcello Martinelli and Franco Pedrotti

Abstract

The objective of this article is to make a reflection upon the methodological question of Environmental Mapping, aiming to reach a systematization proposal. It is assumed that the considerations made regarding the environment are not only directed to nature itself, but also to society. This elaboration begins through the study of the area of interest focusing on its thematic ramifications via analysis maps. After that, a synthesis approach is made which would confirm the delimitation of spatial ensembles, which are groups of unitary areas of analysis characterized by groupings of features or variables—the Types of Environment—also present in relevant literature on Types of Landscape, which would be traced over the synthesis map.

Keywords

Environmental mapping • Graphic representation • Analysis map • Synthesis map • Types of environment • Types of landscape

The current profusion of representations is a social fact par excellence. They are products of human reasoning and they speak to the entire society. It is in this context that one should consider cartography today. Regarded in this sense, it will conduct the elaboration of maps with undeniable participation in the knowledge process.

M. Martinelli (✉)
Postgraduate Program on Human Geography, Department of Geography—FFLCH, University of São Paulo, São Paulo, Brazil
e-mail: marcello.martinelli.3@gmail.com

F. Pedrotti
Department of Botany and Ecology, University of Camerino, Camerino, Italy
e-mail: franco.pedrotti@unicam.it

© Springer International Publishing AG, part of Springer Nature 2018
A.M. Greller et al. (eds.), *Geographical Changes in Vegetation and Plant Functional Types*, Geobotany Studies,
https://doi.org/10.1007/978-3-319-68738-4_16

At the present time, the environmental issue has gained a prominent position and has prompted fierce debates in forums of various levels. Environmental problems are clearly social, because they emerge from society, not from nature.

Inserted in the world of social communication, cartography is distinguished when regarded also as a language, the language of Graphic Representation, which consequently has its own syntax and semiotics (Bertin 1973, 1977).

The Graphical Representation is a visual communication language that can be two-dimensional and timeless, but monosemic, with only one meaning. Its specificity lies essentially in the fact that it is fundamentally linked to the core of relations that can occur between the meaning of symbols. It is therefore interesting to see the relations that exist between symbols that mean relations between objects, facts and phenomena which consists the considered reality, leaving in the background the concern with the relation between the signified and the signifier of symbols, basic characteristic of polysemic semiotic systems with multiple meanings. This is what happens in the communication through photography, painting, drawing, graphics, publicity, advertising, etc., which create figurative or abstract images.

It should be noted that throughout its history, as a result of the flourishing and systematization of the different fields of study constituted of the division of scientific work, carried out in the late eighteenth century and early nineteenth century (Industrial Revolution period), cartography gave way to the definition of Thematic Cartography with greater attention to the representation of known properties than those viewed on objects, facts or phenomena of reality. It is in this context in which Environmental Cartography is inserted (Salichtchev 1979; Robinson 1982; Palsky 1996).

Faced with the complexity of reality, this type of cartography must articulate with different ways of understanding the various scientific branches, each conceiving an adequate temporal scale. It will be a cartography that will take into account, on the one hand, the articulation of different levels of analysis in accordance with the order of magnitude in which the phenomena of reality manifest themselves, and in the other hand, the combinations and contradictions that happen between spatial sets defined by the phenomena under consideration, in a given temporal level.

Today, it cans no longer conceive representations of spatial reality, the space of humanity, in an analytical and fragmented way. One should seek a cartography of integration, one of reconstruction of the whole.

This holistic reality was perceived as landscape by the great naturalists and explorers of the nineteenth century. It was conceived as the relationship between various aspects of nature among themselves. The notion of natural unit was kept, but with a physiognomic characteristic, aesthetic, without history.

Landscape is what one sees before oneself. It is a visible reality. It is a perceived overall view through the surrounding space. Therefore, it does not have an independent existence. It exists through the person who seizes it: each person sees it differently from the other, not only because of the targeting of their observation, but also due to their individual interests (Bertrand 1968; Tuan 1980; Wieber 1985; Pinchemel 1987; Bertrand and Bertrand 2007).

Throughout its history, mankind has had actions that went from simple survival relations within their piece of natural world to a progressive domination, engendering great chances of change through artificiality, culminating in a significant share of techno-science in a globalized space with permanent innovation. Thus, in this new world, it do not count on natural nature anymore, but it do count on an artificial nature, assessed as resource. Therefore, the new expression of space and time becomes the technical-scientific-informational milieu (Santos 1994).

In the Environmental Mapping, the Types of Environment mapping is conceived as environmental synthesis cartography, since it confirms the delimitation of space sets, which are groupings of unitary areas of analysis characterized by groupings of features or variables that the research has made individual (Martinelli 1999; Orsomando et al. 2000, 2007; Martinelli and Pedrotti 2001; García-Abad Alonso 2002; Martinelli 2014a, b).

The organization of the cited grouping within an adequate scientific reasoning will be exposed with transparency through the map legend. This is a necessary theoretical basis. It conveys the meaning of the adopted symbols in the graphical representation in the map, in which the verbalization is essential to overcome the viewing limitations.

The rationale for the development of the Types of Environment cartography follows a methodological reference with stages that establish analysis maps so that later a synthesis can be achieved.

This is a study related to the map "Landscape Systems and Subsystems of the Stelvio National Park" (I), from which a booklet referring to Valle di Trafoi is extracted, originally made in the 1:50.000 scale (Pedrotti et al. 1997) (Fig. 16.1).

With the above basilar posture rooted, it firstly consider the geological and morphological knowledge as fundamental. Thus, the Earth's surface modeled in forms of relief must be related not only to rocks and structural arrangements of different ages and origins that sustain them, but also to the superficial formations and the resulting soils overlying them. No less important is the combined role of the climate flows, considered to be in a constant state of evolution. All of this demonstrates the dynamic characteristic of this important environmental component, which should be taken into consideration in a cartography that pretends to endeavor.

These considerations constitute the foundations that impose certain identifying features for the recognition of environmental categories in wider levels, which would be able to decompose in space sets, coordinating characteristics settings depending on the geological and morphological discontinuities.

Geologically speaking, the territory of Valle de Trafoi occupies much of the northwestern part of the Stelvio National Park dominated on its southern headland by Òrtles, a sharp rise in Horn, reaching 3095 m above sea level surrounded by towering glaciers, modeled in calcareous mountain terrain to the south and land silicate mountains to the north, framed by the Adige river to the north and to the east, and Noce to the south.

For more clarity, the analysis map with generic features of the geology and morphology is shown, highlighting the presence of shapes such as Horn, Glacier and Snowfield, Scarp, Ridge, Col, Sediments and Valley Floors from where rivers

VALLE DI TRAFOI: OVERVIEW

Fig. 16.1 Panoramic set of Valle di Trafoi with a North to South view. In other words, the view of the Trafoi river's mouth towards the background of the summits and the Òrtles glacier

flow, complemented by main climate traits seen in terms of the Endoalpic type of hygric-pluvial continentality (Pedrotti and Martinelli 2010) (Fig. 16.2).

In a second moment, concerning this geological bone and its respective modeled relief, the potential vegetation is assessed, which would be one that would constitute a particular environment from the moment the action operated by human

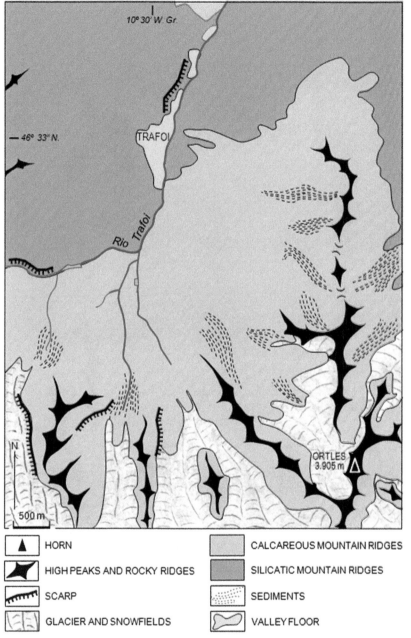

Fig. 16.2 Valle di Trafoi, geology and morphology

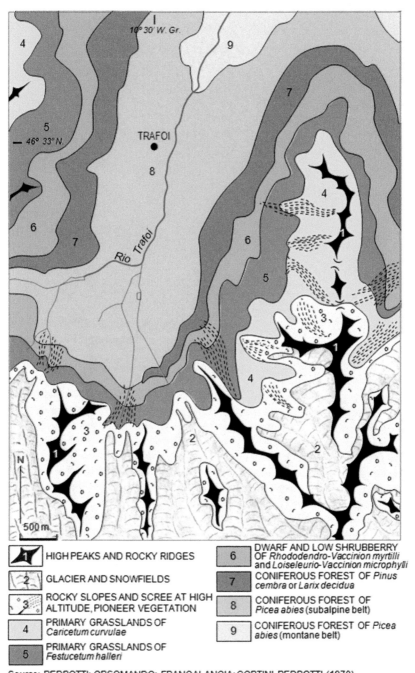

Fig. 16.3 Valle di Trafoi, potential vegetation

16 Environmental Mapping: From Analysis Maps to the Synthesis Map

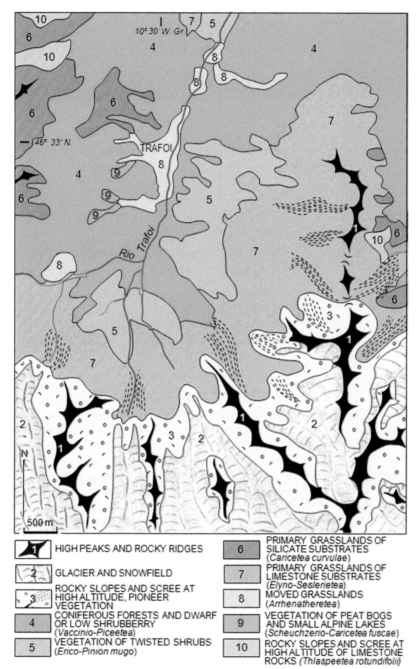

1	HIGH PEAKS AND ROCKY RIDGES	6	PRIMARY GRASSLANDS OF SILICATE SUBSTRATES (*Caricetea curvulae*)
2	GLACIER AND SNOWFIELD	7	PRIMARY GRASSLANDS OF LIMESTONE SUBSTRATES (*Elyno-Seslerietea*)
3	ROCKY SLOPES AND SCREE AT HIGH ALTITUDE, PIONEER VEGETATION	8	MOVED GRASSLANDS (*Arrhenatheretea*)
4	CONIFEROUS FORESTS AND DWARF OR LOW SHRUBBERRY (*Vaccinio-Piceetea*)	9	VEGETATION OF PEAT BOGS AND SMALL ALPINE LAKES (*Scheuchzerio-Caricetea fuscae*)
5	VEGETATION OF TWISTED SHRUBS (*Erico-Pinion mugo*)	10	ROCKY SLOPES AND SCREE AT HIGH ALTITUDE OF LIMESTONE ROCKS (*Thlaspeetea rotundifolii*)

Source: PEDROTTI; ORSOMANDO; CORTINI PEDROTTI (1974).

Fig. 16.4 Valle di Trafoi, actual vegetation

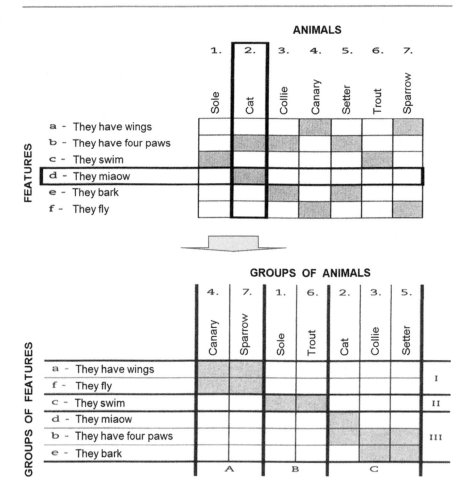

Source: GIMENO (1980).

Fig. 16.5 The reasoning of synthesis

society would cease, considering that the weather would still remain virtually constant. The spatial distribution of these climaxes would happen in a narrow relation with environments that featured favorable ecological conditions. Thus, each set of relief space would correspond to a series of vegetation in which respective plant associations would be met by a dynamic type of bond (Pedrotti 1994; Falinski and Pedrotti 1990; Falinski 1990; Gafta and Pedrotti 1997) (Fig. 16.3).

In the next stage, the reflection would evaluate the actual vegetation in the space produced by dynamic social relations through the succession of the ways of production that mankind has lived and is living throughout its history. In this

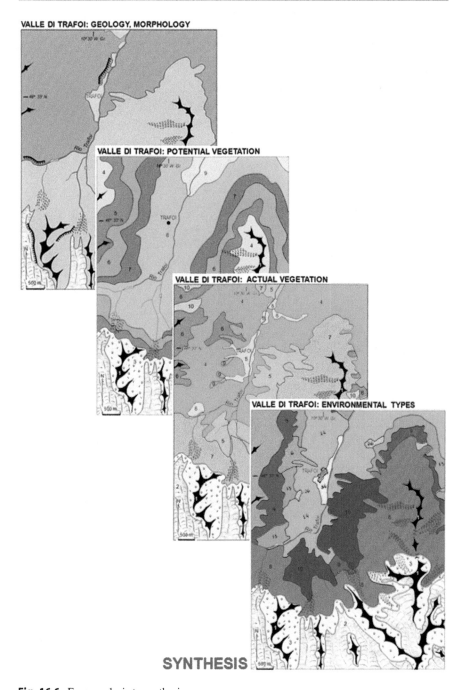

Fig. 16.6 From analysis to synthesis

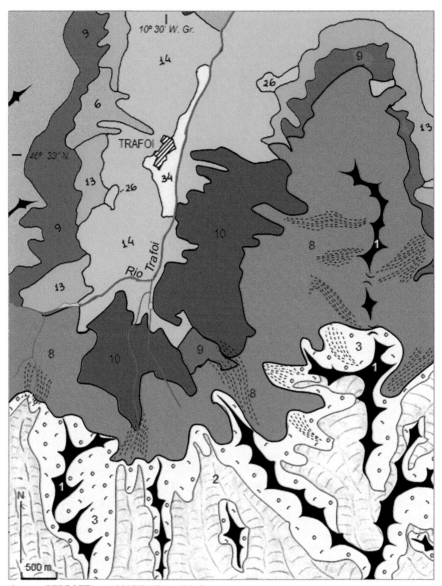

Source: PEDROTTI and MARTINELLI (2015).

Fig. 16.7 Valle di Trafoi, environmental types

operation, there could be spatial sets that would outline a preliminary sketch concerning the articulation of the types of environment that will make up the geographical space (Fig. 16.4).

16 Environmental Mapping: From Analysis Maps to the Synthesis Map

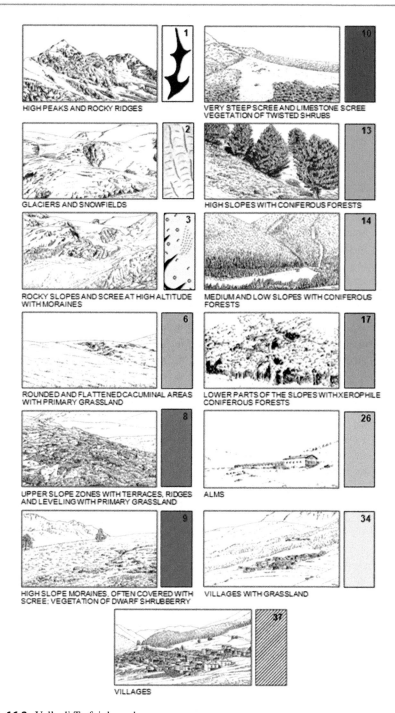

Fig. 16.8 Valle di Trafoi, legend

The last moment of this methodological development considers a reasoning of synthesis, which would confirm the delimitation of groupings of elementary areas of analysis characterized by features or variables groupings.

To expose this thinking it is presented two data matrices, the passage from analysis to synthesis. From the correspondence between animals (7) and its features (6) it moves on to groups of animals (A, B, C) characterized by attribute groups (I, II, III) (Gimeno 1980) (Fig. 16.5).

The groupings mentioned above, in turn, will show combinations of integrated characteristics, which may correspond to a vegetation series fragment or to several fragments of different series, in accordance with its spatial extent and environmental functions, in which, besides nature's dynamics, there would also be the movements of human society activities that would give them specific vital characteristics (Pedrotti et al. 1970) (Fig. 16.6).

These spatial sets thus determined and implemented to the map will find reference in the legend's composition upon concise epithets that will define the respective Types of Environment.

The map legend thus conceived is able to provide full transparency to the resulting composition of the methodological lucubration undertaken in the execution of the map, now a synthesis one, in other words, all the reasoning from taking a position on the interpretation and understanding of the reality to the formalizing of its exposure in a holistic way, by mobilizing for this purpose the rational and correct approach of the guidelines for an adequate syntax of the graphic representation's language.

In this sense, the map legend allows to highlight the Types of Environment, which are noted by symbols defined by their meanings, exposing its main characteristics and accompanied by illustrations (Martinelli and Pedrotti 2016) (Figs. 16.7 and 16.8).

References

Bertin J (1973) Sémiologie graphique: les diagrammes, les réseaux, les cartes. Gauthier-Villars/Mouton, Paris

Bertin J (1977) La cartographie et le traitement graphique de l'information. Flammarion, Paris

Bertrand G (1968) Paysage et géographie physique globale: esquisse méthodologique. Rév Géogr Pyrénées Sud-Ouest 39(3):249–272

Bertrand G, Bertrand C (2007) Uma geografia transversal e de travessias: o meio ambiente através dos territórios e das temporalidades. Massoni, Maringá

Falinski JB (1990) Kartografia geobotaniczna. Varsóvia, PPWK

Falinski JB, Pedrotti F (1990) The vegetation and dynamical tendencies in the vegetation of Bosco Quarto, Promontorio del Gargano, Italy. Braun-Blanquetia 5:1–31

Gafta D, Pedrotti F (1997) Environmental units of the Stelvio National Park as basis for its planning. Oecol Montana 6:17–22

García-Abad Alonso JJ (2002) Cartografía ambiental: Desarrollo y propuestas de sistematización. Obs Medioambiental 5:47–78

Gimeno R (1980) Apprendre à l'école par la graphique. Retz, Paris

Martinelli M (1999) La cartographie environnementale: une cartographie de synthèse. Phytocoenosis 11:123–130

Martinelli M (2014a) Mapas da geografia e cartografia Temática (6ª ed. 2ª reimpressão). Editora Contexto, São Paulo

Martinelli M (2014b) Mapas, gráficos e redes: elabore você mesmo. Oficina de Textos, São Paulo

Martinelli M, Pedrotti F (2001) A cartografia das unidades de paisagem: questões metodológicas. Rev Dep Geogr 14:39–46

Martinelli M, Pedrotti F (2016) Environmental mapping: concepts, methods and case studies. 59th annual symposium of the international association for vegetation science, IAVS, Pirenópolis

Orsomando E, Catorci A, Martinelli M, Raponi M (2000) Carta delle unità ambientali-paesaggistiche dell'Umbria. Firenze, S.E.L.C.A.

Orsomando E, Tardella FM, Martinelli M (2007) Biodiversità forestale e paesaggistica del territorio comunale di Sellano. Camerino, UNICAM

Palsky G (1996) Des chiffres et des cartes: la cartographie quanitative au XIXème siècle. Comité des Travaux Historiques et Scientifiques, Paris

Pedrotti F (1994) Serie di vegetazione e processi dinamici. In: Atti. Seminario la Destinazione Forestale dei Terreni Agricoli, Università di Camerino, Camerino

Pedrotti F, Martinelli M (2010) Carta dei sistemi ambientali (paesaggi) del Trentino-Alto Adige. Università degli Studi di Camerino, Camerino

Pedrotti F, Orsomando E, Cortini Pedrotti C (1970) Parco Nazionale dello Stelvio: Carta della Vegetazione. Firenze, Litografia Artistica Cartografica

Pedrotti F, Gafta D, Martinelli M, Scola AP, Barbieri F (1997) Le unità ambientali del Parco Nazionale dello Stelvio. L'Uomo e l'Ambiente 28:1–103

Pinchemel P (1987) Lire les paysages. Doc Photogr 6088:1–103

Robinson AH (1982) Early thematic mapping in the history of cartography. The University of Chicago Press, Chicago

Salichtchev KA (1979) Cartografía. Pueblo y Educación, La HabanaEditorial

Santos M (1994) Técnica espaço tempo: Globalização e meio técnico-científico informacional. Hucitec, São Paulo

Tuan Y (1980) Topofilia: um estudo de percepção, atitudes e valores do meio ambiente. Difel, São Paulo

Wieber JC (1985) Le paysage visible, objet géographique. Le Courrier du CNRS 57:5–8